Journal of Inherited Metabolic Disease

ISBN 978-0-7923-3855-0   ISBN 978-94-011-9635-2 (eBook)
DOI 10.1007/978-94-011-9635-2
This issue is also available separately
Orders should be sent to: Springer-Science+Business Media, B.V.

ISSN 0141-8955
CODEN = JIMDDP
Vol. 18 Supplement 1, 1995

JOURNAL OF INHERITED METABOLIC DISEASE

Diagnosis of human peroxisomal disorders

A handbook

Frank Roels, Sylvia De Bie, Ruud B.H. Schutgens and Guy T.N. Besley (eds.)

CONTENTS

(Continued on page ii)

Springer-Science+Business Media, B.V.

Contents *(continued from page i)*

J. Inher. Metab. Dis. Suppl. 1 (1995) iii
© SSIEM and Kluwer Academic Publishers.

Acknowledgements

Each year the Federation of European Biochemical Societies (FEBS), through its Advanced Course Committee, finances specialized training in many countries. The publication of this book is the outcome of the Advanced Course 'Diagnostic laboratory methods in peroxisomal disorders', organized in Gent, Belgium, in May 1994, and it greatly benefits from the grant by FEBS. The editors and authors want to express their gratitude to Professor Horst Feldmann (München), chairman of the Advanced Course Committee, and to Professor John Mowbray (London), treasurer of FEBS.

In addition, the European Commission, through its Directorate General XII for Science, Research and Development, has supported the organization of the Course by a sponsorship grant, and the editors are indebted to Dr A. E. Baert, head of the Medical Research Division (Brussels).

Under the Flemish–Spanish Cultural Agreement, these two Governments have funded the invitation of one of the special guest speakers. The Belgian Society for Cell Biology, one of the co-organizers of the Course, also contributed financially.

These grants, and the support of commercial sponsors mentioned below, have enabled the Society for the Study of Inborn Errors of Metabolism and the publisher to offer this illustrated handbook at a price which should make it available throughout the world.

Due to other commitments three specialists in the field were not able to write a chapter for this book, but they contributed actively to the Course, and therefore they are gratefully acknowledged: Dr. Patrick Willems (University of Antwerpen), Dr. François Gadisseux and Dr. Philippe Evrard (Université Catholique de Louvain, Bruxelles).

Finally the editors want to thank those who took care of the many practical tasks in the preparation, printing and distribution of the book, in particular: Mrs Nynke Coutinho and Mr Phil Johnstone of Kluwer Academic Publishers, and Mrs Erika Roosen, secretary of the Department of Human Anatomy, Embryology and Histology at the Faculty of Medicine in Gent.

Frank Roels
Sylvia De Bie
Ruud B.H. Schutgens
Guy T.N. Besley

DEDICATION

This book is dedicated to the patients and their families. Through their motivation and their physical collaboration to diagnostic research and therapeutic trials, they contribute most essentially to our scientific insight into disease and to its prevention.

THE EDITORS

Frank Roels, MD, PhD, is Professor of Human Anatomy and Embryology at the Faculty of Medicine and University Hospital, University of Gent, Belgium.
Sylvia De Bie, MSc, is scientific collaborator, Center for Medical Genetics, University Hospital, Gent, Belgium.
Ruud B.H. Schutgens, PhD, is clinical biochemist, Associate Professor and head of the Clinical Chemistry Laboratory of the Departments of Pediatrics and of Obstetrics and Gynaecology, University Hospital of the University of Amsterdam, The Netherlands.
Guy T.N. Besley, PhD, is clinical biochemist, head of laboratory, Willink Biochemical Genetics Unit, Royal Manchester Children's Hospital, Manchester, United Kingdom.

COMMERCIAL SPONSORS

Amersham
Biocell U.K.
Bioxytech
B.R.L.
Camberra-Packard Benelux N.V.
G.I.B.C.O.
Munc Plastics
Nutricia
N.V. Life Technologies
Pharmacia Biotech
Sanvertech
Shimatsu
Van Hopplinus-Leitz
V.E.L.
World Courier

J. Inher. Metab. Dis. Suppl. 1 (1995) v

Introduction

The last twenty years have witnessed the transformation of Zellweger's cerebro-hepato-renal syndrome from a clinical oddity to a central component of a significant and intriguing group of metabolic diseases which reflect the consequences of structural and enzymatic defects in peroxisomes. This extraordinary progress is the result of a synthesis of classical clinical and histopathological studies linked to those utilizing the most advanced methodology of cellular and molecular biology and biochemistry. These investigations have enriched our understanding of the essential functions of peroxisomes as well as our ability to diagnose (pre- and post-natal) affected individuals, detect carriers and in some instances develop therapeutic approaches to these disorders.

This volume is the most complete collection of laboratory procedures for the diagnosis and investigation of peroxisomal disorders. Beginning with a comprehensive approach to the clinical assessment of peroxisomal diseases, there are chapters on neuropathology, DNA diagnosis, polyunsaturated and very long chain fatty acids, phytanic and pristanic acids, plasmalogens, acyl-CoA oxidases, dihydroxyacetonephosphate acyltransferase, immunocytochemistry, morphometry and ultrastructure.

In addition to providing practical information essential for clinicians responsible for comprehensive diagnostic and counselling services, this Handbook is a provocative guide to future investigations and raises critical, unanswered questions such as clinical variability within a family with the same mutation and the relationship between genetic defect and manifestations of disease. These chapters, which contain numerous, unpublished photographs and photomicrographs (light- and electron microscopy) and detailed laboratory protocols, are both complete and up-to-date, reflecting their authors' expertise and great clinical and research laboratory experience.

Sidney Goldfischer, MD
Professor of Pathology
Associate Dean for Scientific Operations
Albert Einstein College of Medicine of Yeshiva University

J. Inher. Metabl. Dis. 18 Suppl. 1(1995) 1–18
© SSIEM and Kluwer Academic Publishers.

Clinical approach to inherited peroxisomal disorders

F. Poggi-Travert[1], B. Fournier[2], B. T. Poll-The[2] and J.-M. Saudubray[1]
[1]*Department of Pediatrics, Hôpital des Enfants-Malades, Paris, France;* [2]*Department of Metabolic Disorders, Children's Hospital, Wilhelmina Kinderziekenhuis, Utrecht, The Netherlands*

Correspondence: Hôpital des Enfants-Malades, 149 Rue de Sèvres, 75743 Paris Cedex 15, France

Summary: At least 21 genetic disorders have now been found that are linked to peroxisomal dysfunction. Whatever the genetic defect might be, peroxisomal disorders should be considered in various clinical conditions, dependent on the age of onset.

The prototype of peroxisomal disorders is represented by 'classical' Zellweger syndrome (ZS) which is the most severe disorder combining all the characteristic symptoms. ZS is characterized by the association of errors of morphogenesis, severe neurological dysfunction, neurosensory defects, regressive changes, hepatodigestive involvement with failure to thrive, usually early death, and absence of recognizable liver peroxisomes. Other peroxisomal disorders (pseudo-Zellweger syndrome, neonatal adrenoleukodystrophy (NALD), pseudo-neonatal adrenoleukodystrophy, rhizomelic chondrodysplasia punctata (RCDP), and hyperpipecolic acidaemia) share some of these symptoms, but with varying organ involvement, severity of dysfunction, and duration of survival. The diagnosis should not cause difficulty when all the characteristic manifestations are present.

Depending on the main presenting sign, peroxisomal disorders in neonates should be suspected in two categories of circumstances: polymalformative syndrome with craniofacial dysmorphism, and severe neurological dysfunction. During the first 6 months of life, the predominant symptoms may be hepatomegaly, prolonged jaundice, liver failure, anorexia, vomiting and diarrhoea leading to failure to thrive resembling a malabsorption syndrome; severe psychomotor retardation, hearing loss and ocular abnormalities become evident. Beyond 4 years of age, behavioural changes, intellectual deterioration, visual impairment and gait abnormalities may be the presenting symptoms.

Independently of the clinical symptoms and age of onset, most peroxisomal disorders described so far can be clinically screened by recordings of electroretinogram, visual-evoked responses, and brain auditory-evoked responses, which are almost always abnormal. Nine of the 17 peroxisomal disorders with neurological involvement are associated with an accumulation of very long-chain fatty acids (VLCFA), which suggests that assay of plasma VLCFA should be used as a primary test. However, assays of plasma phytanic acid and plasma/urine bile acid intermediates should also be performed in view of the recent reports of atypical chondrodysplasia variants (without rhizomelic shortening) and isolated trihydroxy-cholestanoic aciduria. The differential diagnoses in various clinical conditions and age periods are discussed.

The description of a human inherited peroxisomal disorder was initiated by Goldfischer and colleagues in 1973 by the discovery of an absence of recognizable peroxisomes in liver and kidneys in the cerebrohepatorenal syndrome of Zellweger (ZS) (McKusick 214100). Twenty years later, more than 20 human disorders have been found to be linked to inherited peroxisomal dysfunction, most of them involving neurological defects. By 1994, the nomenclature of these disorders had something of an 'inextricable puzzle' as the amount of information rapidly expanded. This was because most of the 'new' disorders were denominated by reference to the three main original clinical descriptions, namely Zellweger syndrome, neonatal adrenoleukodystrophy (McKusick 202370) (Kelley et al 1986) and infantile Refsum disease (McKusick 266510) (Scotto et al 1982; Poll-The et al 1987). This is not surprising since these three clinical conditions roughly correspond to the three main early-infantile categories of clinical symptoms found in peroxisomal disorders: predominant polymalformative syndrome (as observed in classical ZS or in rhizomelic chondrodysplasia punctata (RCDP) (McKusick 215100) (Heymans et al 1985)), predominant neurological presentation (as in neonatal adrenoleukodystrophy), and predominant hepatodigestive symptoms (as in infantile Refsum disease). Despite this rather simple clinical 'classification' of presenting signs, it is obvious that there is no relationship between clinical and biochemical phenotypes.

Similar clinical phenotypes may correspond to different biochemical lesions, as illustrated by the RCDP 'syndrome', which has been found to be associated either with four biochemical abnormalities (unprocessed peroxisomal thiolase, phytanic acid oxidation defect, acyl-CoA : dihydroxyacetone phosphate acyltransferase (DHAP-AT) and alkyl-dihydroxyacetone phosphate (alkyl-DHAP) synthase defects) (Heikoop et al 1992) or with an isolated enzyme defect (DHAP-AT, alkyl DHAP or phytanic acid oxidation). Conversely, the same biochemical defect(s), or even the same genetic complementation group, can be associated with very dissimilar clinical phenotypes, as illustrated by the classical Zellweger syndrome and infantile Refsum disease (Shimozawa et al 1993). Given the diversity of clinical and biochemical abnormalities, an arbitrary approach has been made of classifying peroxisomal disorders into two groups: those with a deficient assembly of peroxisomes with several or generalized peroxisomal dysfunctions, and those with a single deficient peroxisomal enzyme. This biochemical classification is given in Table 1 but is not useful for physicians who are faced with clinical symptoms and not with biochemical phenotypes. Thus, whatever the genetic defect might be, peroxisomal disorders should be considered in various clinical conditions depending upon the age of the patient and the type and severity of symptoms. The aim of this overview is to give the physician diagnostic clues for the clinical approach to inherited peroxisomal disorders.

Table 1   Classification of peroxisomal disorders

Enzyme defect	Peroxisomes	Disorder
I Peroxisomal assembly deficiencies		
Generalized	Absent	Classical Zellweger syndrome
		Neonatal adrenoleukodystrophy Infantile Refsum disease
Generalized (partial)	Absent	Pseudo-infantile Refsum disease
VLCFA oxidation, THCA oxidation, DHAP-AT	Present	Zellweger-like syndrome
DHAP-AT, alkyl-DHAP synthase phytanic acid oxidase, unprocessed peroxisomal thiolase	Enlarged	Rhizomelic chondrodysplasia punctata (classical/atypical phenotype)
Phytanic acid oxidase, pipecolic acid oxidase	Abnormal	Atypical Refsum disease
II Single peroxisomal enzyme deficiencies		
Isolated DHAPAT or alkyl-DHAP synthase	NA	Rhizomelic chondrodysplasia punctata
VLCFA-CoA synthetase	Normal	X-linked adrenoleukodystrophy
Acyl-CoA oxidase	Enlarged	Pseudo-NALD
Bi(tri)functional enzyme	Normal	Bifunctional enzyme deficiency
Peroxisomal thiolase	Enlarged	Pseudo-Zellweger syndrome
THCA-CoA oxidase	NA	Trihydroxycholestanoic acidaemia
Pipecolic acid oxidase	Abnormal	Isolated pipecolic acidaemia
Mevalonate kinase	NA	Mevalonic aciduria
Phytanic acid oxidase	NA	Classical Refsum disease
Peroxisomal glutaryl-CoA oxidase	Normal	Glutaric aciduria type III
Alanine: glyoxylate aminotransferase	Smaller	Hyperoxaluria type I
Mistargeting	NA	Hyperoxaluria type I
Catalase	Normal	Acatalasaemia

DHAP-AT, dihydroxyacetone phosphate acyltransferase; DHAP, dihydroxyacetone phosphate; VLCFA, very long-chain fatty acids; THCA, trihydroxycholestanoyl; pseudo-NALD, pseudo-neonatal adrenoleukodystrophy; NA = information not available

THE NEONATAL PERIOD

The prototype of peroxisomal disorders is represented by 'classical' ZS which is the most severe condition and combines all symptoms. ZS is characterized by the association of: (1) errors of morphogenesis, (2) severe neurological dysfunction, (3) neurosensory defects, (4) regression, (5) hepatodigestive involvement with failure to thrive and usually early death, and (6) absence of recognizable liver peroxisomes. The other peroxisomal disorders share some of these symptoms, but with varying organ involvement, severity of dysfunction, and duration of survival. The diagnosis should not cause difficulty when all

Table 2 Metabolic processes inducing congenital malformations

Inborn errors affecting the fetus

Peroxisomal disorders
Multiple acyl-CoA dehydrogenase deficiency
Carnitine palmitoyl transferase II deficiency
Respiratory-chain defects
Defect in valine metabolism
Cholesterol biosynthesis (mevalonic aciduria)
Smith–Lemli–Opitz syndrome
Pyruvate dehydrogenase deficiency
Mucopolysaccharidosis, mucolipidosis, sialidosis
Carbohydrate-deficient glycoprotein syndrome

Metabolic disturbances of the mother

Phenylketonuria
Alcohol
Diabetes
Drugs
Vitamin deficiencies (riboflavin)
Abnormal sensitivity to vitamin D (Williams–Beuren syndrome)

these manifestations are present. Depending on the main presenting sign, peroxisomal disorders in neonates should be suspected in two categories of circumstances: polymalformative syndrome with craniofacial dysmorphism, and severe neurological dysfunction.

Polymalformative syndrome and dysmorphism

Craniofacial abnormalities including large fontanelle, high forehead, epicanthus, abnormal ears, and high arched palate may be mistaken for Down syndrome or other chromosomal aberrations (Figures 1, 2, 14). They are frequently associated with other abnormalities such as rhizomelic dwarfism (in RCDP), calcifications of epiphyses (RCDP , ZS) (Figures 5–9), inverted nipples, renal cysts (ZS), neuronal migration defects (gyral abnormalities, neuronal heterotopias), optic nerve dysplasia, choreoretinopathy and cataract. These congenital manifestations point to an abnormal process during the prenatal period, as observed in some inborn errors affecting energy pathways of the fetus and in several conditions of metabolic dysfunction of the mother during pregnancy (Table 2).

Severe neurological dysfunctions

At birth, the predominant symptom is often a severe hypotonia with no reactivity, which can be mistaken for several pathological conditions including congenital neuromuscular disorders, disorders of the central nervous system and autonomic nervous system, and malformation syndromes (Figure 3). Furthermore, there is an increasing number of inborn errors of metabolism without evident biochemical abnormalities by routine laboratory screening that should be considered (Table 3). Severe axial hypotonia may be associated with a neurological distress with hypertonia of the limbs and seizures. Clinically, it may

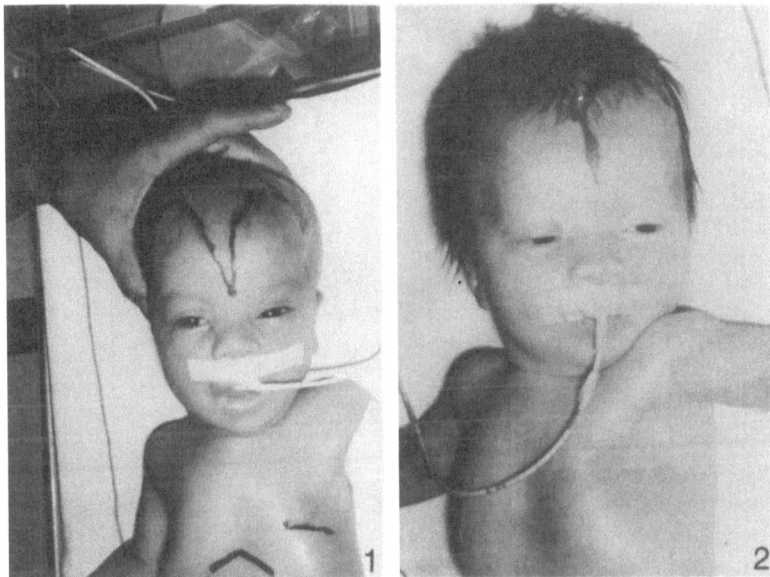

Figure 1 Classical Zellweger syndrome: 2 weeks old, large fontanelle, epicanthal folds, hypertelorism, prominent forehead
Figure 2 Classical Zellweger syndrome: 1 week old, facial dysmorphism similar to Figure 1

be difficult to differentiate between a mitochondrial respiratory-chain disorder and a peroxisomal disease. An important difference is that peroxisomal disorders are not associated with an acute metabolic event or abnormal routine laboratory tests, such as metabolic acidosis or hyperlacticacidaemia.

Peroxisomal disorders in the neonatal period (Table 4)

According to the current nomenclature, peroxisomal disorders with a clinical onset in the neonatal period are: (1) ZS, (2) pseudo-ZS (McKusick 261510) (Goldfischer et al 1986), (3) NALD, (4) pseudo-NALD (McKusick 264470) (Poll-The et al 1988a), (5) RCDP, and (6) hyperpipecolic acidaemia (McKusick 239400). Classical ZS with all the symptoms and RCDP characterized by rhizomelic dwarfism, typical facial appearance, congenital cataracts, and joint contractures (Heymans et al 1985) should be diagnosed easily. Neonatal adrenoleukodystrophy clinically appears as a milder form of ZS. Craniofacial dysmorphism is not always present and NALD patients may show some developmental progress before their progressive deterioration and death, usually within the first 6 years (Figure 4). Cerebral demyelination is more impressive than grey-matter heterotopias, and there is an absence of chondrodysplasia and renal cysts (Kelley et al 1986). Pseudo-ZS and pseudo-NALD have only been observed in a few families. Pseudo-ZS is clinically indistinguishable from the classical ZS (Goldfischer et al 1986); however, liver peroxisomes were abundant and there is a specific defective activity of the peroxisomal enzyme thiolase (Schram et al 1987). Pseudo-NALD displays no dysmorphia and shows a clinical presentation very similar to that observed in NALD (Figure 10); however, liver

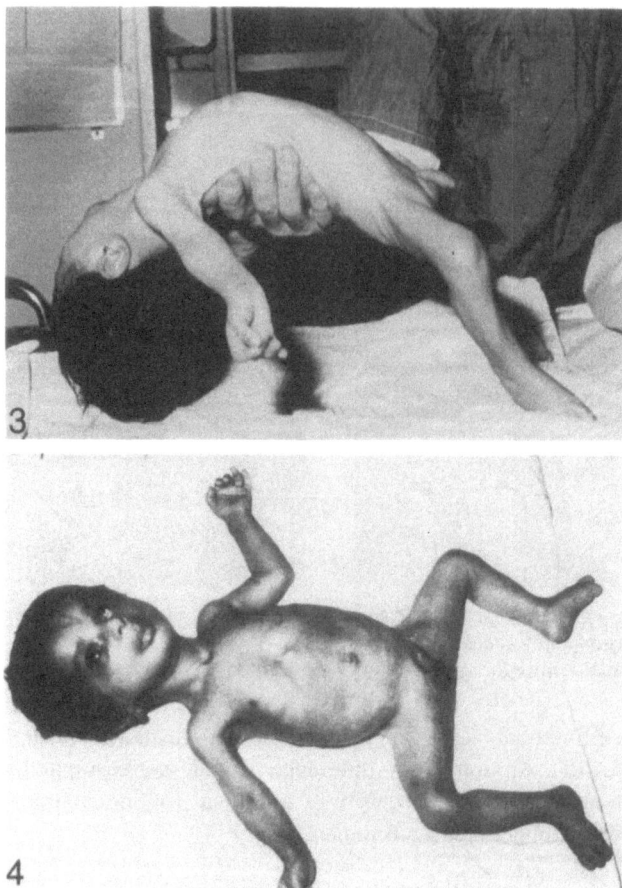

Figure 3 Classical Zellweger syndrome: 2 weeks old, severe hypotonia
Figure 4 Neonatal adrenoleukodystrophy: 3 weeks old, large fontanelle, absence of facial dysmorphia, severe hypotonia

peroxisomes were present and enlarged in size. A specific defect of the peroxisomal acyl-CoA oxidase activity has been found in pseudo-NALD (Poll-The et al 1988a; Fournier et al 1994).

The term hyperpipecolic acidaemia was assigned to patients on the basis of the observed accumulation of pipecolic acid prior to the discovery of the generalized peroxisomal defects (Gatfield et al 1968; Thomas et al 1975; Burton et al 1981). However, hyper-pipecolic acidaemia should be assigned only to patients with only elevated pipecolic acid in body fluids. Hyperpipecolic acidaemia associated with a Joubert syndrome has been observed in three siblings (Poll-The et al 1988b).

Mevalonic aciduria, a disorder with dysmorphic features, cataracts and mental retardation, should be considered as a peroxisomal disorder since mevalonate kinase is predominantly localized in peroxisomes (Hoffmann et al 1986; Biardi et al 1994).

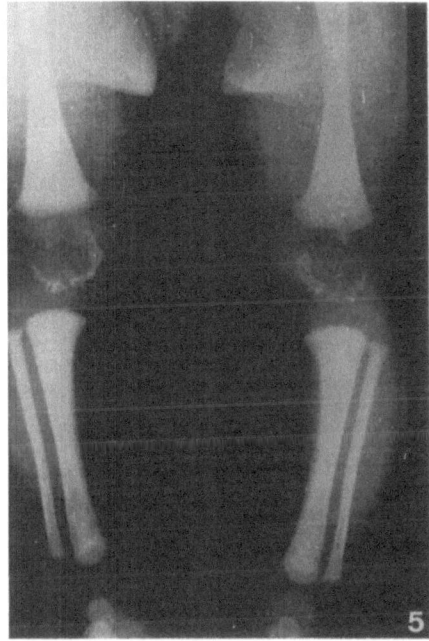 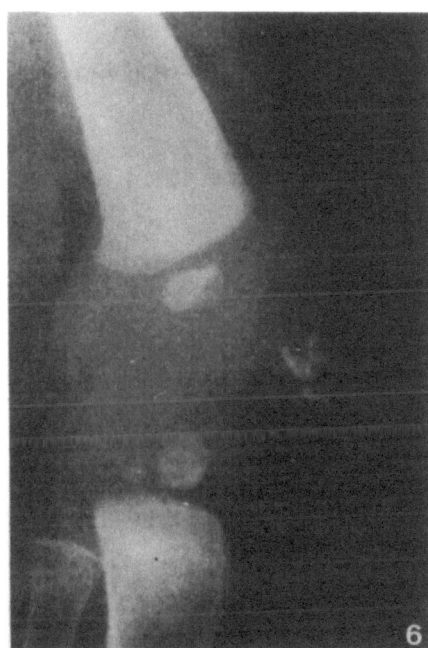

Figure 5   Classical Zellweger syndrome: abnormal calcifications of knees
Figure 6   Classical Zellweger syndrome: abnormal calcifications of knee

Table 3   Differential diagnosis in the neonatal period

Disorders with hypotonia	Inborn errors of metabolism with hypotonia/seizures
• *Neuromuscular disorders* Werdnig–Hoffmann disease Congenital myopathies Congenital muscular dystrophies Congenital myasthenia Congenital polyneuropathy Myotonic dystrophy	Non-ketotic hyperglycinaemia D-Glyceric acidaemia Sulphite oxidase deficiency 3-Methylglutaconic aciduria GABA-transaminase deficiency Cobalamin metabolism defects (Cbl c, Cbl d) Copper metabolism defect (Menkes syndrome)
• *Dysautonomia* Riley–Day syndrome	Biopterin deficiencies Inherited disorders of folate metabolism Respiratory-chain defects
• *Central nervous system disorders* Chromosomal aberrations Prader–Willi syndrome Lowe syndrome	Multiple carboxylase deficiency 3-Methylcrotonyl glycinuria Pompe disease (no seizures) Carbohydrate-deficient glycoprotein syndrome

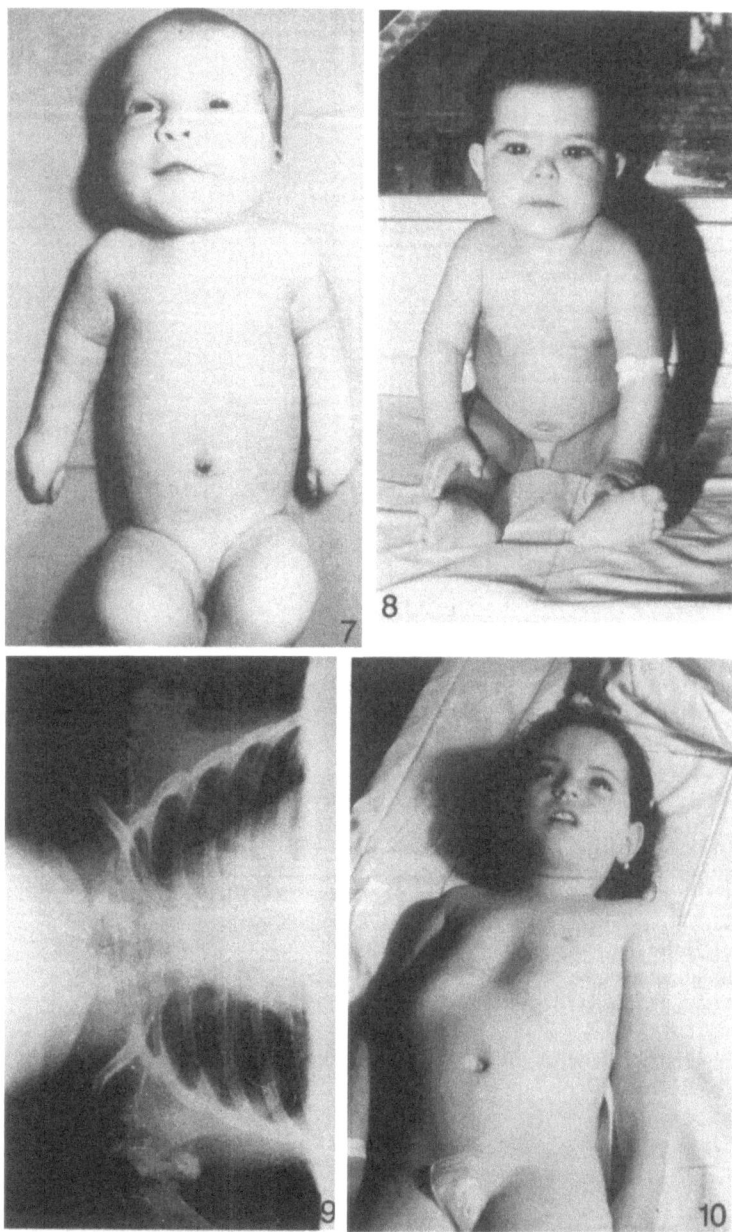

Figure 7 Rhizomelic chondrodysplasia punctata: rhizomelic dwarfism (upper arm and thigh), facial dysmorphia

Figure 8 Atypical RCDP: absence of rhizomelia, absence of facial dysmorphia

Figure 9 Atypical RCDP: abnormal calcifications of shoulders, 2 weeks old

Figure 10 Pseudo-NALD: 4 years old, absence of facial dysmorphia

Table 4   Clinical symptoms of peroxisomal disorders related to age: Part 1

Symptoms	*Disorder*
Neonatal period	
Hypotonia	Zellweger syndrome and variants
A reactivity	Neonatal adrenoleukodystrophy
Seizures	Acyl-CoA oxidase deficiency
Craniofacial dysmorphia	Bifunctional enzyme deficiency
Chondrodysplasia	Rhizomelic chondrodysplasia punctata (typical/atypical)
	Trihydroxycholestanoic acidaemia
	Pipecolic acidaemia
	Mevalonic acidaemia
First 6 months	
Hepatomegaly	Infantile Refsum disease (variants)
Liver failure	Pipecolic acidaemia
Prolonged jaundice	Mevalonic acidaemia
Digestive problems	Neonatal adrenoleukodystrophy
Failure to thrive	Zellweger syndrome (milder forms)
Hypocholesterolaemia	Rhizomelic chondrodysplasia punctata
Osteoporosis	(atypical)
Visual abnormalities	Trihydroxycholestanoic acidaemia

THE FIRST SIX MONTHS OF LIFE

During the first 6 months of life, the predominant symptoms may be hepatomegaly associated or not associated with prolonged jaundice, liver failure and non-specific digestive problems (anorexia, vomiting, diarrhoea) leading to failure to thrive and osteoporosis. Frequently it is associated with hypolipoproteinaemia and decreased plasma values for the fat-soluble vitamins resembling a malabsorption syndrome. In view of this, it is interesting to remember that the three original infantile Refsum disease (IRD) patients described were suspected of having hypo-betalipoproteinaemia, vitamin E deficiency, or cholestatic jaundice (Scotto et al 1982) (Tables 4 and 5).

The peroxisomal disorders clinically expressed in this period of life are IRD, NALD, hyperpipecolic acidaemia, and possibly a few mild ZS variants. In this period, most of the classical ZS patients are developing symptoms such as hepatomegaly and seizures, and die (Table 4). Biochemically, as well as with respect to the absence or significantly decreased number of liver peroxisomes, infantile Refsum disease is similar to ZS (Ogier et al 1985; Poll-The et al 1986; Roels et al 1986). Although IRD patients also share some clinical features with ZS, they differ from ZS with respect to age of onset, initial symptoms, degree of dysfunction and duration of survival (Poll-The et al 1987) (Figures 11–16).

BETWEEN SIX MONTHS AND THREE YEARS OF LIFE

Between 6 months and 3 years, nystagmus, neurosensory deficit and severe psychomotor retardation become evident. Other frequent symptoms are cryptorchidism and clitoris

Table 5   Differential diagnosis in the first 6
months of life

Failure to thrive
• Malabsorption syndromes
  Coeliac disease
  Hypobetalipoproteinaemia
  Cow-milk intolerance
  Hepatic fibrosis with exudative enteropathy
  Respiratory-chain defects
  Carbohydrate-deficient glycoprotein syndrome
• Cholestatic jaundice
  Vitamin E deficiency
  Inborn errors of bile acid metabolism
  Niemann–Pick type C
  Byler disease
  α_1-Antritrypsin deficiency
  Carbohydrate-deficient glycoprotein syndrome
• Hepatic failure
  Respiratory-chain defects
  Tyrosinaemia type I
  Carbohydrate-deficient glycoprotein syndrome

Non-specific failure to thrive
Urea-cycle defects
Organic acidurias
Respiratory-chain defects
Carbohydrate-deficient glycoprotein syndrome

---

hypertrophy. Hearing loss with abnormal brain auditory-evoked responses is consistently present. A number of ocular abnormalities can be observed including cataract, retinitis pigmentosa, optic nerve atrophy or dysplasia, glaucoma and brush-field spots. Electro-retinogram recording and visual-evoked responses are consistently disturbed, sometimes before evidence of severe visual impairment. Retinitis pigmentosa associated with hearing loss and a severe developmental delay, or dysmorphia may be mistaken for several malformation syndromes or inborn errors of metabolism (Tables 6 and 7). In this respect, it can be pointed out that the limits between malformation syndromes and inborn errors are not well delineated, as suggested by the recent finding of a cholesterol biosynthesis defect in the Smith–Lemli–Opitz syndrome (Tint et al 1994). The peroxisomal disorders clinically expressed in this period are IRD, NALD, hyperpipecolic acidaemia, pseudo-ZS, pseudo-NALD, and surviving ZS and RCDP. Most of the NALD patients die in this period after a transient and partial development (Table 6).

BEYOND THE AGE OF FOUR YEARS (Table 6)

X-linked adrenoleukodystrophy should be considered in a prepubertal boy who displays changes in behaviour and deterioration of intellectual functions, vision, and gait. The illness is progressive and will be followed by spastic quadriplegia or hemiplegia evolving to a variable degree of vegetative state and usually causes death in 1–4 years. In a

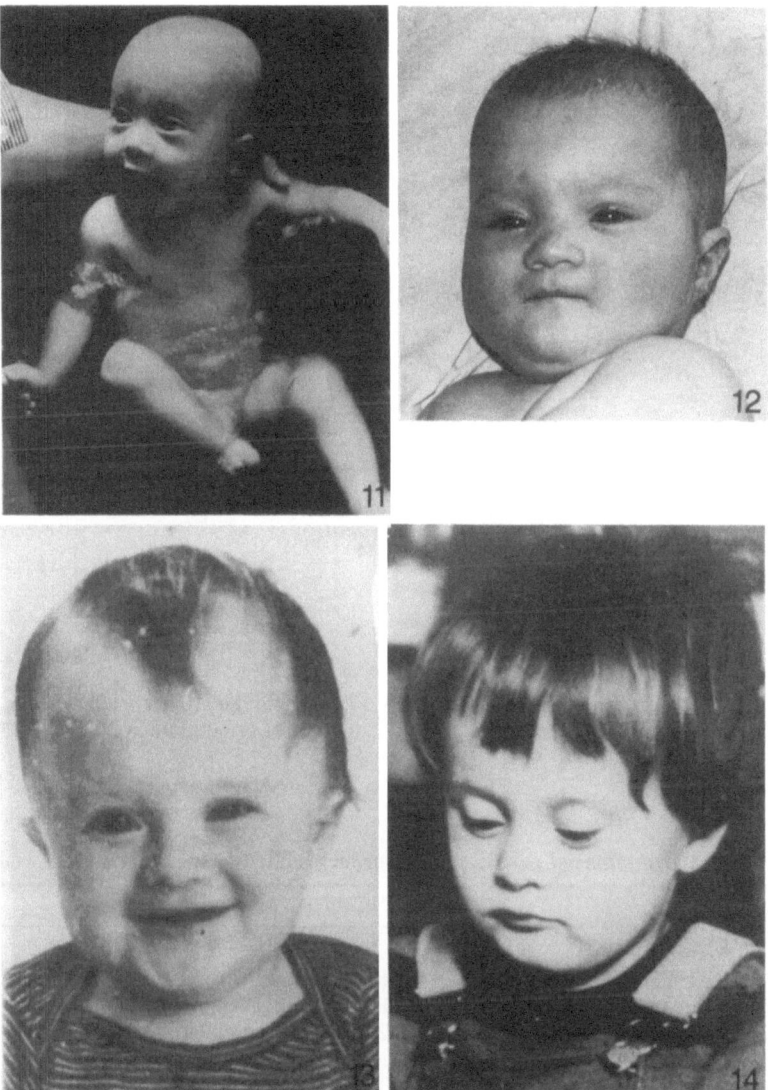

Figure 11 Infantile Refsum disease: 3 months old, normal development
Figure 12 Infantile Refsum disease: 6 months
Figure 13 Infantile Refsum disease: 11 months old, prominent forehead, normal development
Figure 14 Infantile Refsum disease: 2 years old, resembling Down syndrome

minority, signs of adrenal failure accompanied or not by cutaneous pigmentation precede signs of disease in the central nervous system. There is extensive demyelination of cerebral white matter and atrophy of the adrenal cortex (Moser et al 1992; Moser 1995).

Biochemically the disorder is characterized by accumulation of VLCFA in plasma and tissues probably due to a specific defective protein involved in the transport of VLCFA-

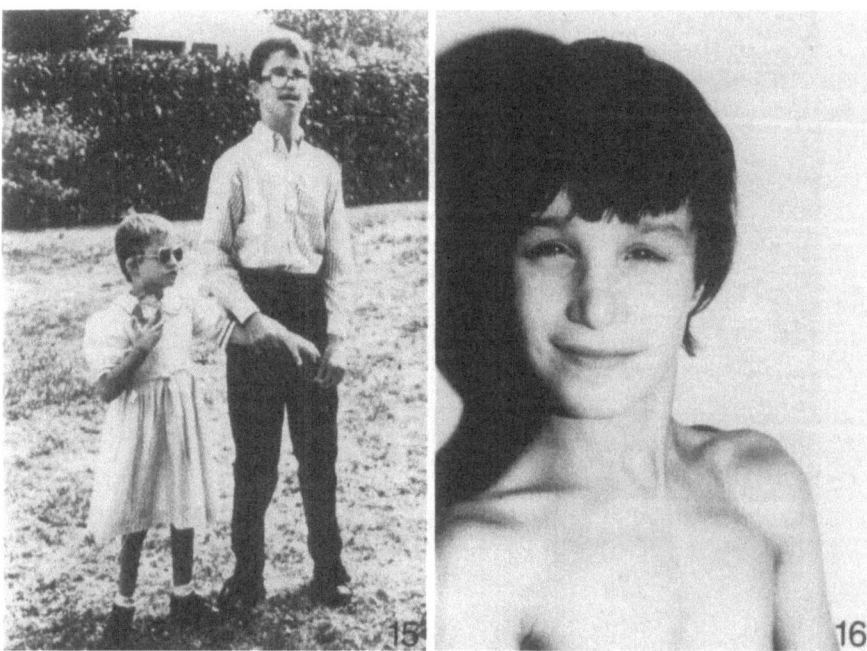

Figure 15 Infantile Refsum disease siblings: boy 17 years old, girl 10 years old
Figure 16 Infantile Refsum disease: 11 years old, absence of facial dysmorphia

Table 6   Clinical symptoms of peroxisomal disorders related to age: Part 2

Symptoms	Disorder
6 months to 4 years	
Neurological presentation	Infantile Refsum disease (pseudo)
Psychomotor retardation	Pipecolic acidaemia
Visual, hearing impairment	Zellweger syndrome (milder)
(electroretinogram,	Rhizomelic chondrodysplasia punctata (atypical)
brain auditory-evoked potentials)	Trihydroxycholestanoic acidaemia
Failure to thrive	
Osteoporosis	
Beyond 4 years	
Behavioural changes	
Intellectual deterioration	X-linked ALD
White-matter demyelination	
Retinitis pigmentosa	
Peripheral neuropathy	
Gait abnormality	Classical Refsum
Ichthyosis	

Table 7 **Differential diagnosis of disorders with retinitis pigmentosa between 6 months and 3 years of life**

Syndrome	Other symptoms
Syndromes with developmental delay	
Joubert	Abnormal respiration, vermis aplasia, hearing deficit, seizures, dysmorphia
Cockayne	Intracranial calcifications, failure to thrive, senile facies, hearing deficit, dermatitis
Sjögren–Larssen (fatty alcohol oxidoreductase deficiency)	Congenital ichthyosis, mental retardation, spastic paraplegia, cataract
Hallgren	Ataxia, hearing deficit
Usher	(Seizures), hearing deficit
Laurence–Moon Biedl	Obesity, hypogenitalism, polydactyly
Flynn–Aird	Muscle wasting, hearing deficit, skin atrophy, baldness, ataxia, seizures
Inborn errors of metabolism	
Abetalipoproteinaemia and vitamin E deficiency	Fat malabsorption, ataxia, peripheral neuropathy (acanthocytosis)
Ceroid lipofuscinosis (infantile types)	Mental regression, seizures, autistic behaviour (microcephaly)
Kearn–Sayre syndrome (respiratory-chain defect)	Ophthalmoplegia, hearing deficit, muscle weakness cardiomyopathy, lactic acidosis
Abnormal purine metabolism (PRPP synthetase overreactivity)	Hearing deficit, mental retardation, congenital disequilibrium
Carbohydrate-deficient glycoprotein syndrome	Mental retardation, hepatic fibrosis, cutaneous signs

CoA synthetase into the peroxisomal membrane or a protein that is associated with the VLCFA-CoA synthase in the membrane (Mosser et al 1993). The number and appearance of liver peroxisomes are normal in childhood ALD.

Changes in behaviour associated with vision impairment may clinically be mistaken for hysteria. Intellectual deterioration in this period of life may be related to many other regressive encephalopathies (Table 8). Diseases such as Sanfilippo (MPS III), Niemann–Pick type C, Wilson disease, subacute sclerosing panencephalitis, and multiple sclerosis are well-known conditions. However, the basic defect of ceroid lipofuscinosis and neuroaxonal dystrophy remains to be elucidated.

'Classical' Refsum disease (McKusick 266500) is another well-known disorder probably involving the peroxisome. The main clinical symptoms are peripheral neuropathy, ataxia, retinitis pigmentosa, ichthyosis, anosmia and nerve deafness, and elevated protein concentration in the CSF, associated with normal intellectual functioning. Biochemically the disorder is characterized by increased plasma phytanic acid concentration, due to a deficiency of phytanic acid α-oxidation, which can be assayed in

Table 8   Differential diagnosis of X-linked childhood adrenoleukodystrophy

Sanfilippo disease (mucopolysaccharidosis type III)
Niemann–Pick disease type C
Metachromatic leukodystrophy
Wilson disease
Juvenile GM2-gangliosidosis
Subacute sclerosing panencephalitis
Multiple sclerosis
Neuroaxonal dystrophy (juvenile form)
Ceroid lipofuscinosis (juvenile type)

cultured skin fibroblasts. A large heterogeneity of features has been described in Refsum patients. The onset of clinical manifestations varies from childhood to the fifth decade. Some patients appear to have a partial deficiency of a peroxisomal enzyme involved in ether-phospholipid biosynthesis and an accumulation of VLCFA, whereas others do not (Van Crugten et al 1986).

Recently, four related patients have been reported, three with classical Refsum disease and the fourth who died from a progressive neurological disorder with clinical and neuropathological abnormalities unusual for classical Refsum disease. Beside the phytanic acid oxidation defect, there was an increase of plasma pipecolic acid in two of these patients (Tranchant et al 1993).

DIAGNOSTIC PROCEDURES

Independently of the clinical symptoms and age of onset, most peroxisomal disorders described so far can be clinically screened by the recordings of electroretinogram, visual-evoked responses, and brain auditory-evoked responses, which are almost always abnormal. Table 9 lists a variety of diagnostic assays that are available for the diagnosis of peroxisomal disorders. Only urinary pipecolic acid excretion, medium- and long-chain dicarboxylic aciduria, hyperoxaluria and mevalonic aciduria can be detected by a general metabolic screening. Nine of the 17 peroxisomal disorders with neurological involvement are associated with an accumulation of VLCFA, which suggests that assay of plasma VLCFA should be used as a primary test. However, assays of plasma phytanic acid and plasma/urine bile acid intermediates should also be performed in view of the recent reports of atypical chondrodysplasia variants (without rhizomelic shortening) and isolated trihydroxycholestanoic aciduria. The clinical presentations of the typical phenotype of RCDP (phytanic acid, plasmalogens) and classical Refsum (phytanic acid) are distinct from the other disorders and should not cause difficulties in their diagnosis. To elucidate whether the accumulation of VLCFA in a patient's plasma results from a defect in peroxisome biogenesis or is caused by a defect in one of the peroxisomal β-oxidation enzyme activities, additional assay procedures must be carried out, in particular plasmalogen levels and immunoblotting of peroxisomal β-oxidation proteins. In fact, in atypical cases, it would be advisable to carry out assays of plasma VLCFA, bile acid intermediates, phytanic, pristanic and pipecolic acid, and DHAP-AT and alkyl-DHAP synthase in cultured skin fibroblasts. However, in some variant forms, the enzymatic

Table 9 Diagnostic assays in peroxisomal disorders

Disease	Material	Type of assay
Classical ZS	Plasma	VLCFA, bile acid, phytanic acid,
Neonatal ALD		pristanic acid, pipecolic acid,
Infantile Refsum		polyunsaturated fatty acids
Zellweger-like	Erythrocytes	Plasmalogens
Pseudo-infantile Refsum	Fibroblasts	Plasmalogens biosynthesis, DHAP-AT,
		alkyl DHAP synthase
		Particle bound catalase
		VLCFA β-oxidation, immunoblotting
		β-oxidation proteins
		Phytanic acid oxidation
	Liver	Cytochemistry
RCDP (classical/	Plasma	Phytanic acid
atypical phenotypes)	Erythrocytes	Plasmalogens
	Fibroblasts	DHAP-AT, alkyl DHAP synthase
		Phytanic acid oxidation
	Liver	Immunoblot thiolase
		Cytochemistry
Isolated peroxisomal	Plasma	VLCFA, bile acids
β-oxidation defects	Fibroblasts	VLCFA, β-oxidation, immunoblotting
		β-oxidation proteins
	Liver	Immunolocalization
Isolated defect of	Plasma	Bile acids
bile acid synthesis	Liver	THCA-CoA oxidase
Isolated pipecolic acidaemia	Plasma	Pipecolic acid
	Liver	Pipecolic acid oxidase
Mevalonic aciduria	Plasma	Organic acids
	Urine	Organic acids
	Fibroblasts	Mevalonate kinase
	Lymphocytes	Mevalonate kinase
Classical Refsum	Plasma	Phytanic acid
	Fibroblasts	Phytanic acid oxidation
Glutaric aciduria type III	Urine	Organic acids
	Liver	Glutaryl-CoA oxidase
Hyperoxaluria type I	Urine	Organic acids
	Liver	AGT; immunocytochemistry
Acatalasaemia	Erythrocytes	Catalase

VLCFA, very long-chain fatty acids; DHAP-AT, dihydroxyacetone phosphate acyltransferase; DHAP, dihydroxyacetone phosphate; THCA, trihydroxycholestanoyl; AGT, alanine glyoxylate aminotransferase;

deficit(s) are expressed only in liver and not in cultured fibroblasts (Espeel et al 1993, 1995; Mandel et al 1994; Roels et al 1993; Schutgens et al 1994b).

Although, in some cases, levels of metabolites in CSF from patients exceed the control range, measurement of VLCFA, bile acids, pristanic acid and phytanic acid does not seem to provide a diagnostic advantage since all measurements can be performed more conveniently in plasma (ten Brink et al 1993a). For some disorders, a retrospective diagnosis can be obtained by analysing stored blood spots collected during neonatal screening (Jakobs et al 1993; ten Brink et al 1993b).

The detection of peroxisomes is facilitated by using the diaminobenzidine staining procedure, which reacts with the peroxisomal marker enzyme catalase, and by immunochemical and immunocytochemical *in situ* techniques with antibodies against matrix and membrane peroxisomal proteins (Roels et al 1991, 1993). The abundance, size and structure of liver peroxisomes should be studied. When peroxisomes are lacking, virtually all of the catalase is present in the cytosolic fraction, instead of the particulate fraction (Roels et al 1993). Hepatic inclusions of fat insoluble in acetone, of polarizing material and of trilamellar structures are diagnostically significant (Roels et al 1993; 1995) and can be detected in autopsy or archival material.

A variety of techniques are available for prenatal diagnosis. Almost all the peroxisomal disorders can be identified prenatally, either by using (cultured) first-trimester chorionic villus samples or amniocytes, or by direct analysis of levels of VLCFA and bile acid intermediates in amniotic fluid (Wanders et al 1988). Today, detection of ZS, NALD, IRD and RCDP is done directly in chorion biopsy material by analysis of DHAP-AT activity and by immunoblotting with anti-peroxisomal thiolase (Schutgens et al 1994a). Another approach is the cytochemical staining of peroxisomes in fresh chorionic villus samples (Roels et al 1987, 1995). In the case of THCA-CoA oxidase deficiency, isolated pipecolic acidaemia, glutaric aciduria type III and hyperoxaluria type I, a fetal liver biopsy is required. Prenatal diagnosis using DNA analysis has been successful in X-ALD and acyl-CoA-oxidase deficiency (Boué et al 1985; Fournier et al 1994; see also Seneca and Lissens 1995).

Heterozygote identification is available for X-linked ALD using VLCFA analysis or restriction fragment polymorphism analyses of DNA (Aubourg et al 1987; Seneca and Lissens 1995).

REFERENCES

Aubourg PR, Sack GH, Meyers DA, Lease JJ, Moser HW (1987) Linkage of adrenoleukodystrophy to a polymorphic DNA probe. *Ann Neurol* **21**: 349–352.

Biardi L, Sreedhar A, Zokaei A et al (1994) Mevalonate kinase is predominantly localised in peroxisomes and is defective in patients with peroxisome deficiency disorders. *J Biol Chem* **269**: 1197–1205.

Boué J, Oberlé I, Heilig R et al (1985) First trimester prenatal diagnosis of adrenoleukodystrophy by determination of very long chain fatty acid levels and by linkage analysis to DNA probe. *Hum Genet* **69**: 272–274.

Burton BK, Reed SP, Remy WT (1981) Hyperpipecolic acidemia: clinical and biochemical observations in two male siblings. *J Pediatr* **99**: 729–734.

Espeel M, Heikoop JC, Smeitink JAM et al (1993) Cytoplasmic catalase and ghostlike peroxisomes in the liver from a child with atypical chondrodysplasia punctata. *Ultrastruct Pathol* **17**: 623–636.

Espeel M, Mandel H, Poggi F, et al. (1995) Peroxisome mosaicism in the liver of peroxisomal deficiency patients. *Hepatology* **22**: 497–504.

Fournier B, Saudubray JM, Benichou B et al (1994) Large deletion of the peroxisomal acyl-CoA oxidase gene in pseudoneonatal adrenoleukodystrophy. *J Clin Invest* **94**: 526–531.

Gatfield PD, Taller E, Hinton GG, Wallace AC, Abdelnour GM, Haust MD (1968) Hyperpipecolatemia: a new metabolic disorder associated with neuropathy and hepatomegaly. *Can Med Assoc J* **99**: 1215–1233.

Goldfischer SL, Moore CL, Johnson AB et al (1973) Peroxisomal and mitochondrial defects in the cerebro-hepato-renal syndrome. *Science* **182**: 62–64.

Goldfischer SL, Collins J, Rapin I et al (1986) Pseudo-Zellweger syndrome: deficiencies in several peroxisomal oxidative activities. *J Pediatr* **108**: 25–32.

Heikoop JC, Wanders RJA, Strijland A, Purvis R, Schutgens RBH, Tager JM (1992) Genetic and biochemical heterogeneity in patients with the rhizomelic form of chondrodysplasia punctata — a complementation study. *Hum Genet* **89**: 439–444.

Heymans HSA, Oorthuys JWE, Nelck G, Wanders RJA, Schutgens RBH (1985) Rhizomelic chondrodysplasia punctata: another peroxisomal disorder. *N Engl J Med* **313**: 187–188.

Hoffmann GF, Gibson KM, Brandt IK, Bader PI, Wappner RS, Sweetman L (1986) Mevalonic aciduria — an inborn error of cholesterol and nonsterol isoprene biosynthesis. *N Engl J Med* **314**: 1610–1614.

Jakobs C, Van den Heuvel CMM, Stellaard F, Largilliere C, Skovby F, Christensen E (1993) Diagnosis of Zellweger syndrome by analysis of very long-chain fatty acids in stored blood spots collected at neonatal screening. *J Inher Metab Dis* **16**: 63–66.

Kelley RI, Datta NS, Dobyns WB et al (1986) Neonatal adrenoleukodystrophy: new cases, biochemical studies and differentiation from Zellweger and related peroxisomal polydystrophy syndromes. *Am J Med Genet* **23**: 869–901.

Mandel H, Espeel M, Roels F et al (1994) A new type of peroxisomal disorder with variable expression in liver and fibroblasts. *J Pediatr* **125**: 549–555.

Moser HW (1995) Adrenoleukodystrophy: natural history, treatment and outcome. *J Inher Metab Dis* **18**: 435–447.

Moser HW, Moser AB, Smith KD et al (1992) Adrenoleukodystrophy: phenotype variability and implications of therapy. *J Inher Metab Dis* **15**: 645–664.

Mosser J, Douar AM, Sarde CO et al (1993) Putative X-linked adrenoleukodystrophy gene shares unexpected homology with ABC transporters. *Nature* **361**: 726–730.

Ogier H, Roels F, Cornelis A et al (1985) Absence of hepatic peroxisomes in a case of infantile Refsum's disease. *Scand J Clin Lab Invest* **45**: 767–768.

Poll-The BT, Saudubray JM, Ogier H et al (1986) Infantile Refsum's disease: biochemical findings suggesting multiple peroxisomal dysfunction. *J Inher Metab Dis* **9**: 169–174.

Poll-The BT, Saudubray JM, Ogier H et al (1987) Infantile Refsum disease: an inherited peroxisomal disorder. Comparison with Zellweger syndrome and neonatal adrenoleukodystrophy. *Eur J Pediatr* **146**: 477–483.

Poll-The BT, Roels F, Ogier H et al (1988a) A new peroxisomal disorder with enlarged peroxisomes and a specific deficiency of acyl-CoA oxidase (pseudo-neonatal adrenoleukodystrophy). *Am J Hum Genet* **42**: 422–434.

Poll-The BT, Lombes A, Lenoir G et al (1988b) *Joubert's syndrome associated with hyper-pipecolatemia. Three siblings.* PhD thesis, University of Amsterdam, 201–219.

Roels F, Cornelis A, Poll-The BT et al (1986) Hepatic peroxisomes are deficient in infantile Refsum disease: a cytochemical study of 4 cases. *Am J Med Genet* **25**: 257–271.

Roels F, Verdonck V, Pauwels M et al (1987) Light microscopic visualization of peroxisomes and plasmalogens in first trimester chorionic villi. *Prenat Diagn* **7**: 525–530.

Roels F, Espeel M, De Craemer D (1991) Liver pathology and immunocytochemistry in peroxisomal disorders: A review. *J Inher Metab Dis* **14**: 853–875.

Roels F, Espeel M, Poggi F et al (1993) Human liver pathology in peroxisomal diseases: A review including novel data. *Biochimie* **75**: 281–292.

Roels F, De Prest B, De Pestel G (1995) Liver and chorion cytochemistry. *J Inher Metab Dis* **18**: (**Suppl. 1**): 155–171.

Schram AW, Goldfischer SL, Van Roermund CWT et al (1987) Human peroxisomal 3-oxoacyl-coenzyme A thiolase deficiency. *Proc Natl Acad Sci USA* **84**: 2494–2497.

Schutgens RBH, Dekker C, Mooijer P, Wanders RJA (1994a) First trimester prenatal diagnosis of peroxisome deficiency disorders (PDD) and rhizomelic chondrodysplasia punctata (RCDP). *Abstracts of the 32nd SSIEM Symposium.* Society for the Study of Inborn Errors of Metabolism. Kluwer Academic Publishers (ISBN 1 870617), 238.

Schutgens RBH, Wanders RJA, Jakobs C et al (1994b) A new variant of Zellweger syndrome with normal peroxisomal functions in cultured fibroblasts. *J Inher Metab Dis* **17**: 319–322.

Scotto J, Hadchouel M, Odièvre M et al (1982) Infantile phytanic acid storage disease, a possible variant of Refsum's disease: three cases, including ultrastructural studies of liver. *J Inher Metab Dis* **5**: 83–90.

Seneca S, Lissens W (1995) DNA diagnosis of X-linked adrenoleukodystrophy. *J Inher Metab Dis* **18 (Suppl. 1)**: 34–44.

Shimozawa N, Suzuki Y, Orii T, Moser AB, Moser HW, Wanders RJA (1993) Standardization of complementation grouping of peroxisome-deficient disorders and the second Zellweger patient with peroxisomal assembly factor-I (PAF-I) defect. *Am J Hum Genet* **52**: 843–844.

ten Brink HJ, Van den Heuvel CMM, Poll-The BT, Wanders RJA, Jakobs C (1993a) Peroxisomal disorders: concentrations of metabolites in cerebrospinal fluid compared with plasma. *J Inher Metab Dis* **16**: 587–590.

ten Brink HJ, Van den Heuvel CMM, Christensen E, Largilliere C, Jakobs C (1993b) Diagnosis of peroxisomal disorders by analysis of phytanic and pristanic acids in stored blood spots collected at neonatal screening. *Clin Chem* **39**: 1904–1906.

Thomas GH, Haslam RHA, Bashaw ML, Capute AJ, Neidengard L, Ransom JL (1975) Hyper-pipecolic acidemia associated with hepatomegaly, mental retardation, optic nerve dysplasia and progressive neurological disease. *Clin Genet* **8**: 376–382.

Tint GS, Irons M, Elias ER et al (1994) Defective cholesterol biosynthesis associated with the Smith–Lemli–Opitz syndrome. *N Engl J Med* **330**: 107–113.

Tranchant C, Aubourg PR, Mohr M, Rocchiccioli F, Zaenker C, Warter JM (1993) A new peroxisomal disease with impaired phytanic and pipecolic acid oxidation. *Neurology* **43**: 2044–2048.

Van Crugten JT, Paton B, Poulos A (1986) Partial deficiency of dihydroxyacetone phosphateacyltransferase activity in both classical and infantile Refsum's disease. *J Inher Metab Dis* **9**: 163–168.

Wanders RJA, Heymans HSA, Schutgens RBH et al (1988) Peroxisomal disorders in neurology. *J Neurol Sci* **88**: 1–39.

J. Inher. Metab. Dis. 18 Suppl. 1 (1995) 19–33

Neuropathology of peroxisomal diseases

J.-J. MARTIN
Born-Bunge Foundation and University of Antwerp, Universiteitsplein 1, B-2610 Wilrijk, Antwerp, Belgium

Summary: This paper gives a description of the essential neuropathological techniques applied to the study of metabolic disorders affecting the nervous system. Subsequently, the neuropathological features of a series of peroxisomal disorders are described with special attention being paid to adrenoleukodystrophy.

Peroxisomes contain enzymes for essential reactions in a number of different metabolic pathways; they thus play an indispensable role in intermediary metabolism. Many different human disease states may be associated with abnormal peroxisomal functions. The Zellweger cerebrohepatorenal syndrome (McKusick 214100) represents, for example, one of the most serious peroxisomal diseases since it is associated with malfunction of virtually every organ. Children with the disease usually do not survive beyond the fourth month of life (Moser et al 1991).

Peroxisomal diseases include: (1) those in which peroxisomes are virtually absent, leading to a generalized impairment of peroxisomal functions (the cerebrohepatorenal syndrome of Zellweger, neonatal adrenoleukodystrophy (McKusick 202370), infantile Refsum disease (McKusick 266510), hyperpipecolic acidaemia (McKusick 239400); (2) those in which peroxisomes are present and several peroxisomal functions are impaired (the rhizomelic form of chondrodysplasia punctata (McKusick 215100), combined peroxisomal β-oxidation enzyme protein deficiency, etc.); and (3) those in which peroxisomes are present and only a single peroxisomal function is impaired, such as X-linked adrenoleukodystrophy (McKusick 300100), peroxisomal thiolase deficiency or pseudo-Zellweger syndrome (McKusick 261510), acyl-CoA oxidase deficiency or pseudo-neonatal adrenoleukodystrophy (McKusick 264470), and, possibly, the classical form of Refsum disease (McKusick 266500) (Wanders et al 1988; Tager et al 1990).

Biochemical, immunohistochemical and complementation studies (Tager et al 1990) are, of course, of crucial importance. Neuropathology illustrates the terminal stages of these disorders, while magnetic resonance imaging (MRI) offers the unique opportunity to study the progression of the damage in the central nervous system. Many peroxisomal disorders affect the central nervous system, some of them producing demyelination while others induce neuronal migration disturbances. The MRI pattern of X-linked adrenoleukodystrophy is now well known, but this cannot be said of other peroxisomal disorders (van der Knaap and Valk 1991).

Recent progress made in molecular biology will improve our knowledge concerning the defective gene products (Mosser et al 1993).

TECHNICAL FEATURES

Biopsies to study conditions affecting the central and peripheral nervous system include sampling of skin, conjunctiva, peripheral nerve, skeletal muscle, rectum and brain. Brain biopsies have been largely used in the past but have lost much of their importance except for the diagnosis of brain tumours. Indeed, in neurological disorders the sampling of easily available tissues such as skin, conjunctiva, peripheral nerve and skeletal muscle has progressively replaced more aggressive procedures. Also, the enzymatic and biochemical assays on one hand, and the recent developments of molecular biology on the other, have considerably alleviated the need and usefulness of brain biopsies for metabolic disorders. However, modern stereotaxic surgery using sophisticated neuroradiological procedures makes it possible to reach nearly every part of the central nervous system and to take small specimens of nervous tissue. As most of the procedures used for the handling of biopsies are similar to the ones used for the examination of post-mortem material, the techniques will be discussed later.

Post-mortem examination must be thorough and the results of the autopsy of the visceral organs must be available. Autopsy of the nervous system should include removal of the whole brain with brainstem, cerebellum, spinal cord, spinal ganglia, cauda equina, peripheral nerves and muscles. It is no longer acceptable to put all these different pieces into formalin. In one hemisphere, small pieces of cortex and white matter must be fixed in buffered glutaraldehyde for electron microscopy, other specimens must be fixed in methacarn to allow for special immunohistochemical techniques. The rest of the same cerebral hemisphere as well as one cerebellar hemisphere, pieces of nerve and muscle must be kept deep frozen to allow further biochemical studies or isolation of brain proteins. It is advisable to cut the hemisphere in coronal (verticofrontal) sections starting from the frontal lobe to the occipital lobe and to keep these brain slices apart. It is mandatory to identify the fragments and their exact anatomical localization. The correct identification of brain areas is no longer possible if the hemisphere is kept in one piece in the deep-freezer at $-70°C$.

Specific neuropathological techniques are used for the study of the nervous system. They can be subdivided into two main groups as follows.

Techniques making use of light microscopy (LM)

Techniques on fragments fixed in methacarn: Methacarn is a Carnoy's fixative containing absolute methanol instead of absolute ethanol. After paraffin embedding, the slices can be stained for immunohistochemistry of amyloid precursor protein, amyloid A4, microglia, prion protein and various brain-specific proteins using the peroxidase–antiperoxidase (PAP) method for the polyclonal antibodies (Sternberger et al 1970) and the avidin–biotin complex (ABC) technique for the monoclonal antibodies (Hsu et al 1981).

Fixation in neutral formalin: After a period of 4 weeks in neutral formalin, one cerebral hemisphere, the brainstem with the cerebellum and the rest of the nervous system are cut in coronal sections. If necessary, black-and-white photographs and colour slides are made.

Sections for light-microscopic examination are prepared with two purposes in mind. The first is to describe the topography and the progression of the morphological alterations. These data are useful for understanding anatomoclinical correlations and reconstructing the evolution of the disease. To obtain this information, large sections are of course to be preferred in order to give an accurate anatomical localization of the neuropathological lesions. The second purpose is to study the structure of the lesions to better understand their physiopathogenesis.

To fulfil such purposes, frozen sections including large slices of the brain are cut on a freeze microtome and stained according to various techniques in order to illustrate different cell types, myelin, reactive glia and the destruction of the tissue (all techniques are to be found in Romeis (1948) and Gasser (1961)). These are relatively thick sections of ±30 μm. The best techniques are the following:

(1) Spielmeyer's method with haematoxylin to show myelin; the colour of normal myelin is deep blue to black; demyelination or dysmyelination is characterized by a less dense coloration or a discoloration going from light blue to white.

(2) Staining with cresyl-violet, which shows the nuclei of neurones and glial cells, the ribosomes in the neuronal perikaryon (known as the Nissl bodies). This technique enables the neuropathologist to examine in great detail the distribution and the organization of the neurones and to determine whether there is neuronal loss, increase in the number of glial cells, presence of macrophages, etc.

(3) Staining according to Holzer's method with phosphomolybdic acid. This staining shows the filaments in the astrocytes. In normal circumstances, the colour of the Holzer slides is light blue. When there is an increase in the number of glial cells and intracellular filaments, the colour becomes deep blue. If there is a diffuse homogeneous staining one speaks of isomorphic gliosis (e.g. the diffuse involvement of the white matter such as in leukodystrophies), but if there is an irregular increase of fibrillary astrocytes one speaks of anisomorphic gliosis (e.g. when there is a focal necrosis).

(4) The Sudan III method shows neutral fats and is used to detect recent damage to nervous tissue. The neutral fats present in macrophages and also in astrocytes are stained red, contrasting with the yellow coloration of the normal tissues.

(5) The PAS (periodic acid–Schiff) method is used to demonstrate the presence of lipopigments, lipofuscin, corpora amylacea, etc.

(6) Bielschowsky's silver impregnation shows neurofibrillar degeneration or senile plaques. The major advantage of frozen sections is that there is no embedding and therefore the same specimens can be employed later for other purposes such as electron microscopy.

Celloidin sections are prepared from blocks dehydrated in alcohol and embedded in nitrocellulose. This is a classical neuropathological technique. Because the specimens are passed through different concentrations of alcohol, there is considerable shrinkage of the celloidin slices. Frozen sections and celloidin sections from the same patient will differ in size. On celloidin sections, which are 25 μm thick but represent more tissue because of the ~30% shrinkage, two main techniques are applied: one is Woelcke's haematoxylin staining for myelin, which gives about the same information as Spielmeyer's staining on frozen sections but with fewer artefacts and a much denser blue-black colour, and the other is Nissl's method for cytology. The use of thionine in Nissl's method will show the neurones, their perikarya and the glial nuclei as well.

Finally paraffin sections are made from smaller representative areas of the nervous system (isocortex, allocortex, basal ganglia, thalamus, brainstem, cerebellum, etc.). Besides cresyl-violet and haemtoxylin–eosin stainings for cytology and Bodian silver protein impregnation for the axons, paraffin sections can be stained for immuno-histochemistry. PAP or ABC techniques are used with a whole panel of polyclonal and monoclonal antibodies against glial fibrillary acidic protein, myelin basic protein, S100 protein, microtubules, neuron-specific enolase, microtubule-associated protein tau, neurofilaments, microglia, macrophages, lysosomes, etc. (Sternberger et al 1970).

Techniques making use of electron microscopy (EM)

Standard electron-microscopic techniques are used, including a 4-h fixation in 4% buffered glutaraldehyde solution followed by 2h post-fixation in buffered osmium tetroxide. The embedding is in Araldite. The only special precaution with skin biopsies is to preserve, during sectioning and flat embedding of the skin specimens, such an orientation of the fragments that a full-thickness section of epidermis, dermis and hypodermis can be examined. Semi-thin sections are prepared and stained with 2% buffered toluidine blue. Thin sections are contrasted with 2% uranyl acetate and lead citrate and examined at 60 kV with an electron microscope. Pre-embedding according to the PAP or ABC method and post-embedding using the immunogold staining technique can also be applied for the same monoclonal and polyclonal antibodies used on sections for light microscopy. Of course, the pre-embedding technique does not ensure optimal preservation of the tissues and more fixation artefacts are to be expected than with the classical EM fixation. More specific staining and better tissue preservation are obtained with the immunogold post-embedding method. These techniques depend on the availability of appropriate monoclonal antibodies.

PEROXISOMAL DISORDERS

Adrenoleukodystrophy (ALD) and adrenomyeloneuropathy (AMN)

ALD is an X-linked disorder characterized, when fully expressed, by a combination of a progressive demyelination of the central nervous system and adrenal insufficiency. Lesions are also found in the peripheral nerves and in the cutaneous or conjunctival nerve twigs.

Figure 1 (opposite)   Classical example of adrenoleukodystrophy with occipitotemporal demyelination, progressing in rostral direction. (**A**) Frontotemporal lobe and neostriatum: normal myelination in the frontal lobe (black areas) and myelin pallor in the temporal lobe. (Celloidin section; Woelcke's haematoxylin staining for myelin; magnification×0.96.) (**B**) Frontotemporal lobe with thalamus and basal ganglia: severe demyelination (white areas) but for some subcortical zones, part of the cerebral peduncle and the thalamus (in black on the photograph). Note the severe involvement of the optic tracts (arrow). (Celloidin section; Woelcke's haematoxylin staining for myelin; magnification×0.96.) (**C**) Parietotemporal lobe with pulvinar: diffuse isomorphic fibrillary gliosis in the white matter and the thalamus as shown by the greyish appearance on the photograph. (Frozen section; Holzer's staining for the fibrillary glia; magnification×0.8.) (**D**) Parietotemporal lobe: huge amounts of neutral fats (black areas on the photograph as shown by the arrow) are accumulated in the white matter and correspond to a rather recent myelin breakdown. (Frozen section; Sudan III staining for neutral fats; magnification×0.8.)

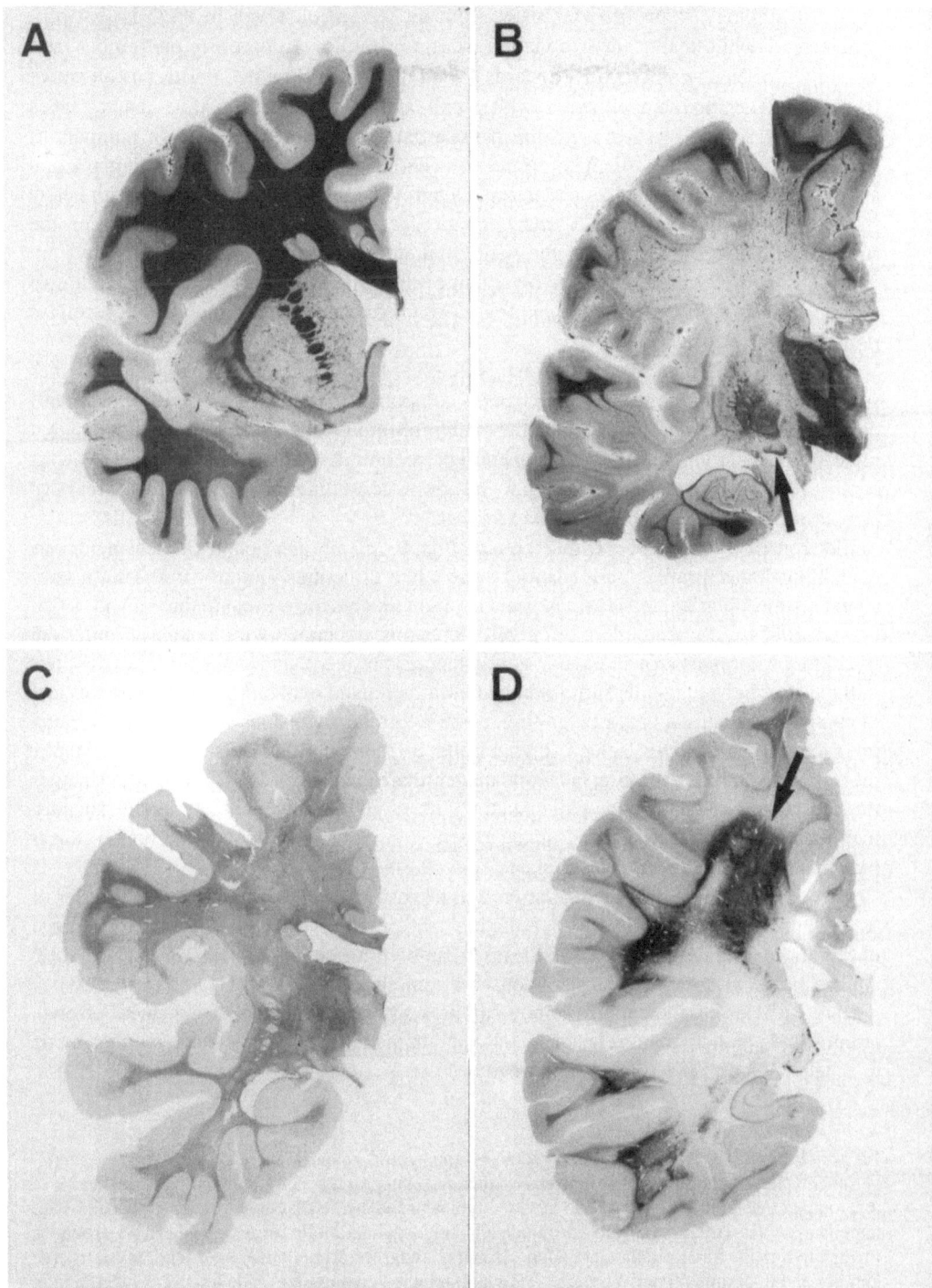

Neuropathology shows that the more frequent pattern observed in ALD is that of a progressive white-matter demyelination starting in the temporo-occipital lobes and extending, often asymmetrically, towards the frontal areas (Figure 1). There is an active front of advancing demyelination with small amounts of PAS-positive macrophages followed by a zone of severe myelin loss, partial axonal destruction, large numbers of lipid-filled macrophages, lymphoplasmocytic perivascular infiltrates and gemistiocytic astrocytes. The more caudal area represents a burnt-out demyelinated territory with heavy isomorphic fibrillary gliosis. A characteristic feature is the irregular sparing of the subcortical arcuate fibres and of the white matter in the gyri. The visual pathways are damaged in ALD with constant involvement of the geniculocalcarine radiations but also with focal lesions in optic chiasm and tracts. The reported loss of retinal ganglion cells is probably of a retrograde nature.

Prominent inflammation in the demyelinative lesion of adrenoleukodystrophy has suggested an immune-mediated pathogenic component. A study by Boutin et al (1989) demonstrated that macrophages were the main infiltrating cells in the white matter and that there was a wide variability in their phenotype according to their location. Data (Powers et al 1992) support a natural immune response consisting predominantly of reactive astrocytes, macrophages, T-cells and cytokines.

Electron microscopy of the macrophages shows membrane-bound or free inclusions containing linear profiles and trilaminar sheet-like structures (Figure 2A). They correspond to the polarizing inclusions seen by light microscopy (Schaumburg et al 1975; Powers 1985). Free needle-like or curved inclusions associated with lamellar profiles are found in the adrenals (zona reticularis and inner part of the zona fasciculata), Leydig cells in the testes, Schwann cells surrounding small myelinated axons in peripheral nerves and in cutaneoconjunctival nerve twigs (Figure 2B). Similar inclusions have also been found in macrophages in liver, spleen, thymus and in renal tubular cells. Beside the many macrophages found in the affected white matter, there are also early ultrastructural changes in the seemingly unaffected white matter with slight involvement of myelin sheaths, membrane-free intracytoplasmic inclusions in oligoglia and membrane-bound linear inclusions in swollen astrocytes (Takeda et al 1989).

In another example of ALD, the onset will be rostral, affecting first the white matter of frontal lobes (Figure 3). The demyelination will progress in caudal direction (personal observation; Shiga et al 1992). Such lesions can be found in adolescent males and also in adult patients with ALD. A presenile onset is sometimes observed (Uyama et al 1993).

Rare cases of adrenoleukodystrophy can present as spinocerebellar degeneration with atrophy and parenchymatous lesions of cerebellum and brain stem, detectable by magnetic resonance imaging (Kusaka and Imai 1992).

Figure 2 *(opposite)* Electron microscopy of the stored material in (A) the central and (B) peripheral nervous system. (**A**) Central nervous system: presence of trilaminar lipid leaflets in a macrophage. (**B**) Peripheral nervous system: typical needle-like or curved clefts associated with lamellar profiles (see arrow) are surrounded by glycogen granules in a Schwann cell around a myelinated axon. Such pictures are found in nerve twigs of conjunctiva, skin, endomysial nerve bundles and in peripheral nerve biopsies. (Glutaraldehyde – osmium tetroxide fixation, embedding in Araldite, contrast with uranylacetate and lead citrate; bar = 1 μm)

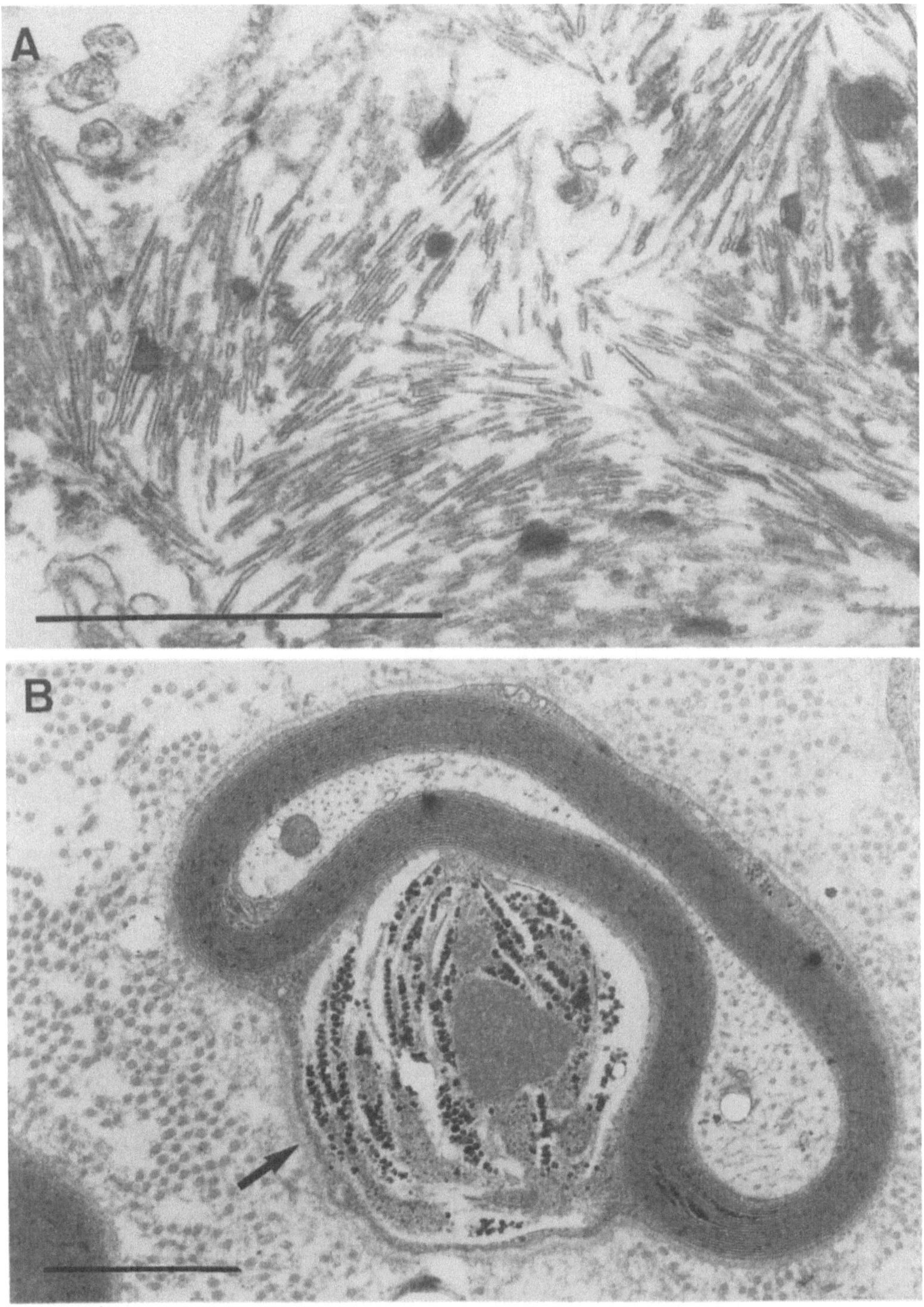
A
B

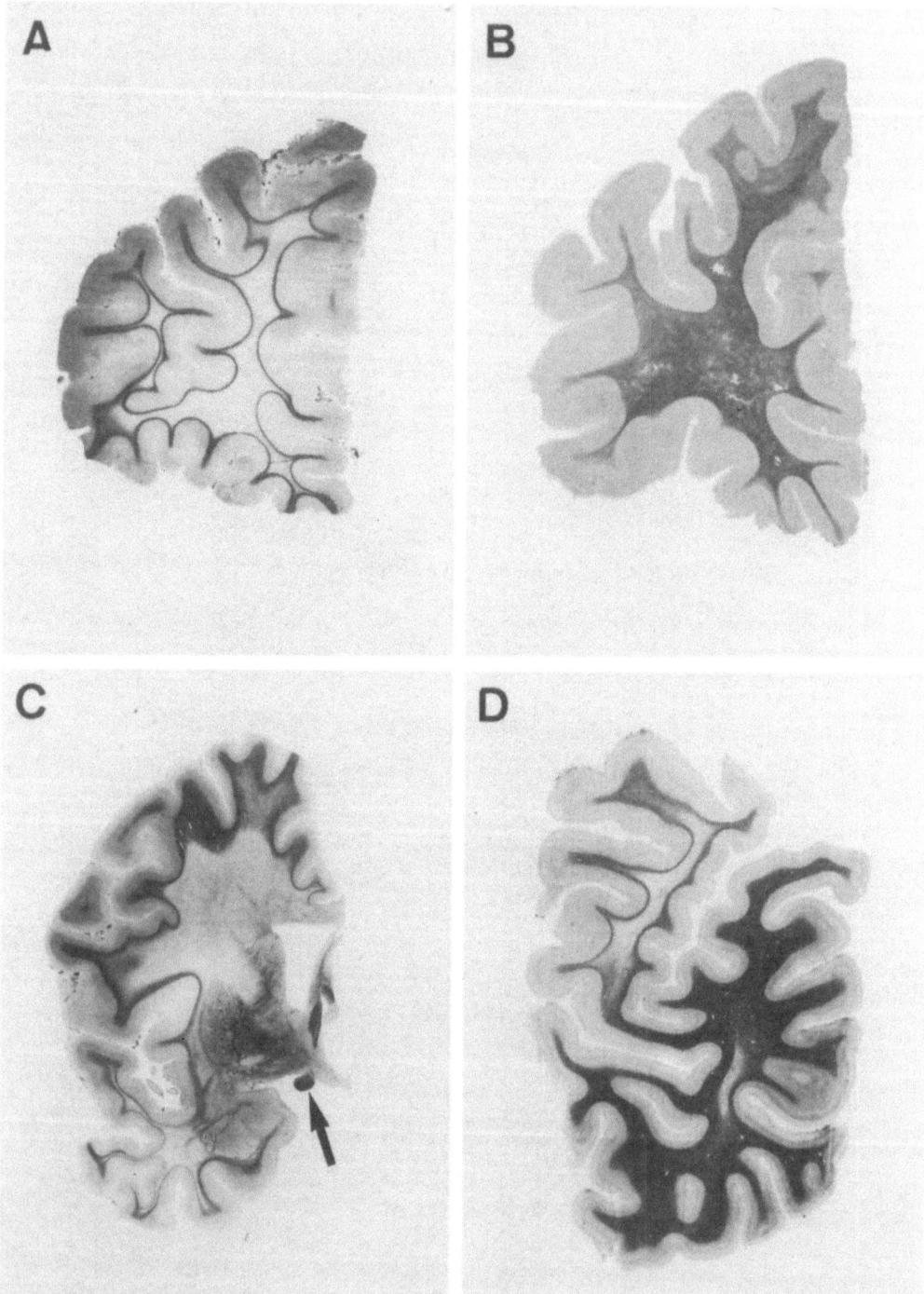

There is a good correlation between the clinical signs and the progression of the demyelinating lesions. The behavioural disturbances, the delayed comprehension of speech, the decrease of visual acuity with homonymous hemianopia, cortical blindness or optic atrophy correspond to the temporo-occipital onset of the demyelination. The pyramidal signs and the (initially) unilateral hemiplegic attitude reflect the asymmetry of the rostral progression of the demyelination. The evolution proceeds towards dementia, blindness and spastic diplegia in a few years. It can be interrupted by death due to a fast progression of the disease or complicating conditions such as pneumonia. In the examples with frontal onset and progression towards the caudal areas, the first symptoms are those found in a frontal syndrome with behavioural disturbances, followed later by asymmetrical pyramidal features.

Magnetic resonance imaging (MRI) and positron emission tomography can be used in correlation with neuropathology to follow the progression of the disease by showing the onset and the development of demyelination (Iinuma et al 1989). Gadolinium-MRI may show, on T_1-weighted image, an enhancement of the rim of the lesions (Murai et al 1992).

In AMN (McKusick 300100), there exists a multilevel demyelination of the pyramidal tracts in brainstem and spinal cord, and of posterior columns and medial and lateral lemnisci. In some AMN cases, a diffuse myelin pallor has been found in the hemispheric white matter. MRI of the spinal cord in subjects with ALD-AMN complex can also deliver useful information by showing decreased spinal cord diameter and also possibly intracranial white matter changes (Kurihara et al 1991; Snyder et al 1991). Intermediate cases between adrenoleukodystrophy and adrenomyeloneuropathy have been documented by neuropathological examinations (Kanda et al 1989).

Evidence for a peripheral neuropathy with segmental demyelination, remyelination and onion-bulb formations has been shown in AMN (Vercruyssen et al 1982). Major diagnostic advances in the fields of biochemistry and molecular biology have been preceded by the demonstration of specific inclusions in adrenal-, testis- and cutaneo-conjunctival tissues (Martin et al 1980). Spicular, rectilinear or slightly curved electron-lucent clefts are indeed demonstrated in Schwann cells surrounding normal myelinated axons in adrenoleukodystrophy (Figure 2B). The deposits are not membrane-bound but they are lined by free glycogen particles, osmiophilic lamellar structures and mito-chondria. Prominent lesions occur in clusters. Serial sections are sometimes necessary. Unmyelinated axons are normal. Similar but less numerous lesions are found in myelinated nerve twigs in adrenomyeloneuropathy.

Figure 3 *(opposite)* Less classical adrenoleukodystrophy with frontal onset of the demyelination and progression towards the temporo-occipital white matter. (**A**) Frontal lobe: demyelination in the white matter with sparing of the subcortical fibres (black on the photograph). (Celloidin section; Woelcke's haematoxylin staining for myelin; magnification×1.2.) (**B**) Frontal lobe: diffuse isomorphic fibrillary gliosis including the subcortical fibres. (Frozen section; Holzer's staining for fibrillary glia; magnification×1.2.) (**C**) Frontotemporal lobe with basal ganglia: severe demyelination but for some relatively preserved subcortical areas and the better stained optic tract (arrow). (Celloidin section; Woelcke's haematoxylin staining for myelin; magnification×0.8.) (**D**) Occipital lobe: demyelination in the white matter of the upper occipital gyri while the rest of the occipital lobe is adequately myelinated. The whitish area in the centre corresponds to the most caudal part of the occipital horn. (Celloidin section; Woelcke's haematoxylin staining for myelin; magnification×0.96.)

Zellweger cerebrohepatorenal syndrome

In the brain, which may have increased weight, different alterations have been described: cerebral malformations with pachygyria, polymicrogyria, deep parietal clefts and neuronal migration defects. Microscopy has shown dysplastic dentate and olivary nuclei as well as neuronal heterotopias. Golgi studies have revealed irregular neuronal arrangements, immature neurones, poor dendritic arborization and poor spine development, all of which suggest abnormal morphogenesis and delayed maturation (Takashima et al 1991). There are also demyelinating lesions in the white matter with evidence of myelin breakdown and lipid deposition in macrophages, and also in glial cells. Developmental immuno-histochemistry of catalase, acyl-CoA oxidase and ketoacyl-CoA thiolase has revealed that positive reaction appeared with normal neuronal and glial maturation. Abnormal peroxisomal membrane or peroxisomal metabolites might cause a migration disorder in intrauterine development and myelination disturbance in perinatal maturation. Diffuse dysmyelination might be related to maldevelopment of oligodendroglia, and migration disorder to abnormality of endothelial cells or radial glia.

The cerebra of four abortuses of gestational ages between 14 and 22 weeks, diagnosed as cerebrohepatorenal (Zellweger) syndrome *in utero*, showed centrosylvian architectonic abnormalities with thin cortical plates and broad subcortical heterotopic zones. Astrocytes, neuroblasts, immature neurones and radial glia contained abnormal pleomorphic cytosomes, presumably of variable lipid composition (Powers et al 1989).

Immunohistochemical studies with antisera against peroxisomal enzymes, catalase and β-oxidation enzymes (acyl-CoA oxidase, bifunctional protein, and 3-ketoacyl-CoA thiolase) should be applied to brain sections from patients with peroxisomal disorders such as Zellweger syndrome or neonatal adrenoleukodystrophy to check the results of Kamei et al (1993). These authors reported a weak or negative reaction of neurons in the cerebral cortex and a weak reaction of glial cells in the white matter of patients affected by peroxisomal disorders. However, the same authors found only a *diffuse* staining in normal brain tissue, which casts doubt on their findings in pathological conditions. Repeat studies are therefore needed to validate the application of immunohistochemical methods to the study of peroxisomal enzymes in the central nervous system. The importance of immunohistochemistry has already been adequately documented in the liver by Roels et al (1991, 1993).

Electron microscopy discloses the presence of macrophages with lipid clefts, lamellae and lamellar lipid profiles. Trilaminar and lamellar lipid profiles are also found in striated cells in the adrenal cortex. Light and electron microscopy in first-trimester chorionic villi show that normal cytotrophoblast cells contain small peroxisomes in which a cytochemical reaction for catalase may be demonstrated (Roels et al 1987; Takashima et al 1992). Zellweger amniocytes appear to lack these organelles, although some cells have rare structures that might be residual or abnormal peroxisomes (Lazarow et al 1988; see also the paper by Roels et al in this issue).

Neonatal adrenoleukodystrophy

The migration defects are less prominent than those in Zellweger syndrome. Demyelin-ation occurs as in classical adrenoleukodystrophy.

Other cases are more difficult to classify, such as the one reported by Naidu et al (1988) of a girl with neonatal seizures and severe mental retardation, malformative and destructive lesions of central grey and white matter, atrophy of adrenal cortex with striated adrenocortical cells, hepatic fibrosis, PAS-positive macrophages in several organs and biochemical features of X-linked adrenoleukodystrophy. Assignment to the recently reported entity of peroxisomal acyl-CoA oxidase deficiency was excluded by the normal amounts of this enzyme in the liver.

Pathological findings in a sural nerve biopsy in a case of neonatal adrenoleukodystrophy (Mito et al 1989) were not shown by light microscopy since the density and total number of myelinated fibres were not significantly different from controls. However, electron microscopy revealed thinner myelin sheaths than in controls. Some linear or trilamellar inclusion bodies were found in Schwann cells and fibroblasts, as in X-linked adrenoleukodystrophy. Büngner's bands were also seen by electron microscopy, and myelin ovoids or balls were seen in teased fibres.

Infantile Refsum disease

Atrophic cerebellar cortex with Purkinje cells in the molecular layer is found. No clear evidence is found for active demyelination. Macrophages can be present but they do not contain sudanophilic material.

Autopsy findings by Chow et al (1992) in two sisters with probable infantile Refsum disease who died at 8 months and 3.5 years, respectively, showed adrenal atrophy, cirrhosis and foamy histiocytes in multiple organs. The brain showed no demyelination, little cytoarchitectural abnormality, occasional perivascular histiocytes in the grey matter and meninges, and prominent Purkinje cells in the molecular layer of the cerebellum. In the younger patient the changes were very subtle in spite of the marked clinical similarity.

The main gross and microscopic findings in the first autopsy report of infantile Refsum disease (Torvik et al 1988) showed liver cirrhosis, hypoplastic adrenals, and large groups of lipid macrophages in liver, lymph nodes and the cerebral white matter. The brain exhibited no malformations except for a severe hypoplasia of the cerebellar granule layer and ectopic location of the Purkinje cells in the molecular layer. A mild and diffuse reduction of axons and myelin was found in the corpus callosum and periventricular white matter, the corticospinal tracts and the optic nerves. Large numbers of perivascular macrophages were present in the same areas but there was no active demyelination. The retina and cochlea showed severe degenerative changes. Peripheral nerves, skeletal system and kidneys were normal. Electron microscopy showed characteristic cytoplasmic inclusions with bilamellar profiles in macrophages in liver, lymph nodes and brain but not in the adrenals. Similar inclusions were found in liver cells and astrocytes.

The findings differed from those observed in Zellweger syndrome and in neonatal adrenoleukodystrophy.

Pseudo-infantile Refsum disease (Aubourg et al 1993) has also been described, but no morphological data are available as yet.

Hyperpipecolic acidaemia

Differences from patient to patient are noted as far as the cerebral lesions are concerned. In one case, there was a demyelination resembling sudanophilic leukodystrophy, demyelination in pons and cerebellum and also presence of trilaminar inclusions.

Pseudo-neonatal adrenoleukodystrophy

A new peroxisomal disorder with enlarged peroxisomes and a specific deficiency of acyl-CoA oxidase has been described in a pseudo-neonatal adrenoleukodystrophy (Poll-The et al 1988). In contrast to neonatal adrenoleukodystrophy patients, hepatic peroxisomes were enlarged and not decreased in number. Accumulation of very long-chain fatty acids (VLCFA) was associated with an isolated deficiency of the fatty acyl-CoA oxidase, the enzyme that catalyses the first step of the peroxisomal β-oxidation.

Central nervous system malformations and white matter changes have been diagnosed by neuroradiology (Kyllerman et al 1990) since a CT scan showed low density of cerebral white matter and a MRI examination demonstrated white matter changes, a thin corpus callosum, cerebellar malformation and dorsal displacement of the brainstem. There are no autopsy data on the central nervous system.

Other conditions

Other patients belong to a heterogeneous group of early-onset peroxisomal disorders distinct from the Zellweger syndrome and other generalized peroxisomal disorders such as the one reported by Van Maldergem et al (1992). In this male newborn with hypotonia and seizures, other clinical features were glaucoma, absence of facial dysmorphism and of liver enlargement, and renal cysts. The patient died at the age of 3 months. Neuropathological findings included pachygyria of the olivary nuclei (Figure 4) and cerebellar neuronal heterotopias. There was no evidence for a demyelinating process. Trilamellar inclusions typical of a peroxisomal fatty acid oxidation defect were present in macrophages.

In the ill-defined group of orthochromatic leukodystrophies, some cases may belong to the group of peroxisomal disorders although the biochemical identification is often lacking or insufficiently accurate. There is, for example, a sporadic case of leukodystrophy with early onset and rapid clinical course (Taniike et al 1992). Classical metabolic leukodystrophies like metachromatic leukodystrophy (McKusick 205100), globoid cell leukodystrophy (McKusick 245200) and adrenoleukodystrophy were excluded by biochemical assays. Autopsy findings were compatible with the diagnosis of the pigmentary type of orthochromatic leukodystrophy. However, there was a severe neuronal loss and collection of globoid-like cells at the interface of the grey matter and the white matter. The lack of adequate biochemical and enzymatic studies renders this case unclassifiable at the present time.

An unusual orthochromatic leukodystrophy with epithelioid cells (Norman–Gullotta) has been shown to belong to the group of peroxisomal disorders because of an increase of very long-chain fatty acids in formalin-fixed brain white matter (Molzer et al 1993). Ultrastructural findings in brain showed typical lamellar inclusions. The particular type of

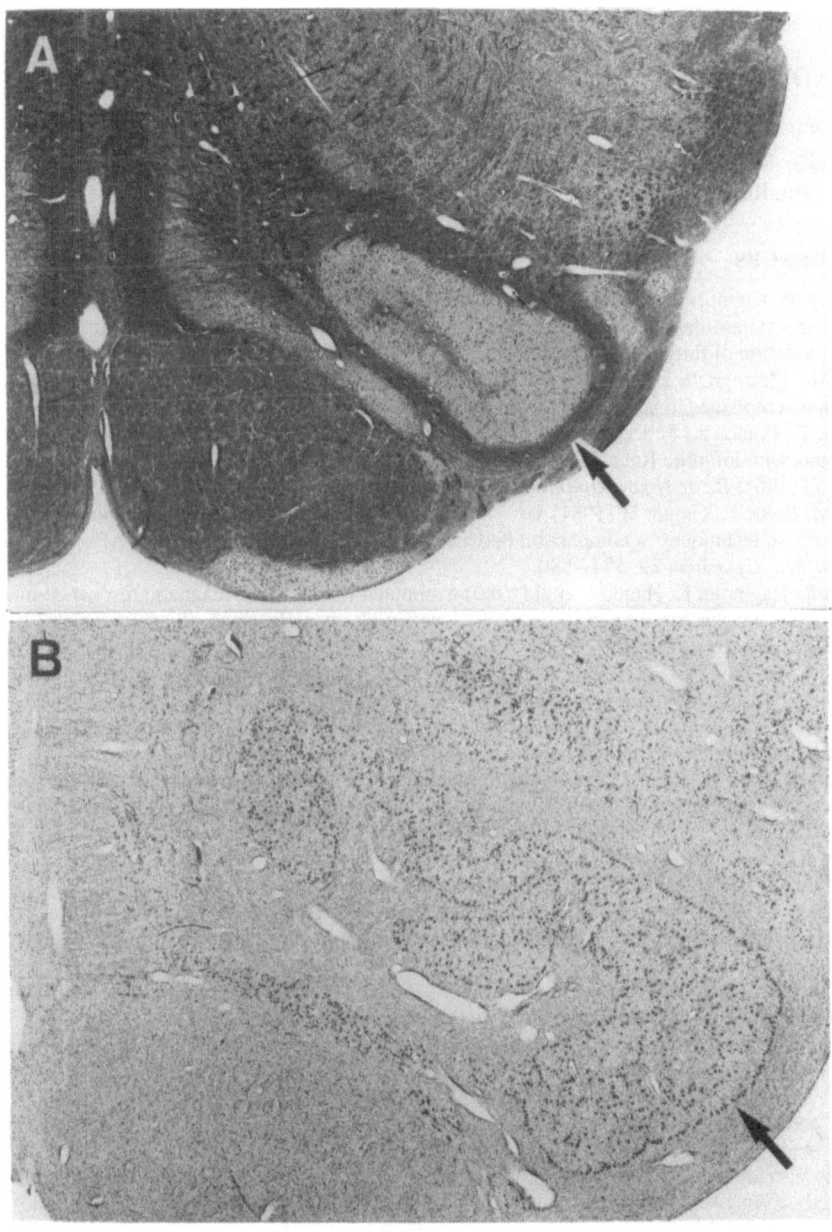

Figure 4 Patient reported by Van Maldergem et al (1992). (**A**) Medulla oblongata: pachygyria of the olivary nucleus on a myelin stain. (Frozen section; Spielmeyer's staining for myelin; magnification×1.2.) (**B**) Medulla oblongata: pachygyria of the olivary nucleus with the typical distribution of neurons at the margins of the pachygyric nucleus as shown on this cytological preparation. (Frozen section; cresyl violet staining for cytology; magnification×16)

peroxisomal disorder present in this case (heterozygote of X-linked adrenoleukodystrophy?) remains speculative.

ACKNOWLEDGEMENTS

The contributions of Dr C. Ceuterick, of Miss E. De Leenheir and of Mrs U. Lübke are gratefully acknowledged. This work has been realized with the help of the Fonds voor Geneeskundig Wetenschappelijk Onderzoek (grant no. 3.0020.94).

REFERENCES

Aubourg P, Kremser K, Roland MO, Rocchiccioli F, Singh I (1993) Pseudo-infantile Refsum disease: catalase-deficient peroxisomal particles with partial deficiency of plasmalogen synthesis and oxidation of fatty acids. *Pediatr Res* **34**: 270–276.
Boutin B, Matsuguchi L, Lebon P, Ponsot G, Arthuis C (1989) Immunohistochemical analysis of brain macrophages in adrenoleukodystrophy. *Neuropediatrics* **20**: 202–206.
Chow CW, Poulos A, Fellenberg AJ, Christodoulou J, Danks DM (1992) Autopsy findings in two siblings with infantile Refsum disease. *Acta Neuropathol (Berl)* **83**: 190–195.
Gasser G (1961) *Basic Neuropathological Technique*. Oxford: Blackwell Scientific Publ.
Hsu SM, Raine L, Fanger H (1981) Use of avidin–biotin peroxidase (ABC) complex in immunoperoxidase techniques: a comparison between ABC and unlabeled antibody (PAP) procedures. *J Histochem Cytochem* **29**: 577–580.
Iinuma K, Haginoya K, Handa I, et al (1980) Computed tomography, magnetic resonance imaging, positron emission tomography and evoked potentials at early stage of adrenoleukodystrophy. *Tohoku J Exp Med* **159**: 195–203.
Kamei A, Houdou S, Takashima S, Suzuki Y, Becker LE, Armstrong DL (1993) Peroxisomal disorders in children: immunohistochemistry and neuropathology. *J Pediatr* **122**: 573–579.
Kanda F, Inoue K, Jinnai K, Takahashi K, Mannen T (1989) A case of adrenomyeloneuropathy with localized cerebral white matter degeneration. *Rinsho Shinkeigaku* **29**: 483–487
Kurihara K, Kim H, Tamura N, Iwasaki S, Hamaguchi K (1991) A case of adrenoleukodystrophy presenting large lesion of the cerebellar white matter and dentate nuclei on brain CT and MRI. *Rinsho Shinkeigaku* **31**: 72–78.
Kusaka H, Imai T (1992) Ataxic variant of adrenoleukodystrophy: MRI and CT findings. *J Neurol* **239**: 307–310.
Kyllerman M, Blomstrand S, Mansson JE, Conradi NG, Hindmarsh T (1990) Central nervous system malformations and white matter changes in pseudo-neonatal adrenoleukodystrophy. *Neuropediatrics* **21**: 199–201.
Lazarow PB, Small GM, Santos M et al (1988) Zellweger syndrome amniocytes: morphological appearance and a simple sedimentation method for prenatal diagnosis. *J Pediatr Res* **24**: 63–67.
Martin J-J, Ceuterick C, Libert J (1980) Skin and conjunctival nerve biopsies in adrenoleukodystrophy and its variants. *Ann Neurol* **8**: 291–295.
Mito T, Takada K, Akaboshi S, Takashima S, Takeshita K, Origuchi Y (1989) A pathological study of a peripheral nerve in a case of neonatal adrenoleukodystrophy. *Acta Neuropathol (Berl)* **77**: 437–440.
Molzer B, Gullotta F, Harzer K, Poulos A, Bernheimer H (1993) Unusual orthochromatic leukodystrophy with epitheloid cells (Norman–Gullotta): increase of very long chain fatty acids in brain discloses a peroxisomal disorder. *Acta Neuropathol (Berl)* **86**: 187–189.
Moser HW, Bergin A, Cornblath D (1991) Peroxisomal disorders. *Biochem Cell Biol* **69**: 463–474.
Mosser J, Douar AM, Sarde CO, et al (1993) Putative X-linked adrenoleukodystrophy gene shares unexpected homology with ABC transporters. *Nature* **361**: 726–730.
Murai H, Sasagasako N, Yoshimura T, Kira J, Goto I (1992) Gadolinium-MRI findings of two adrenoleukodystrophy cases treated with gamma-globulin. *Rinsho Shinkeigaku* **32**: 416–420.
Naidu S, Hoefler G, Watkins PA, et al (1988) Neonatal seizures and retardation in a girl with

biochemical features of X-linked adrenoleukodystrophy: a possible new peroxisomal disease entity. *Neurology* **38**: 1100–1107.

Poll-The BT, Roels F, Ogier H, et al (1988) A new peroxisomal disorder with enlarged peroxisomes and a specific deficiency of acyl-CoA oxidase (pseudo-neonatal adrenoleukodystrophy). *Am J Hum Genet* **42**: 422–434.

Powers JM (1985) Adreno-leukodystrophy (adreno-testiculo-leukomyelo-neuropathic-complex) *Clin Neuropathol* **4**: 181–199.

Powers JM, Tummons RC, Caviness VS Jr, Moser AB, Moser HW (1989) Structural and chemical alterations in the cerebral maldevelopment of fetal cerebro-hepato-renal (Zellweger) syndrome. *J Neuropathol Exp Neurol* **48**: 270–289.

Powers JM, Liu Y, Moser AB, Moser HW (1992) The inflammatory myelinopathy of adreno-leukodystrophy: cells, effector molecules, and pathogenetic implications. *J Neuropathol Exp Neurol* **51**: 630–643.

Roels F, Verdonck V, Pauwels M, et al (1987) Light microscopic visualization of peroxisomes and plasmalogens in first trimester chorionic villi. *Prenat Diagn* **7**: 525–530.

Roels F, Espeel M, De Craemer D (1991) Liver pathology and immunocytochemistry in congenital peroxisomal diseases: a review. *J Inher Metab Dis* **14**: 853–875.

Roels F, Espeel M, Poggi F, Mandel H, Van Maldergem L, Saudubray JM (1993) Human liver pathology in peroxisomal diseases: a review including novel data. *Biochimie* **75**: 281–292.

Romeis B (1948) *Mikroskopische Technik.* Munich: Leibniz Verlag.

Schaumburg HH, Powers JM, Raine CS, Suzuki K, Richardson EP Jr (1975) Adrenoleukodystrophy. A clinical and pathological study of 17 cases. *Arch Neurol* **32**: 577–591.

Shiga Y, Saito H, Mochizuki H, Chida K, Tsuburaya K (1992) A case of adrenoleukodystrophy having progressed from the frontal lobes. *Rinsho Shinkeigaku* **32**: 600–605.

Snyder RD, King JN, Keck GM, Orrison WW (1991) MR imaging of the spinal cord in 23 subjects with ALD-AMN complex. *Am J Neuroradiol* **12**: 1095–1098.

Sternberger LA, Hardy PH, Cuculis JJ, Meyer HG (1970) The unlabeled antibody enzyme method of immunohistochemistry: preparation and properties of soluble antigen–antibody complex (horseradish peroxidase–antihorseradish peroxidase) and its use in the identification of spirochetes. *J Histochem Cytochem* **18**: 315–333.

Tager JM, Brul S, Wiemer EA, et al (1990) Genetic relationship between the Zellweger syndrome and other peroxisomal disorders characterized by an impairment in the assembly of peroxisomes. *Prog Clin Biol Res* **321**: 545–558.

Takashima S, Chan F, Becker LE, Houdou S, Suzuki Y (1991) Cortical cytoarchitectural and immunohistochemical studies on Zellweger syndrome. *Brain Dev* **13**: 158–162.

Takashima S, Houdou S, Kamei J, et al (1992) Neuropathology of peroxisomal disorders, Zellweger syndrome and neonatal adrenoleukodystrophy. *No To Hattatsu* **24**: 186–193.

Takeda S, Ohama E, Ikuta F (1989) Adrenoleukodystrophy. Early ultrastructural changes in the brain. *Acta Neuropathol (Berl)* **178**: 124–130.

Taniike M, Fujimura H, Kogaki S, et al (1992) A case of pigmentary type of orthochromatic leukodystrophy with early onset and globoid cells. *Acta Neuropathol (Berl)* **83**: 427–433.

Torvik A, Torp S, Kase BF, Ek J, Skjeldal O, Stokke O (1988) Infantile Refsum's disease: a generalized peroxisomal disorder. Case report with postmortem examination. *J Neurol Sci* **85**: 39–53.

Uyama E, Iwagoe H, Maeda J, Nakamura M, Terasaki T, Ando M (1993) Presenile-onset cerebral adrenoleukodystrophy presenting as Balint's syndrome and dementia. *Neurology* **43**: 1249–1251.

Van Maldergem L, Espeel M, Wanders RJA, et al (1992) Neonatal seizures and severe hypotonia in a male infant suffering from a defect in peroxisomal beta-oxidation. *Neuromuscular Dis* **2**: 217–224.

van der Knaap MS, Valk J (1991) The MR spectrum of peroxisomal disorders. *Neuroradiology* **33**: 30–37.

Vercruyssen A, Martin J-J, Mercelis R (1982) Neurophysiological studies in adrenomyeloneuropathy. A report on five cases. *J Neurol Sci* **56**: 327–336.

Wanders RJ, Heymans HS, Schutgens RB, Barth PG, van den Bosch H, Tager JM (1988) Peroxisomal disorders in neurology. *J Neurol Sci* **88**: 1–39.

J. Inher. Metab. Dis. 18 Suppl. 1 (1995) 34–44
© SSIEM and Kluwer Academic Publishers.

DNA diagnosis of X-linked adrenoleukodystrophy

S. Seneca and W. Lissens*
Department of Medical Genetics, University Hospital-Vrije Universiteit Brussel, Brussels, Belgium

*Correspondence: Department of Medical Genetics, University Hospital-Vrije Universiteit Brussel, Laarbeeklaan 101, 1090 Brussels, Belgium

Summary: The X-linked adrenoleukodystrophy (ALD) gene was identified recently and is predicted to encode a 745-amino-acid peroxisomal membrane protein. Strategies have been designed for the search for mutations in the ALD gene in patients. Several mutations have now been found and it seems that many different mutations are responsible for ALD. There is no straightforward correlation between genotype and phenotype since the same mutation can cause different ALD phenotypes in the same family. However, once a mutation has been found in a family, it can be traced in all at-risk individuals of that family, both post- and prenatally, without the need for very long-chain fatty acid (VLCFA) analysis. Segregation analysis with extragenic and intragenic polymorphisms may remain useful in families where mutation analysis is not possible for practical reasons; VLCFA analysis and measurement of the peroxisomal β-oxidation with $C_{26:0}$ fatty acid as a substrate will remain the alternative. We also briefly discuss the possibilities of DNA diagnosis for other peroxisomal disorders.

Adrenoleukodystrophy (ALD; McKusick 300100) is an X-linked neurodegenerative disorder affecting about one in 20 000 males (Moser and Moser 1989). The disease shows great variation in clinical phenotype (Moser et al 1992). The commonest and most severe phenotype is cerebral childhood ALD (50% of cases). Affected boys initially develop normally and the first clinical signs appear between the ages of 4 and 10 years. Behavioural abnormalities, hearing loss and visual problems are usually the first symptoms of the disease, followed by motor disturbances and reflex anomalies. Progressive degeneration of the brain white matter leads to dementia, blindness and quadriplegia, and death usually occurs within 5 years from the onset of the symptoms. Adrenomyeloneuropathy (AMN) represents a milder form of ALD progressing more slowly and involving mainly the spinal cord and the peripheral nerves (30% of cases; in contrast with all other reports concerning the phenotypic distribution in ALD patients, AMN is more frequent than childhood cerebral ALD in the Netherlands (van Geel et al 1994)). Male patients with AMN present in the second or third decade of life with symptoms of peripheral neuropathy and progressive spastic paraparesis. Almost 90% of males with cerebral childhood ALD or AMN also have varying degrees of adrenal insufficiency. Isolated adrenal insufficiency (Addison disease) can be an early sign of childhood ALD or AMN, but can also remain the only clinical expression of ALD. A diagnostic workup for ALD is therefore necessary in each patient with Addison disease.

Childhood cerebral ALD and AMN frequently occur in the same family, even in sibships (Moser et al 1984, 1992; Moser and Moser 1988; Fournier et al 1994a). Also, at least 30% of females who are heterozygous for ALD develop a mild AMN picture after approximately age 30 years. Several hypotheses have been proposed to explain this clinical variability in heterozygous females and hemizygous males. These include an X-linked dominant inheritance pattern, immunological and environmental factors, skewing of X-inactivation or the presence of autosomal modifier genes (Maestri and Beaty 1992; Moser et al 1992; Watkiss et al 1993). So far, no satisfactory explanation has been found.

ALD is the most frequent peroxisomal disorder; it is characterized biochemically by the accumulation of very long-chain fatty acids (VLCFAs) in brain white matter and the adrenal glands, and to a lesser extent in other tissues and body fluids (Moser et al 1984; Moser and Moser 1989; Wanders et al 1992). The accumulation is due to a defect in the degradation of VLCFAs in the peroxisomes. Biogenesis of the peroxisomes is not impaired. It has been suggested that the deficiency of the peroxisomal VLCFA-CoA synthetase (also called lignoceroyl-CoA ligase), the enzyme necessary for the activation of VLCFAs to their CoA esters before entering the β-oxidation pathway, is the primary defect in ALD since affected males have diminished activity of this enzyme. Alternatively, another gene product, necessary for the proper expression, functioning or positioning of the VLCFA-CoA synthetase, could be deficient.

The discovery of impaired VLCFA degradation in ALD has led to the development of reliable diagnostic tests based on the measurement of the saturated $C_{22:0}$, $C_{24:0}$ and $C_{26:0}$ acid concentrations and ratios in serum, plasma and fibroblasts for postnatal diagnosis and chorionic villi or cultured amniotic fluid cells for prenatal diagnosis of ALD (for review see Wanders et al (1992)). Determination of these biological parameters, however, is not always reliable in female carriers as not all carriers have elevated values (Moser et al 1983; Moser and Moser 1989).

In this paper we discuss the use of DNA technology for the localization and identification of the ALD gene, and how this information can be used for the diagnosis of ALD. A short overview of DNA diagnosis for other peroxisomal diseases is also presented.

LOCALIZATION AND ISOLATION OF THE ALD GENE

The first evidence for the physical localization of the ALD gene to the X-chromosome came from the study of cultured skin fibroblasts of female carriers of ALD (Migeon et al 1981). Clones derived from individual fibroblast cells were of two types: some of the clones showed elevated VLCFA patterns similar to hemizygous male patients while the others had normal VLCFA levels. This effect of Lyonization (i.e. random inactivation of one X-chromosome in females) is only observed in relation to the X-chromosomes of females; these results, therefore, indicated that the gene for ALD was indeed on the X-chromosome. Moreover, the disease in the families of heterozygous females was shown to co-segregate with electrophoretic variants of the enzyme glucose-6-phosphate dehydrogenase. The gene coding for this enzyme was later shown to be localized on the X-chromosome at Xq28 (Szabo et al 1984). A further confirmation of the localization of the ALD gene came from studies with the recombinant DNA probe St14 which detects highly polymorphic variable number tandem repeat (VNTR) sequences at locus DXS52 in Xq28 (Oberlé et al 1985; Aubourg et al 1987). The consistent co-segregation of the disease with one of the polymorphic DXS52 alleles in individual ALD families indicated that the gene causing ALD and DXS52 should be close together; only exceptionally was a recombinant event between the two loci observed (van Oost et al 1991). Close genetic linkage between the ALD gene and other polymorphic markers in the Xq28 region has since been demonstrated. The chromosome Xq28 band contains many disease-causing genes and, among others, the genes responsible for haemophilia A (Gitschier et al 1984), nephrogenic diabetes insipidus (Pan et al 1992), X-linked hydrocephalus, X-linked spastic paraplegia and MASA syndrome (Jouet et al 1994; Vits et al 1994), red/green colour blindness (Nathans et al 1986a,b) and Emery–Dreifuss muscular dystrophy (Bione et al 1994) have been isolated from this region.

The ALD gene was finally obtained by using a positional cloning approach (Mosser et al 1993). The high incidence of abnormal colour vision in AMN patients (accurate assessment of colour vision in childhood ALD is difficult or impossible owing to neurological degeneration in these patients) suggested that these genes could be close together and that deletions comprising both the ALD gene and the red/green colour genes could be involved (Aubourg et al 1988; Sack et al 1989). This latter idea involving large deletions finally proved to be wrong and the colour vision abnormalities in ALD are most likely secondary to cerebral degeneration in these patients (Aubourg et al 1990; Feil et al 1991; Mosser et al 1993; Sack and Morrell 1993; Sack et al 1993). Nevertheless, an AMN patient with blue-monochromatic colour vision was the clue to the identification of the ALD gene (Feil et al 1991; Mosser et al 1993). In this particular patient, two deletions were found separated by a large inversion (at least 110 kb of DNA) of the DNA region between the two deletions. In the first deletion resides the red colour gene in normal individuals and the AMN patient was thus missing this gene, which explained his colour vision abnormality. However, no other deletions were found among 81 ALD patients in this region, thus making it unlikely that the ALD gene resides here. The region corresponding to the second deletion (19.2 kb) and flanking DNA was isolated from normal individuals and used to analyse the DNA of other ALD patients. In 5 out of 84 unrelated ALD patients, deletions varying in length between 1.6 kb and 15.3 kb were also found, but 82 control males had no deletions in this region. It was thus likely that at least part of the ALD gene was present in the regions deleted in these patients. The complete gene was finally isolated by a combination of molecular techniques (Mosser et al 1993). The gene is expressed as mRNA of about 3.7 kb in normal tissues, including adrenal gland and brain, and has an open reading frame of 2235 bases encoding a protein of 745 amino acids. The genomic structure has recently been determined (Sarde et al 1994); the gene extends over a 21 kb DNA region and consists of ten protein coding regions (exons) ranging in length from 85 bp to about 1300 bp interrupted by nine non-coding regions (introns). The DNA sequence of at least 150 bp of each intron flanking the exons has been determined.

From the coding DNA sequence the amino acid sequence of the adrenoleukodystrophy protein (ALDP) was deduced and was shown to exhibit significant homology with the peroxisomal membrane protein (PMP) of 70 kDa relative molecular mass (PMP70) that is involved in peroxisome biogenesis and is defective in a subset of Zellweger syndrome patients (Gärtner et al 1992; Mosser et al 1993). The PMP70 protein belongs to a super-family of ATP-dependent membrane transporters (ATP binding cassette transporters or ABC transporters) (Hyde et al 1990; Kamijo et al 1990). Likewise, the putative ALDP belongs to this family of transporters and it was recently demonstrated to be associated with the peroxisomal membrane (Contreras et al 1994; Mosser et al 1994). The ALDP was also shown to be absent in lymphoblasts and fibroblasts from seven ALD patients, five of whom had partial deletions of the ALD gene. The ALDP shows no homology with the deduced amino acid sequence of the rat VLCFA-CoA synthetase (Aubourg 1994) and, although the exact biological function of the ALDP is still unknown, it is most likely involved in the transport or activation of the latter enzyme, since the activity of this enzyme is deficient in ALD (Contreras et al 1994; Mosser et al 1994).

From all the data available today, it seems that less than 10% of the ALD patients (6 out of 85 patients, or 7%) have detectable deletions in the ALD gene (Mosser et al 1993). In 3 out of 4 patients with deletions, no mRNA expression was found. The fourth patient with a deletion of 1.6 kb had a mRNA 200 bp shorter than normal (Cartier et al 1993). In one family, two brothers with a 5.7 kb deletion had different clinical phenotypes, indicating that there is no straightforward correlation between genotype and phenotype. A first mis-sense mutation (G to A at nucleotide position 1258; numbering according to Mosser et al 1993) was identified in exon 1, leading to a Glu to Lys substitution at amino acid position 291 of the ALDP (Cartier et al 1993).

Since then, several groups have identified mutations in the ALD gene starting from either mRNA or genomic DNA of patients or obligate carriers (Barcelo et al 1994; Fanen et al 1994; Fuchs et al 1994; Kemp et al 1994; Uchiyama et al 1994; Ligtenberg et al 1995). Since the flanking intronic sequences for each exon are known, each exon can be specifically amplified by the polymerase chain reaction (PCR) using intron-specific primers. Likewise, mRNA can be transcribed *in vitro* in cDNA (reverse transcription) and the resulting cDNA can serve as template for PCR with primers specific for the ALD gene. These amplified fragments can then either be screened for sequence variations by estab-lished technology followed by DNA sequencing of variant fragments or be directly sequenced (Grompe 1993). The mutations identified in unrelated families using these strategies comprise mostly missense mutations (50% or more), but also small frameshifts (deletion of one or two bases from the coding region), nonsense mutations and splice site mutations affecting the correct splicing out of the introns from the pre-mRNA.

The identification of the mutations also confirms that there is little, if any, correlation between the genotype (mutation) and the clinical phenotype of the patients. Also, many different mutations have been found in unrelated ALD families. The only exception is an AG deletion at cDNA positions 1801/1802 causing a shift in the translation reading frame of the gene which has been found in several unrelated patients of different ethnic origins (Barcelo et al 1994; Fuchs et al 1994; Kemp et al 1994; Ligtenberg et al 1995). Therefore, it will be necessary in most families to screen the whole gene for mutations. However, once the mutation is found, its presence or absence can be ascertained in each member in the family at risk. Female carriers of an ALD allele, who are asymptomatic and have normal VLCFA levels, will also be identified.

DNA DIAGNOSIS OF ALD

The vast majority of our body cells are nucleated, with only few exceptions (for instance erythrocytes). All of these cells can thus be used for the isolation of DNA for diagnosis. In practice, white cells isolated from whole blood will most often be used for postnatal DNA diagnosis since this sample can easily be obtained. Cultured skin fibroblasts, lymphoblastoid cell lines or, in the case of deceased patients, frozen stored tissues or blood spots on Guthrie cards (Rubin et al 1989) are also valuable tools for DNA isolation. For prenatal diagnosis, chorionic villi obtained in the first trimester of pregnancy can be used directly without culturing. In the second trimester of pregnancy, cultured (or uncultured) amniotic fluid cells can also be used. In those cases where PCR-based methods are available for diagnosis (for instance for mutation analysis) preimplantation diagnosis of embryos at the 6- to 8-cell stage of development could be considered.

The most direct way of DNA diagnosis in ALD families is by searching for mutations in the ALD gene. This study can start with Southern blot analysis of restriction enzyme-digested DNA of the index case of the family to detect large deletions or rearrangements in the gene. When no deletions or rearrangements are found, the individual exons of the gene and/or the cDNA can then be amplified by PCR, either as one fragment or as several overlapping fragments where the fragments would be too long for proper analysis (for example, exons 1 and 10, which are almost 1300 bp long). Direct sequencing of the resulting fragments or detection of sequence variations followed by sequencing of variant fragments would then lead to the identification of the mutation causing ALD. As mentioned before, the index case is preferentially used for this analysis. If no index case is available, an obligate carrier female should be used. However, since these females also carry a normal ALD gene, the absence of an individual exon in the mutated allele could be masked by the normal one when using PCR amplification. Therefore, absence of that exon will not be detected owing to the presence of the normal exon. In mothers of sporadic ALD patients, the mutation may also not be detectable since the mutations in these families may result from a *de novo* mutation or from germ-line mosaicism in the mother (Graham et al 1992). Prenatal diagnosis by VLCFA analysis and measurement of the activity of the peroxisomal β-oxidation remains the option in these cases. Once a mutation has been found in a family, it can be traced in all individuals at risk of that family, both post- and prenatally, without the need for VLCFA analysis.

The technical facilities for mutation analysis in the ALD gene may not be available in all laboratories. Therefore, segregation analysis of the disease with polymorphic markers in the Xq28 region may be used. Many polymorphic markers have been isolated in this region and can potentially be used for segregation analysis. An example of segregation analysis in an ALD family is shown in Figure 1. Several authors have described carrier detection and prenatal diagnosis in this way. The markers used include DXS52 (Boué et al 1985; Aubourg et al 1987; Willems et al 1990; Van Oost et al 1991), DXS15 (Willems et al 1990), polymorphisms in the factor VIII gene, and the probe F814 which detects polymorphisms both at DXS52 and in the factor VIII gene (Willems et al 1990; Lalloz et al 1991; Van Oost et al 1991). Although very few recombinations between the ALD gene and these markers have been reported, it should be advised to use markers on both sides of the ALD gene, thereby allowing the identification of possible recombinations. The risk

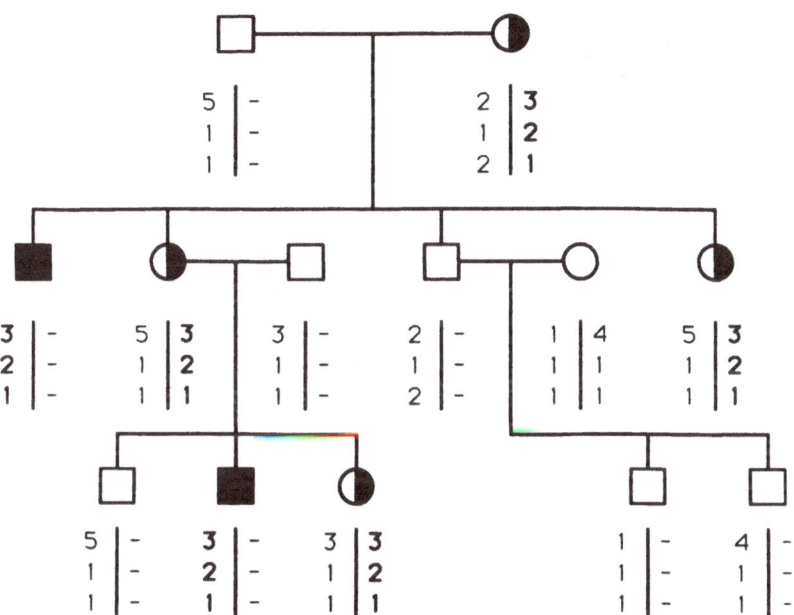

Figure 1 Segregation analysis in a pedigree with two male patients affected with adrenoleukodystrophy (closed squares). Half-closed circles refer to carrier females. The DNA haplotypes are shown under each individual and were obtained by Southern blot analysis with probe F814 at the DXS52 locus (alleles 1−5) and at the factor VIII locus (alleles 1 or 2), and with probe DX13 at the DXS15 locus (alleles 1 or 2). The results demonstrate that, in this family, ALD segregates with the allele 3−2−1 (shown in bold)

of recombination would be further reduced when intragenic polymorphisms were available. Two polymorphisms in the ALD gene have been described (G or A at nucleotide position 1934, predicting no change of the leucine at amino acid position 516, and C or G at position 2632 just behind the translation stop codon) and these polymorphisms could potentially be useful for segregation analysis (Fuchs et al 1994; Kemp et al 1994). The disadvantage of segregation analysis is that family studies are necessary and that a minimum number of family members is necessary for the reliable determination of the polymorphic fragments segregating with the disease. However, in the majority of families this will be no problem, and also sporadic cases of ALD are rather uncommon. The highly polymorphic nature of DNA markers in the Xq28 region also means that only in a minority of families will DNA diagnosis not be possible. In these families, VLCFA analysis remains the alternative.

PRACTICAL ASPECTS OF MUTATION ANALYSIS

Once the normal cDNA sequence and the genomic structure of a disease-causing gene are known, mutation analysis in the genes of patients can be considered, starting from either RNA or genomic DNA. Genomic DNA and RNA can be isolated from patient tissues, either directly or after culture (for instance skin fibroblasts or lymphoblastoid cell lines), by standard laboratory techniques or by using commercial reagent kits. The RNA has to be reverse-transcribed into cDNA before use and this can also be done with commercial reagent kits.

Several mutation detection techniques have been developed in recent years (for review see Grompe (1993)) and most of these could potentially be used for analysis of the ALD gene. Many of these techniques rely on the primer-specific PCR amplification of the target DNA or cDNA prior to mutation analysis. So far, the most complete and straightforward mutation analysis of the ALD gene has been conducted by Ligtenberg and colleagues (1995). These authors isolated RNA from cultured fibroblasts from 28 ALD patients or female carriers. After reverse transcription of the RNA, the whole coding region of the ALD gene was PCR-amplified in five partially overlapping fragments with ALD gene-specific PCR primers (RT-PCR). The fragments were then directly sequenced using the PCR primers and newly designed internal primers. This approach is very effective in mutation detection and the mutations in all 28 unrelated families were identified. However, RNA of patients may not always be available. Another drawback of this technique may present in patients and carriers either not expressing or showing strongly reduced expression of the mutant gene at the RNA level. However, this is probably only a minor problem for the analysis of the ALD gene, since all mutations could be identified in the study of Ligtenberg et al (1995).

A similar approach was used by other groups starting from genomic DNA of patients. In these studies, primers were designed from the intronic sequences of the ALD gene as published by Sarde et al (1994). In contrast to RT-PCR, at least 10 genomic fragments have to be amplified to cover the 10 exons of the ALD gene. Another problem in genomic DNA analysis is that the human genome contains sequences with high homology (up to 90%) to the 3' end of the ALD gene (exons 7–10 region). Some of these sequences interfere with PCR amplification of the ALD gene and render mutation analysis difficult (Sarde et al 1994). Nevertheless, several mutations have been identified using this approach. Barcelo et al (1994) used PCR amplification and direct sequencing to detect the AG deletion at cDNA positions 1801/1802 in exon 5. Other groups have included well-established PCR-based mutation screening methods such as single-strand conformation polymorphism analysis (SSCP; Orita et al 1989) and denaturing gradient gel electrophoresis (DGGE; Lerman and Silverstein 1987). These screening methods have the advantage that, from all fragments studied, only those that show aberrant migration (relative to fragments from normal individuals, in well-defined conditions in polyacrylamide gels) have to be sequenced. However, the sensitivity of these methods is not absolute and not all mutations will be detected (Hayashi 1992; Grompe 1993).

On the basis of the distribution of the mutations that have been found so far, a strategy could be developed for mutation analysis of other ALD families. The PCR amplification and analysis of exons 1 (the largest exon) and 5 will probably identify the mutation in more than 50% of the families. Both exons can be amplified from genomic DNA. When no mutations are found in these exons, the remaining exons must be analysed either in genomic DNA or in RNA. The exons 2–4 and 6 can be analysed starting from genomic DNA, while the exons 7–10 region is probably best studied through cDNA analysis.

DNA DIAGNOSIS IN OTHER PEROXISOMAL DISORDERS

The genes for several enzymes involved in the loss of a single peroxisomal function have been isolated, but so far mutations have only been identified in the genes for peroxisomal acyl-CoA oxidase (Fournier et al 1994b; Varanasi et al 1994) and alanine:glyoxylate aminotransferase (AGT) (Takada et al 1990; Purdue et al 1991). In two siblings with pseudoneonatal adrenoleukodystrophy, a large deletion encompassing almost the whole peroxisomal acyl-CoA oxidase gene has been identified recently (Fournier et al 1994b). In about one-third of patients with primary hyperoxaluria type 1 (PH1), the intracellular distribution of the AGT enzyme is impaired owing to a glycine to arginine substitution at amino acid position 170 (Purdue et al 1990). The substitution leads to a mistargeting of the liver-specific AGT to the mitochondria. In several patients, a glycine to glutamate substitution at position 82 has been identified leading to a complete loss of AGT catalytic activity (Purdue et al 1992). Mutations involved in the intraperoxisomal aggregation of AGT or leading to non-expression of AGT have also been identified (for review see Danpure et al (1994)).

The disorders Zellweger syndrome, neonatal adrenoleukodystrophy and infantile Refsum disease lead to a loss of multiple peroxisomal functions. These disorders belong to at least nine complementation groups, implying that at least nine genes are required for peroxisome assembly. In one patient with Zellweger syndrome, a nonsense mutation resulting in premature translation termination of the peroxisome assembly factor-1 protein has been demonstrated to be the underlying defect (Shimozawa et al 1992). The patient was homozygous for the mutation and her parents were shown to be heterozygous carriers of this mutation. This information was used for DNA diagnosis in a fetus in this family and resulted in the detection of a healthy carrier fetus (Shimozawa et al 1993a). Recently, the same mutation was identified in a Zellweger patient belonging to the same complementation group (Shimozawa et al 1993b). As mentioned before, mutations in the PMP70 have been found in two Zellweger syndrome families of complementation group 1 (Gärtner et al 1992). The first patient is a heterozygous carrier of a *de novo* splice site mutation resulting in the insertion of 23 bp of intronic DNA between exons 1 and 2. In the second family, two patients are carriers of a G to A transition, predicting the substitution of glycine at amino acid position 17 of the PMP70 protein by aspartic acid. The father and an asymptomatic female sibling of the patients also carried this mutation. However, a second mutation could not be found in these two families, and no other mutations have been found in the PMP70 gene of 19 other complementation group 1 patients. These results indicate that the deficiency in complementation group 1 patients is heterogeneous.

ACKNOWLEDGEMENTS

We thank Mrs L. Dumoulin and Mr E. Van Emelen for help in the preparation of the manuscript.

REFERENCES

Aubourg P (1994) Adrenoleukodystrophy and other peroxisomal diseases. *Current Opinion in Genetics and Development* **4**: 407–411.

Aubourg PR, Sack GH, Meyers DA, Lease JJ, Moser HW (1987) Linkage of adrenoleukodystrophy to a polymorphic DNA probe. *Ann Neurol* **21**: 349–352.

Aubourg PR, Sack GH, Moser HW (1988) Frequent alterations of visual pigment genes in adreno-leukodystrophy. *Am J Hum Genet* **42**: 408–413.

Aubourg P, Feil R, Guidoux S, et al (1990) The red-green visual pigment gene region in adreno-leukodystrophy. *Am J Hum Genet* **46**: 459–469.

Barcelo A, Giros M, Sarde CO, et al (1994) Identification of a new frameshift mutation (1801delAG) in the ALD gene. *Hum Mol Genet* **10**: 1889–1890.

Bione S, Maestrini E, Rivella S, et al (1994) Identification of a novel X-linked gene responsible for Emery–Dreifuss muscular dystrophy. *Nature Genetics* **8**: 323–327.

Boué J, Oberlé I, Heilig R, et al (1985) First trimester prenatal diagnosis of adrenoleukodystrophy by determination of very long chain fatty acid levels and by linkage analysis to a DNA probe. *Hum Genet* **69**: 272–274.

Cartier N, Sarde CO, Douar AM, Mosser J, Mandel JL, Aubourg P (1993) Abnormal messenger RNA expression and a missense mutation in patients with X-linked adrenoleukodystrophy. *Hum Mol Genet* **2**: 1949–1951.

Contreras M, Mosser J, Mandel JL, Aubourg P, Singh I (1994) The protein coded by the X-adrenoleukodystrophy gene is a peroxisomal integral membrane protein. *FEBS Lett* **344**: 211–215.

Danpure CJ, Jennings PR, Fryer P, Purdue PE, Allsop J (1994) Primary hyperoxaluria type 1: genotypic and phenotypic heterogeneity. *J Inher Metab Dis* **17**: 487–499.

Fanen P, Guidoux S, Sarde CO, Mandel JL, Goossens M, Aubourg P (1994) Identification of mutations in the putative ATP-binding domain of the adrenoleukodystrophy gene. *J Clin Invest* **94**: 516–520.

Feil R, Aubourg P, Mosser J, et al (1991) Adrenoleukodystrophy: a complex chromosomal rearrangement in the Xq28 red/green-color-pigment gene region indicates two possible gene localizations. *Am J Hum Genet* **49**: 1361–1371.

Fournier B, Smeitink JAM, Dorland L, Berger, Saudubray JM, Poll-The BT (1994a) Peroxisomal disorders: a review. *J Inher Metab Dis* **17**: 470–486.

Fournier B, Saudubray JM, Benichou B, et al (1994b) Large deletion of the peroxisomal acyl-CoA oxidase gene in pseudoneonatal adrenoleukodystrophy. *J Clin Invest* **94**: 526–531.

Fuchs S, Sarde CO, Wedemann H, Schwinger E, Mandel JL, Gal A (1994) Missense mutations are frequent in the gene for X-chromosomal adrenoleukodystrophy (ALD). *Hum Mol Genet* **3**: 1903–1905.

Gärtner J, Moser H, Valle D (1992) Mutations in the 70K peroxisomal membrane protein gene in Zellweger syndrome. *Nature Genetics* **1**: 16–23.

Gitschier J, Wood WI, Goralka TM, et al (1984) Characterization of the human factor VIII gene. *Nature* **312**: 326–330.

Graham GE, MacLeod PM, Lillicrap DP, Bridge PJ (1992) Gonadal mosaicism in a family with adrenoleukodystrophy: molecular diagnosis of carrier status among daughters of a gonadal mosaic when direct detection of the mutation is not possible. *J Inher Metab Dis* **15**: 68–74.

Grompe M (1993) The rapid detection of unknown mutations in nucleic acids. *Nature Genetics* **5**: 111–117.

Hayashi K (1992) PCR-SSCP: a method for detection of mutations. *Genet Anal Techn Appl* **9**: 73–79.

Hyde SC, Emsley P, Hartshorn MJ, et al (1990) Structural model of ATP-binding proteins associated with cystic fibrosis, multidrug resistance and bacterial transport. *Nature* **346**: 362–365.

Jouet M, Rosenthal A, Armstrong G, et al (1994) X-linked spastic paraplegia (SPG1), MASA syndrome and X-linked hydrocephalus result from mutations in the L1 gene. *Nature Genetics* **7**: 402–407.

Kamijo K, Taketani S, Yokota S, Osumi T, Hashimoto T (1990) The 70-kDa peroxisomal membrane protein is a member of the Mdr(P-glycoprotein)-related ATP-binding protein superfamily. *J Biol Chem* **265**: 4534–4540.

Kemp S, Ligtenberg MJL, Van Geel BM, et al (1994) Identification of a two base pair deletion in five unrelated families with adrenoleukodystrophy: a possible hotspot for mutations. *Biochem Biophys Res Commun* **202**: 647–653.

Lalloz MRA, McVey JH, Pattinson JK, Tuddenham EGD (1991) Haemophilia A diagnosis by analysis of a hypervariable dinucleotide repeat within the factor VIII gene. *Lancet* **338**: 207–211.

Lerman LS, Silverstein K (1987) Computational simulation of DNA melting and its application to denaturing gel electrophoresis. *Methods Enzymol* **155**: 482–501.

Ligtenberg MJL, Kemp S, Sarde CO, et al (1995) Spectrum of mutations in the gene encoding the adrenoleukodystrophy protein. *Am J Hum Genet* **56**: 44–50.

Maestri NE, Beaty TH (1992) Predictions of a 2-locus model for disease heterogeneity: application to adrenoleukodystrophy. *Am J Med Genet* **44**: 576–582.

Migeon BR, Moser HW, Moser B, Axelman J, Silence D, Norum RA (1981) Adrenoleukodystrophy: evidence for X-linkage, inactivation, and selection favoring the mutant allele in heterozygous cells. *Proc Natl Acad Sci USA* **78**: 5066–5070.

Moser HW, Moser AB (1989) Adrenoleukodystrophy (X-linked). In Scriver CR, Beaudet AL, Sly WS, Valle D, eds. *The Metabolic Basis of Inherited Disease*, 6th edn. New York: McGraw-Hill, 1511–1532.

Moser HW, Moser AE, Trojak JE, Supplee SW (1983) Identification of female carriers of adreno-leukodystrophy. *J Pediatr* **103**: 54–59.

Moser HW, Moser AE, Singh I, O'Neill BP (1984) Adrenoleukodystrophy: survey of 303 cases: biochemistry, diagnosis and therapy. *Ann Neurol* **16**: 628–641.

Moser HW, Moser AB, Smith KD, et al (1992) Adrenoleukodystrophy: phenotypic variability and implications for therapy. *J Inher Metab Dis* **15**: 645–664.

Mosser J, Douar AM, Sarde CO, et al (1993) Putative X-linked adrenoleukodystrophy gene shares unexpected homology with ABC transporters. *Nature* **361**: 726–730.

Mosser J, Lutz Y, Stoeckel ME, et al (1994) The gene responsible for adrenoleukodystrophy encodes a peroxisomal membrane protein. *Hum Mol Genet* **3**: 265–271.

Nathans J, Thomas D, Hogness DS (1986a) Molecular genetics of human color vision: the genes encoding blue, green, and red fragments. *Science* **232**: 193–202.

Nathans J, Piantanida TP, Eddy RL, Shows TB, Hogness DS (1986b) Molecular genetics of inherited variation in human color vision. *Science* **232**: 203–210.

Oberlé I, Drayna D, Camerino G, White R, Mandel JL (1985) The telomeric region of the human X-chromosome long arm: presence of a highly polymorphic DNA marker and analysis of recombination frequency. *Proc Natl Acad Sci USA* **82**: 2824–2828.

Orita M, Iwahana H, Kanazawa H, Hayashi K, Sekiya T (1989) Detection of polymorphisms of human DNA by gel electrophoresis as single-strand conformation polymorphisms. *Proc Natl Acad Sci USA* **86**: 2766 –2770.

Pan Y, Metzenberg A, Das S, Jing B, Gitschier J (1992) Mutations in the V2 vasopressin receptor gene are associated with X-linked nephrogenic diabetes insipidus. *Nature Genetics* **2**: 103–106.

Purdue PE, Takada Y, Danpure CJ (1990) Identification of mutations associated with peroxisome-to-mitochondrion mistargeting of alanine·glyoxylate amino transferase in primary hyperoxaluria type 1. *J Cell Biol* **111**: 2341–2351.

Purdue PE, Lumb MJ, Fox M, et al (1991) Characterization and chromosomal mapping of a genomic clone encoding human alanine: glyoxylate amino transferase. *Genomics* **10**: 34–42.

Purdue PE, Lumb MJ, Allsop J, Minatogawa Y, Danpure C (1992) A glycine-to-glutamate substitution abolishes alanine: glyoxylate aminotransferase catalytic activity in a subset of patients with primary hyperoxaluria type 1. *Genomics* **13**: 215–218.

Rubin EM, Andrews KA, Kan YW (1989) Newborn screening by DNA analysis of dried blood spots. *Hum Genet* **82**: 134–136.

Sack GH, Morrell JC (1993) Adrenoleukodystrophy: overlapping deletions point to a gene location in Xq28. *Biochem Biophys Res Commun* **191**: 955–960.

Sack GH, Raven MB, Moser HW (1989) Color vision defects in adrenomyeloneuropathy. *Am J Hum Genet* **44**: 794–798.

Sack GH, Alpern M, Webster T, et al (1993) Chromosomal rearrangement segregating with adreno-leukodystrophy: a molecular analysis. *Proc Natl Acad Sci USA* **90**: 9489–9493.

Sarde CO, Mosser J, Kioschis P, et al (1994) Genomic organization of the adrenoleukodystrophy gene. *Genomics* **22**: 13–20.

Shimozawa N, Tsukamoto T, Suzuki Y, et al (1992) A human gene responsible for Zellweger syndrome that affects peroxisome assembly. *Science* **255**: 1132–1134.

Shimozawa N, Suzuki Y, Orii T, Tsukamoto T, Fujiki Y (1993a) Prenatal diagnosis of Zellweger syndrome using DNA analysis. *Prenat Diagn* **13**: 149.

Shimozawa N, Suzuki Y, Orii T, Moser AB, Moser HW, Wanders RJA (1993b) Standardization of complementation grouping of peroxisome-deficient disorders and the second Zellweger patient with peroxisomal assembly factor (PAF-1) defect. *Am J Hum Genet* **52**: 843–844.

Szabo P, Purrello M, Rocchi M, et al (1984) Cytological mapping of the human glucose-6-phosphate dehydrogenase gene distal to the fragile-X site suggests a high rate of meiotic recombination across this site. *Proc Natl Acad Sci USA* **81**: 7855–7859.

Takada Y, Kanedo N, Esumi H, Purdue PE, Danpure CJ (1990) Human peroxisomal L-alanine: glyoxylate amino transferase. Evolutionary loss of mitochondrial targeting signal by point mutation of the initiation codon. *Biochem J* **268**: 517–520.

Uchiyama A, Suzuki Y, Song X-Q, et al (1994) Identification of a nonsense mutation in ALD protein cDNA from a patient with adrenoleukodystrophy. *Biochem Biophys Res Commun* **198**: 632–636.

Van Geel BM, Assies J, Weverling GJ, Barth PG (1994) Predominance of the adrenomyeloneuropathy phenotype of X-linked adrenoleukodystrophy in the Netherlands: a survey of 30 kindreds. *Neurology* **44**: 2343–2346.

Van Oost BA, Van Zandvoort PM, Tünte W, et al (1991) Linkage analysis in X-linked adrenoleukodystrophy and application in post- and prenatal diagnosis. *Hum Genet* **86**: 404–407.

Varanasi U, Chu R, Chu S, Espinosa R, LeBeau MM, Reddy JK (1994) Isolation of the human peroxisomal acyl-CoA oxidase gene: organization, promoter analysis, and chromosomal localization. *Proc Natl Acad Sci USA* **91**: 3107–3111.

Vits L, Van Camp G, Coucke P, et al (1994) MASA syndrome is due to mutations in the neural cell adhesion gene L1CAM. *Nature Genetics* **7**: 408–413.

Wanders RJA, Van Roermund CWT, Lageweg W, et al (1992) X-linked adrenoleukodystrophy: biochemical diagnosis and enzyme defect. *J Inher Metab Dis* **15**: 634–644.

Watkiss E, Webb T, Bundey S (1993) Is skewed X inactivation responsible for symptoms in female carriers for adrenoleukodystrophy? *J Med Genet* **30**: 651–654.

Willems PJ, Vits L, Wanders RJA, et al (1990) Linkage of DNA markers at Xq28 to adrenoleukodystrophy and adrenomyeloneuropathy present within the same family. *Arch Neurol* **47**: 665–669.

J. Inher. Metab. Dis. 18 Suppl. 1 (1995) 45–60

Pre- and postnatal diagnosis of peroxisomal disorders using stable-isotope dilution gas chromatography–mass spectrometry

N. M. Verhoeven, W. Kulik, C.M.M. van den Heuvel and C. Jakobs*
Department of Pediatrics, Free University Hospital Amsterdam, De Boelelaan 1117, 1081 HV Amsterdam, The Netherlands
*Correspondence

Summary: Quantitative analysis of the following peroxisomal metabolites is reported: very long-chain fatty acids (VLCFA), pipecolic acid, bile acid intermediates, phytanic and pristanic acid, in plasma, urine, cerebrospinal fluid (CSF), blood spots collected at neonatal screening and amniotic fluid. An overview is given of the concentrations of these metabolites in body fluids from control subjects and all patients investigated so far in this laboratory. The method of choice is gas chromatography–mass spectrometry (GC-MS) with electron capture detection, combined with the use of stable-isotope-labelled internal standards

The peroxisome is a cellular organelle whose importance is evident from the severity of clinical signs and symptoms in patients with disturbed peroxisomal function.

The peroxisomal disorders are biochemically characterized by accumulation in tissues and body fluids of the metabolites that are normally degraded in the peroxisome. These metabolites include saturated and unsaturated VLCFA and the branched-chain fatty acids, phytanic and pristanic acid, of which oxidation is primarily a peroxisomal process. Defects in bile acid biosynthesis caused by deficiency or absence of peroxisomes result in elevated levels of the bile acids dihydroxy- and trihydroxycoprostanic acid (DHCA and THCA). Also pipecolic acid, an intermediate in the degradation of the amino acid lysine, is found to accumulate in patients with general peroxisomal dysfunction. Peroxisomal disorders can be classified into three groups depending upon the extent of peroxisomal dysfunction (see Wanders et al (1988) for a review). The first group is characterized by the absence of morphologically distinguishable peroxisomes, resulting in a virtually complete loss of peroxisomal functions. One of the diseases belonging to this group is the Zellweger syndrome (McKusick 214100), a severe disease with death usually occurring before the first year of life. The general deficiency of peroxisomal functions causes accumulations of VLCFA, pipecolic acid, bile acid intermediates, phytanic and pristanic acids. In the second group, although peroxisomes are present, several but not all peroxisomal functions are impaired. In this group is rhizomelic chondrodysplasia punctata (RCDP) (McKusick 215100), in which there are deficiencies in plasmalogen synthesis and phytanic acid oxidation, resulting in accumulation of phytanic acid and deficient plasmalogen levels in tissues and blood cells. Group three consists of disorders with loss of a single peroxisomal function: e.g., X-linked adrenoleukodystrophy (X-ALD) (McKusick 300100), peroxisomal acyl-CoA oxidase deficiency (McKusick 264470), bifunctional protein deficiency and peroxisomal 3-oxoacyl-CoA thiolase deficiency (McKusick 261510).

Classical Refsum disease (McKusick 266500) is biochemically characterized by accumulation of phytanic acid, caused by defective oxidation of this fatty acid. Because it is still not clear whether α-oxidation is a peroxisomal or a mitochondrial process or a process taking place in both organelles, Refsum disease is not yet definitively classified as a peroxisomal disorder (Steinberg 1989; Singh et al 1993).

The fact that a broad range of clinical symptoms can be found within one particular type of peroxisomal disorder (e.g. Zellweger syndrome) together with the similarity of clinical abnormalities among different peroxisomal disorders makes correct diagnosis on clinical grounds difficult. Virtually all disorders from groups 1 to 3 present with severe neurological dysfunction, cranial dysmorphism, hearing loss, visual problems and severe psychomotor retardation (Fournier et al 1994; Heymans et al 1990; Poggi-Travert et al, this issue). This similarity in clinical signs and symptoms among the different peroxisomal disorders makes biochemical differential diagnosis very important. Thus analysis of the peroxisomal metabolites VLCFA, phytanic and pristanic acid, bile acids and pipecolic acid is of great value.

In this chapter is summarized determination of these metabolites in plasma, urine, CSF, blood spots collected at neonatal screening and, for prenatal diagnosis, in amniotic fluid. An overview is given of the concentrations of these peroxisomal metabolites in body fluids from control subjects and from all patients investigated in our laboratory. The method of choice for quantitative analysis is GC-MS with electron capture chemical ionization, combined with the use of stable-isotope-labelled internal standards.

In the following section we explain the principal features of this technique.

Gas chromatography–mass spectrometry (McCloskey 1990)

In order to determine low concentrations (up to fmol/μl) of compounds in complex mixtures accurately, it is necessary to use a technique that is both highly selective and sensitive. Therefore, quantification of peroxisomal metabolites in body fluids is performed by GC combined with electron capture (EC) MS. The required accuracy is derived from the use of internal standards labelled with stable (non-radioactive) isotopes, referred to as stable-isotope dilution.

After clean-up and derivatization, extracts of body fluids are still extremely complex, containing a large variety of compounds. The GC is used for the separation of these compounds into different fractions prior to their introduction into the MS. In the ideal case, the compounds elute one after another from the GC column, entering the ion source of the MS at different times.

In the ion source, ions are produced proportionately to the amount of the compound entering from the GC. These charged particles move into the high vacuum of the mass analyser where they are exposed to the applied electric and/or magnetic fields, thus allowing separation on the basis of their mass-to-charge ratio. After mass separation, the ions are measured by the detector as an electric current proportional to the number of ions

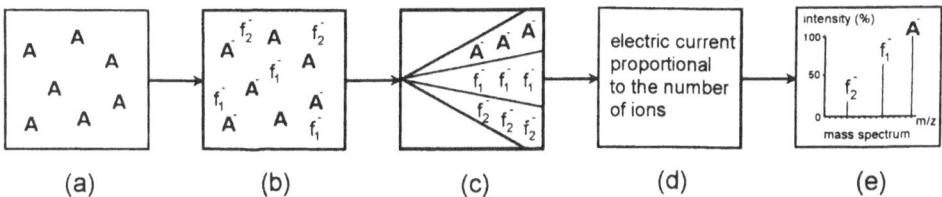

a. sample inlet system
b. ion source (ion formation)
c. mass analyzer with ion (m/z)-separating fields
d. ion detector
e. data system

Figure 1 Schematic representation of a mass spectrometer. The molecules of a compound (A) enter the source. In the source, ions (A⁻, f_1^-, f_2^-) are formed. The ions are separated on their mass-to-charge ratio in the electric and/or magnetic field, after which they reach the detector. The data system transforms the signals into a mass spectrum

carrying the same mass (Figure 1). By varying the electric and/or magnetic field, the ions for a range of mass values can be focused consecutively on the detector. This continuous scanning over a range of mass values produces the mass spectrum: a graph of 'mass' versus 'relative intensity'. Since the different compounds produce their own characteristic patterns, the recorded signals for these masses are characteristic for the identity and amount of these compounds.

For quantitative analysis, ion-chromatograms with Gaussian peak shapes (graphs of response of a selected ion vs time) are recorded. With this method, the 'scan speed' within the mass range of interest should be chosen in such a way that the response for all masses of interest can be measured at least 10–12 times (Figure 2). This can be achieved by monitoring only a few characteristic ions of the analyte(s) and the internal standard(s) and by ignoring the ions at all other mass values. This latter technique is known as selected-ion monitoring (SIM). Approximately 100–1000 times more sensitivity can be achieved by using SIM as compared to repetitive scanning of complete mass spectra.

For the analysis of known peroxisomal metabolites it is only necessary to confirm the identity of the compound of interest by one properly selected characteristic ion. Furthermore, an ionization technique (electron capture chemical ionization) is chosen that ensures minimal fragmentation of the ionized compounds. This leads to maximal efficiency of ion formation.

Electron capture chemical ionization (Harrison, 1992)

In electron capture chemical ionization (ECCI, also referred to as negative ion chemical ionization, NICI), the reagent gas acts as a moderating gas that 'thermalizes' the electrons. The most common moderating gases are CH_4, NH_3 and i-C_4H_{10}. The sensitivity for ionization of the analyte depends on (a) the efficiency of the moderating gas to thermalize the electrons; (b) the efficiency of the moderating gas in collisionally stabilizing the molecular anions formed in the EC process.

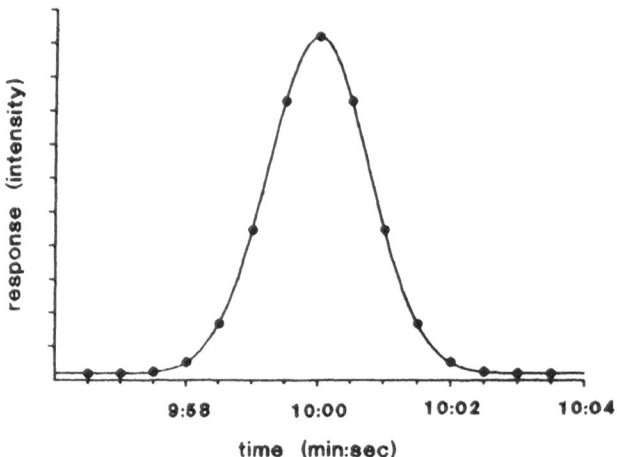

Figure 2 Ion chromatogram. The dots in the curve represent the data points. At least 10–12 data points are required to construct a proper curve

The major advantage of ECCI is the potentially high sensitivity (up to attomol/μl). It is, however, a prerequisite that the analyte has a substantial electron affinity. For this reason, analytes with a low electron affinity are derivatized with reagents that introduce these particular groups.

It is, however, not sufficient to have a facile EC only; it is also necessary that characteristic ions from the analyte are formed. For example, pentafluorobenzyl bromide (PFB-Br) is a derivatizing agent that can form ethers and/or esters that satisfy both requirements of high electron affinity and formation of characteristic ions by loss of the PFB moiety with retention of the negative charge on the analyte. In the same way it is a practical advantage of ECCI that all compounds that do not possess the required electron affinity will not be ionized and thus will not be detected.

It should be noted that the technique of ECCI requires specialized technical and instrumental expertise.

Stable-isotope dilution

The use of an internal standard is based on the assumption that a known amount of this standard, added at the start of the analytical procedure, will behave in a similar way to the analyte throughout the complete analysis. The ratio between the final detected response for the internal standard and the response for the analyte enables quantification of the metabolite. In this way, there is a complete correction for unforeseen and/or uncontrollable losses during the analytical process since this ratio does not change. The compounds most suited as internal standards are compounds with the same extraction coefficient, derivatization efficiency, retention time on the GC column and electron affinity as the analyte to be detected. Compounds that meet these requirements are usually analogues of the metabolites, labelled with stable isotopes. If, for example, four hydrogen atoms in a VLCFA are replaced by four deuterium atoms, the behaviour of the new stable-isotope-labelled compound will be virtually identical to that of the unlabelled analyte upon extraction and derivatization. Only the retention time on the GC column will be slightly shorter (which does not affect the ratio between unlabelled/labelled compound). Owing to the labelling, the masses of the ions of the internal standard are 4 atomic mass units higher than the masses of the analyte ions. This results in two distinct responses for the analyte, or unlabelled compound (unknown quantity), and the labelled compound (known quantity). The use of a calibration curve (ratio vs concentration) gives accurate values for the original concentrations of the analyte.

The presence of the signal of the labelled internal standard identifies when and where the signal of the analyte should occur and indirectly confirms its structure. It is inherent to this method of stable-isotope dilution (SID) that every analyte requires its own labelled analogue.

In summary, SID GC-MS with ECCI owes its selectivity to: (a) the extraction and derivatization procedure, (b) GC separation, (c) ECCI as ionization method, (d) mass-selective detection, (e) stable-isotope-labelled internal standards. Its sensitivity results from its selectivity and the high ionization efficiency.

MATERIALS AND METHODS

Body fluids

Control plasma, urine and CSF samples and control blood spots collected at neonatal screening were obtained from individuals without any symptoms of metabolic disease or impaired renal and liver function. Control amniotic fluid samples were obtained from healthy pregnant women by amniocentesis between the 16th and 18th weeks of pregnancy. Amniocentesis was done on indication of advanced maternal age. Plasma, urine, CSF and blood spot samples were obtained from patients with various peroxisomal disorders. Determination of the exact type of peroxisomal disorder was based on multiple characteristic clinical, metabolic and enzymatic abnormalities. Amniotic fluid samples from pregnancies 'at risk' (index case mostly classified) were also obtained between the 16th and 18th weeks of gestational age. Most samples were obtained from foreign countries and were received lyophilized or frozen in dry ice and stored at −20°C prior to analysis.

Methods

Determination of VLCFA in plasma, urine, amniotic fluid, and CSF was performed according to a method described by Stellaard et al (1990) and of phytanic acid and pristanic acid according to ten Brink et al (1992). Analysis of VLCFA in blood spots was performed according to Jakobs et al (1993), and of phytanic acid and pristanic acid in blood spots according to ten Brink et al (1993a). Analysis of pipecolic acid in body fluids was performed either according to Kok et al (1987) or to Zee et al (1992). The methods used to quantify bile acids in plasma and in amniotic fluid were described by Stellaard et al (1989, 1991).

In Figure 3, a schematic representation of the sample preparations for quantitative analysis of peroxisomal metabolites is given.

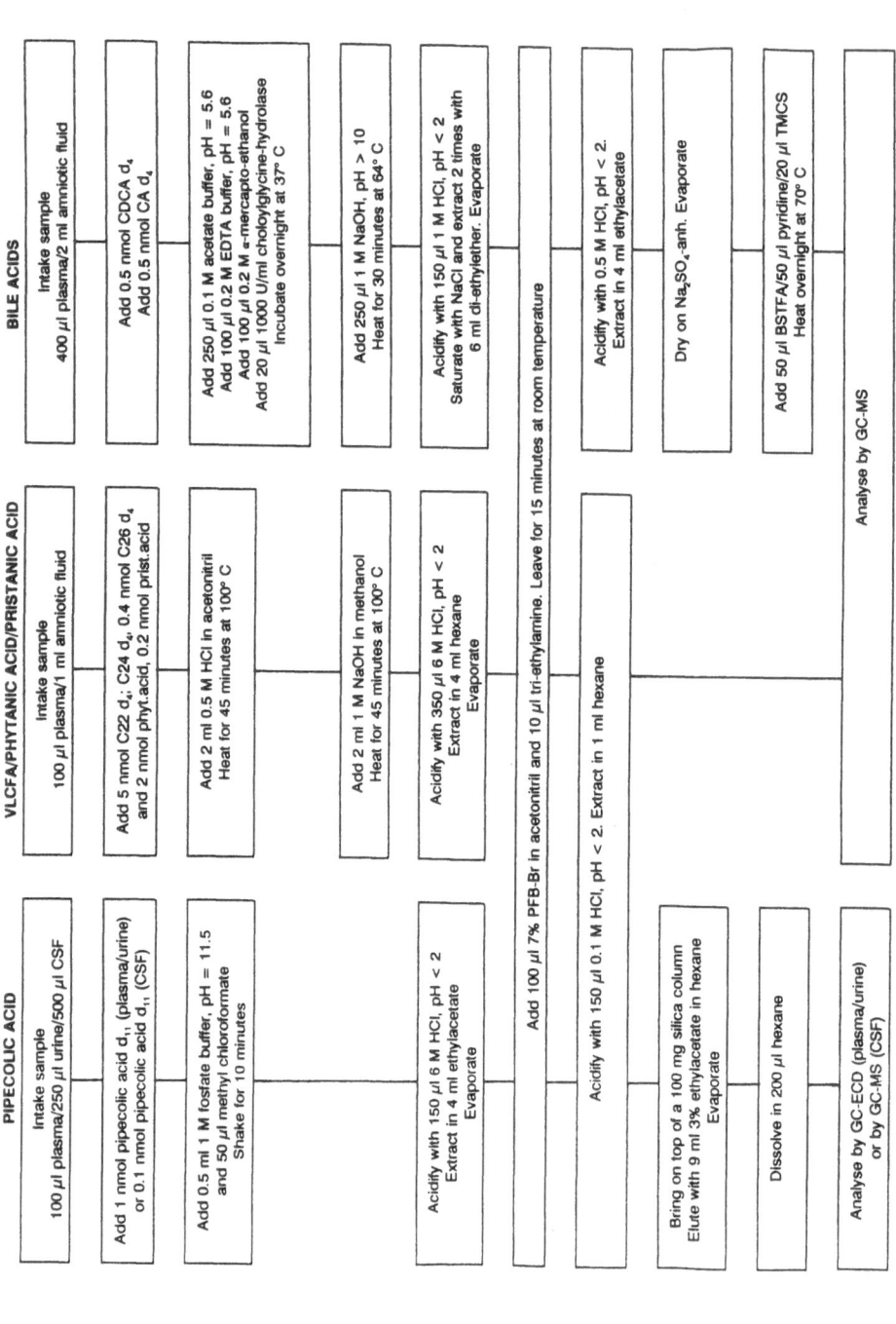

Figure 3 Sample preparation for quantitative analysis of pipecolic acid, VLCFA, phytanic acid and pristanic acid and bile acids

Table 1 **Origin of unlabelled authentic compounds and stable-isotope-labelled internal standards**

Compound	Origin	Reference
Docosanoic acid ($C_{22:0}$)	Sigma Chemicals, St Louis, MO, USA	
Tetracosanoic acid ($C_{24:0}$)	Sigma Chemicals, St Louis, MO, USA	
Hexacosanoic acid ($C_{26:0}$)	Sigma Chemicals, St Louis, MO, USA	
(3,3,4,4-D_4)Docosanoic acid	Own synthesis	ten Brink (1989b)
(3,3,5,5-D_4)Tetracosanoic acid	Own synthesis	ten Brink (1989b)
(3,3,7,7-D_4)Hexacosanoic acid	Own synthesis	ten Brink (1989b)
Phytanic acid	Own synthesis	ten Brink (1989a)
Pristanic acid	Own synthesis	ten Brink (1989a)
(3-Methyl-D_3)phytanic acid	Own synthesis	ten Brink (1989a)
(2-Methyl-D_3)pristanic acid	Own synthesis	ten Brink (1989a)
Cholic acid (CA)	Supelco Inc, Bellefonte, PA, USA	
Chenodeoxycholic acid (CDCA)	Supelco Inc, Bellefonte, PA, USA	
Trihydroxycoprostanic acid (THCA)	Private gift	Stellaard (1989)
Dihydroxycoprostanic acid (DHCA)	Private gift	Stellaard (1989)
(2,2,4,4-D_4)Cholic acid	MSD-Isotopes, Montreal, Canada	
(2,2,4,4-D_4)Chenodeoxycholic acid	MSD-Isotopes, Montreal, Canada	
L-Pipecolic acid	Sigma Chemicals, St Louis, MO, USA	
DL-(1,2,2,3,3,4,4,5,5,6,6-D_{11})Pipecolic acid	MSD-Isotopes, Montreal Canada	

Table 1 summarizes the origin of the authentic compounds and internal standards used for analysis of VLCFA, phytanic and pristanic acid, bile acids and pipecolic acid in body fluids and blood spots.

RESULTS AND DISCUSSION

Measurement of peroxisomal metabolites, of which the concentrations in body fluids are low with only marginally elevated levels in some peroxisomal disorders (e.g. X-ALD), places great demands on the technique used. We therefore selected GC-MS analysis with ECCI and stable-isotope-labelled internal standards as the method of choice. Although this method is in general analytically superior to other methods, it has two main disadvantages. The first is the necessity to synthesize internal standards that are often not commercially available. The second disadvantage arises from the use of the GC-MS equipment, which has high costs of acquisition and servicing. Therefore, for pipecolic acid determination we now frequently use a GC method with EC detection (Zee et al 1992). This method is less expensive but still yields high accuracy and sensitivity, owing to the use of the [$^2H_{11}$]pipecolic acid internal standard.

VLCFA

Table 2 summarizes the concentrations of VLCFA in plasma, blood spots, CSF, urine and amniotic fluid in controls and patients affected by different peroxisomal disorders.

Among the VLCFA, $C_{26:0}$ and $C_{26:1}$ (both absolute and in proportion to $C_{22:0}$) are the most

Table 2   Concentrations of VLCFA and VLCFA ratios in different body fluids of controls and individuals affected by different peroxisomal disorders

Body fluid	$C_{22:0}$	$C_{26:1}$	$C_{26:0}$	$C_{26:1}/C_{22:0}$	$C_{26:0}/C_{22:0}$
Plasma/serum (μmol/L)					
Controls ($n=50$)	17–96	0.05–1.30	0.22–1.31	0.001–0.021	0.003–0.021
Per. Def. Dis					
($n=20$)	6–31	1.23–31	1.18–1.77	0.070–1.14	0.056–0.76
RCDP ($n=3$)	30–54	0.08–0.70	0.40–0.79	0.003–0.018	0.013–0.020
X-ALD ($n=4$)	18–87	1.67–2.29	2.26–5.77	0.019–0.035	0.026–0.10
Acyl-CoA ox.					
Def. ($n=2$)	40, 60	1.62, 1.82	2.70, 3.53	0.027, 0.046	0.059, 0.068
Bifunct/Thiolase					
Def. ($n=5$)	23–37	4.35–17.8	5.59–18.4	0.19–0.65	0.25–0.68
Blood spots (nmol/g paper)					
Controls ($n=40$)	45–135	10–33	8–35	0.11–0.31	0.10–0.33
Per. Def. Dis.					
($n=4$)	44–94	20–54	19–49	0.45–0.57	0.42–0.54
RCDP ($n=1$)	96	31	31	0.32	0.32
X-ALD ($n=1$)	54	18	19	0.33	0.35
CSF (μmol/L)					
Controls ($n=13$)	n.d.–0.170	n.d.–0.004	n.d.–0.014	n.d.–0.058	n.d.–0.187
Per. Def. Dis.					
($n=1$)	n.d.	0.019	0.014		
X-ALD ($n=1$)	n.d.	0.002	0.017		
Acyl-CoA ox.					
Def. ($n=1$)	0.104	0.444	0.298	4.44	2.98
Bifunct/Thiolase					
Def. ($n=1$)	n.d.	0.065	0.060		
Urine (mmol/mol creatinine)					
Controls ($n=23$)	0.008–0.286	n.d.	n.d.–0.059	n.d.	n.d.–0.68
Per. Def. Dis.					
($n=4$)	0.011–0.043	n.d.	0.007–0.024	n.d.	0.27–0.73
X-ALD ($n=1$)	0.012	n.d.	0.001	n.d.	0.10
Acyl-CoA ox.					
Def. ($n=1$)	0.046	n.d.	0.013	n.d.	0.27
Amniotic fluid (μmol/L)					
Controls ($n=50$)	0.33–1.18	0.011–0.098	0.036–0.19	0.022–0.14	0.052–0.21
Per. Def. Dis.					
($n=10$)	0.16–0.89	0.11–0.53	0.062–0.46	0.21–0.71	0.18–0.60
X-ALD ($n=1$)	0.76	0.13	0.29	0.17	0.38
Acyl-CoA ox.					
Def. ($n=3$)	0.40–0.92	0.21–0.37	0.24–0.43	0.29–0.53	0.36–0.59
Bifunct/Thiolase					
Def. ($n=3$)	0.59–0.83	0.15–0.35	0.16–0.32	0.18–0.59	0.20–0.54

Per. Def. Dis. = peroxisome deficiency disorders; Acyl-CoA ox. Def = peroxisomal acyl-CoA oxidase deficiency; Bifunct/Thiolase Def. = bifunctional protein or peroxisomal 3-oxoacyl-CoA thiolase deficiency; n.d. = not detectable

important parameters. $C_{26:0}$ and $C_{26:1}$ were found to be elevated in plasma in all cases but one of peroxisome deficiency disorders and peroxisomal β-oxidation defects. However, the ratio $C_{26:0}/C_{22:0}$ was elevated in all cases, which makes this parameter the most reliable. In patients with a peroxisomal deficiency disorder and in patients with bifunctional protein or thiolase deficiency, the $C_{26:1}$ concentration and the $C_{26:1}/C_{22:0}$ ratio are as high as or even higher than the corresponding $C_{26:0}$ concentration and $C_{26:0}/C_{22:0}$ ratio. In X-ALD and peroxisomal acyl-CoA oxidase deficiency, the concentration of the unsaturated $C_{26:1}$ as well as the $C_{26:1}/C_{22:0}$ ratio are generally lower than the corresponding $C_{26:0}$ concentration and $C_{26:0}/C_{22:0}$ ratio.

If no plasma is available, blood spots collected for neonatal screening can be used for quantification of VLCFA (Jakobs et al 1993). Comparison of the concentrations of VLCFA in blood spots and plasma is difficult, because the amount of blood spotted is not exactly known. The VLCFA concentrations are expressed in nmol per gram filter paper, to correct for differences in thickness and size of the paper used in different screening programmes. Absolute values of $C_{26:0}$ and $C_{26:1}$ are of limited importance, owing to overlap in the ranges for controls and patients. In blood spots of patients affected by a peroxisome deficiency disorder, the ratios $C_{26:0}/C_{22:0}$ and $C_{26:1}/C_{22:0}$, however, are elevated, making ratios diagnostically useful.

It appears that for all metabolites investigated, concentrations in control CSF are very low, and often not detectable ($<0.001\,\mu mol/L$), suggesting that these substances do not have a major physiological function as brain metabolites. In some cases, accumulation of peroxisomal metabolites in plasma from patients with peroxisomal disorders is associated with elevated levels in CSF. For VLCFA, accumulation in plasma was found to be associated with slightly elevated levels in CSF in three cases with CSF/plasma concentration ratios <0.01. The VLCFA concentration in the CSF sample from the patient with acyl-CoA oxidase deficiency, however, was exceptionally high. It is obvious that even small amounts of blood present in the CSF sample will significantly influence metabolite concentrations in CSF. However, blood cells were microscopically undetectable in the CSF from this patient. Accordingly, there is no apparent explanation for the relatively high CSF/plasma ratio observed (ten Brink et al 1993b). In general, measurement of VLCFA in CSF does not seem to offer a diagnostic approach, since it can be done more conveniently in plasma.

Urinary concentrations of VLCFA are much lower than in plasma, indicating that renal excretion of VLCFA is negligible, which is no surprise considering the apolar structure of the compounds (Stellaard et al 1990). As a consequence, patients with peroxisomal disorders and elevated plasma VLCFA levels do not exhibit elevated urinary concentrations of $C_{26:0}$ and/or $C_{26:1}$.

Quantification of VLCFA in amniotic fluid may be a diagnostic tool for prenatal diagnosis of peroxisome deficiency disorders and peroxisomal β-oxidation defects (Jakobs 1994). In all affected cases an elevation of $C_{26:1}$ was found, whereas $C_{26:0}$ was only elevated in 11 of the 17 cases. The $C_{26:0}$ and $C_{26:1}$ levels in amniotic fluid from unaffected fetuses at risk were within the control ranges. The observation that $C_{26:0}$ in amniotic fluid was normal in 5 cases of Zellweger syndrome and one case of thiolase deficiency may imply that $C_{26:0}$ is not a reliable marker for the prenatal diagnosis of peroxisomal disease in amniotic fluid. On the other hand, $C_{26:1}$ (absolute and in proportion to $C_{22:0}$) in amniotic fluid was

Table 3 Concentrations of pristanic acid and phytanic acid in different body fluids of controls and individuals affected by different peroxisomal disorders

Body fluid	Pristanic	Phytanic	Pristanic/Phytanic
Plasma/serum (μmol/L)			
Controls ($n=90$)	0.01–2.98	0.04–9.88	0.05–0.40
Per. Def. Dis. ($n=22$)	0.10–34.7	0.52–118	0.08–0.44
RCDP ($n=3$)	0.05–3.18	58–941	0.0008–0.003
X-ALD ($n=3$)	0.42–2.40	3.02–12	0.11–0.45
Acyl-CoA ox. Def. ($n=2$)	0.74, 1.67	3.72, 4.36	0.20, 0.38
Bifunct/Thiolase Def. ($n=5$)	8.95–77	2.61–50	1.36–6.08
Refsum disease ($n=4$)	n.d.–0.76	350–1054	n.d.–0.0007
Blood spots (nmol/g paper)			
Controls ($n=30$)	0.03–0.47	0.21–3.7	0.09–0.47
Per. Def. Dis ($n=4$)	1.4–7.1	3.7–22	0.25–0.44
RCDP ($n=1$)	0.16	20	0.008
X-ALD ($n=1$)	0.07	0.66	0.11
CSF (μmol/L)			
Controls ($n=13$)	n.d.–0.001	n.d.–0.012	n.d.
Per. Def. Dis. ($n=1$)	n.d.	0.006	n.d.
X-ALD ($n=1$)	n.d.	0.008	n.d.
Acyl-CoA ox. Def. ($n=1$)	n.d.	0.012	n.d.
Bifunct/Thiolase Def. ($n=1$)	0.034	0.016	2.13
Amniotic fluid (μmol/L)			
Controls ($n=37$)	n.d.–0.014	n.d.–0.040	n.d.–0.40
Per. Def. Dis. ($n=8$)	0.001–0.014	0.004–0.035	0.10–0.48
Bifunct/Thiolase Def. ($n=2$)	0.005, 0.009	0.019, 0.040	0.23, 0.26

Per. Def. Dis. = peroxisome deficiency disorders; Acyl-CoA ox. Def. = peroxisomal acyl-CoA oxidase deficiency; Bifunct/Thiolase Def. = bifunctional protein or peroxisomal 3-oxoacyl-CoA thiolase deficiency; n.d. = not detectable

consistently elevated and seems to be a valuable parameter. However, an authentic standard and a labelled internal standard will be needed for this compound to ensure accurate quantification. The VLCFA levels in amniotic fluid are strikingly higher as compared to urine. This points to a source of VLCFA other than fetal urine.

Because of this and the other reasons mentioned above, we do not recommend the use of the $C_{26:0}$ and $C_{26:1}$ levels as a sole marker in prenatal diagnosis of disorders with defective peroxisomal β-oxidation.

Phytanic/pristanic acid

Phytanic and pristanic acid levels in plasma, blood spots, CSF, and amniotic fluid in controls and peroxisomal disorders are summarized in Table 3.

Phytanic acid is a branched-chain fatty acid present in plasma of healthy individuals. As all phytanic acid in the body is from dietary origin, its concentration is age-dependent. Accumulation of phytanic acid in plasma is found in patients with peroxisome deficiency disorders, RCDP, classical Refsum disease and some isolated peroxisomal β-oxidation disorders. When a peroxisomal disorder is suspected early in a child's life, phytanic acid may be within the control range, as is observed in some cases of Zellweger syndrome. Pristanic acid, originating from phytanic acid breakdown and from the diet, accumulates in both peroxisomal deficiency disorders and some disorders with disturbed peroxisomal β-oxidation. Pristanic acid accumulation is always accompanied by elevated phytanic acid levels, probably owing to inhibition of phytanic acid oxidation by pristanic acid. Determination of pristanic acid along with phytanic acid is meaningful, because the pristanic/phytanic acid ratio contributes to the precise identification of the disease (ten Brink et al 1992). Where breakdown of both pristanic and phytanic acid is impaired (peroxisome deficiency disorders), a normal ratio between these metabolites is found. However, in a single β-oxidation defect (bifunctional protein/thiolase deficiency), the ratio is found to be elevated, while in a single α-oxidation defect (RCDP or classical Refsum disease), the ratio is extremely low.

Next to plasma, blood spots can be used for quantification of phytanic and pristanic acid (ten Brink et al 1993a). Whether or not the phytanic and pristanic acid concentrations in blood spots and plasma are similar is difficult to establish, the exact amount of blood per blood spot being unknown. In analogy with VLCFA, phytanic and pristanic acid concentrations are expressed in nmol per gram filter paper. A surprising finding was the observation of clearly elevated levels of phytanic acid in blood spots from 3 patients with peroxisome deficiency disorders and one patient with RCDP (ten Brink et al 1993a). Since phytanic acid accumulation is age-dependent, early-diagnosed patients do not always show elevated levels of phytanic acid in plasma. Nevertheless, in blood spots, which are collected at days 6–10 of life, phytanic acid was found to be elevated in all cases but one. This may suggest that the phytanic acid partly originates from blood cells. In peroxisome deficiency disorders, pristanic acid was elevated in all cases, whereas the pristanic acid/phytanic acid ratio, as determined in a blood spot from a RCDP patient, was below the control range.

In CSF of patients, phytanic acid concentration was not significantly raised (ten Brink et al 1993b). On the contrary, the CSF pristanic acid level in a patient with disturbed pristanic β-oxidation (bifunctional protein/thiolase deficiency) was elevated. The increased pristanic/phytanic acid ratio, as seen in plasma of this patient, was also observed in CSF.

In analogy with VLCFA, phytanic and pristanic acid excretion in urine is negligible and elevated levels in plasma are not reflected in urine (data not shown).

Also, in amniotic fluid of affected fetuses both phytanic and pristanic acids were found to be normal, and therefore not diagnostically useful.

Bile acids

In Table 4 concentrations of the bile acids cholic acid (CA), chenodeoxycholic acid (CDCA), DHCA and THCA in controls and patients with peroxisomal disorders are given.

Table 4 Concentrations of CDCA, CA, THCA, DHCA and THCA/CA, DHCA/CDCA ratios in different body fluids of controls and individuals affected by different peroxisomal disorders

Body fluid	CDCA	CA	THCA	DHCA	THCA/CA	DHCA/CDCA
Plasma/serum (μmol/L)						
Controls (n=25)	0.46–5.98	0.061–7.42	n.d.–0.035	n.d.–0.12	n.d.–0.030	n.d.–0.029
Per. Def. Dis. (n=19)	0.34–41.2	0.072–19.9	0.11–47.2	0.32–44.1	0.21–63.1	0.052–21.6
RCDP (n=2)	1.13, 1.22	0.16, 0.39	n.d., 0.001	n.d., 0.008	n.d., 0.018	n.d., 0.016
X-ALD (n=3)		0.20–1.21	n.d.–0.006		n.d.–0.008	
Bifunct/Thiolase Def. (n=3)	0.43–2.17	0.23–7.18	0.45–1.15	1.29–5.00	0.39–5.71	0.59–11.0
Refsum disease (n=2)	0.94, 1.73	0.44, 0.63	n.d., 0.002	0.005, 0.017	n.d., 0.012	0.010, 0.012
CSF (μmol/L)						
Controls (n=3)	n.d.–0.017	n.d.–0.021	n.d.	n.d.	n.d.	n.d.
Per. Def. Dis. (n=1)	0.016	0.016	0.015	0.030	0.94	1.88
Bifunct/Thiolase Def. (n=1)	0.015	0.012	0.010	0.006	0.83	0.40
Amniotic fluid (μmol/L)						
Controls (n=40)	0.20–1.37	0.09–0.60	n.d.–0.007	n.d.–0.010	n.d.–0.034	n.d.–0.021
Per. Def. Dis. (n=10)	0.11–0.26	0.09–0.14	0.018–0.21	0.005–0.059	0.13–1.47	0.024–0.24
Acyl-CoA ox. Def. (n=3)	0.28–0.42	0.08–0.28	n.d.–0.002	n.d.–0.001	n.d.–0.024	n.d.–0.002
Bifunct/Thiolase Def. (n=3)	0.08–0.18	0.03–0.04	0.018–0.061	0.008–0.016	0.54–1.47	0.041–0.098

Per. Def. Dis.=peroxisome deficiency disorders; Acyl-CoA ox. Def.=peroxisomal acyl-CoA oxidase deficiency; Bifunct/Thiolase Def.=bifunctional protein or peroxisomal 3-oxoacyl-CoA thiolase deficiency; n.d.=not detectable

The best indicator for abnormalities in bile acid synthesis is plasma concentrations of THCA and DHCA and the ratios THCA/CA and DHCA/CDCA. All patients affected by peroxisomal disorders involving bile acid synthesis display elevated values for each of these parameters (Stellaard et al 1989).

In blood spots collected at neonatal screening, elevated levels of bile acid intermediates are found in peroxisomal disease, as has been described by Gustafsson et al (1987).

CSF levels of DHCA and THCA were found to be elevated in one case of a peroxisome deficiency disorder and one case of bifunctional protein deficiency (Stellaard et al 1989). Plasma levels of these metabolites, however, were considerably higher and thus were more reliable diagnostic parameters.

Urinary levels of bile acids can be used for diagnosis of peroxisomal disorders (Eyssen et al 1985), but concentrations in urine are less reliable than those in plasma because bile acid excretion may be influenced by renal discrimination (Stellaard et al 1989).

Concentrations of the bile acid intermediates THCA and DHCA in amniotic fluid were found to be very low, even in affected cases (Stellaard et al 1991). The ratios of THCA/CA and DHCA/CDCA in amniotic fluid were elevated in all cases in which there was either a deficiency of peroxisomes or defective enzymes in the β-oxidation of THCA and DHCA. In all other cases at risk these ratios were within the control range, implying that a peroxisomal disorder could be ruled out or did not involve bile acid synthesis. It can be concluded that the determination of bile acid intermediates in amniotic fluid appears to be an important tool in the prenatal diagnosis of those peroxisomal disorders in which bile acid biosynthesis is impaired.

Pipecolic acid

Levels of pipecolic acid in plasma, CSF, urine and amniotic fluid are summarized in Table 5. Concentrations of pipecolic acid in both plasma and urine of a healthy population were found to be age-dependent (Kok et al 1987; Zee et al 1992). Pipecolic acid is an intermediate in lysine degradation, and is generally found to be elevated in plasma and urine of patients affected by a peroxisomal deficiency disorder. However, in 2 cases with a generalized disorder of peroxisome biogenesis (mild Zellweger variants), pipecolic acid concentrations in urine and in plasma were normal. Another cause of elevated pipecolic acid levels, although relatively minor, may be liver failure, which suggests that the finding of only moderately elevated pipecolic acid is not diagnostic for peroxisomal disorders.

A good diagnostic criterion may be determination of pipecolic acid in CSF, because pipecolic acid is elevated in CSF from all patients affected by a peroxisome deficiency disorder studied so far (Kok et al 1987; Zee et al 1992).

The concentration range of pipecolic acid in amniotic fluid was less than expected on the basis of the urinary values. The normal pipecolic acid levels found in the amniotic fluids of 3 fetuses affected by a peroxisome deficiency disorder indicate that pipecolic acid levels cannot be used for prenatal diagnosis of this disorder (Kok et al 1987).

CONCLUSION

For postnatal diagnosis of peroxisomal disorders we advise determination of VLCFA, phytanic and pristanic acid, THCA and DHCA and pipecolic acid in plasma. In addition,

Table 5 Concentration of pipecolic acid in different body fluids of controls and individuals affected by different peroxisomal disorders

Body fluid	Pipecolic acid	
Plasma/serum (μmol/L)	*Age < 1 week*	*Age > 1 week*
Controls ($n=54$)	0.55–10.8 ($n=18$)	0.54–2.46 ($n=35$)
Per. Def. Dis ($n=24$)	5.28–22.0 ($n=4$)	7.53–391 ($n=20$)
RCDP ($n=1$)		1.08
X-ALD ($n=1$)		1.04
Acyl-CoA ox. Def. ($n=2$)		1.38–1.66
Bifunct/Thiolase Def. ($n=2$)		7.51–9.22
CSF (μmol/L)		
Controls ($n=42$)	0.009–0.12	
Per. Def. Dis. ($n=7$)	0.28–4.53	
Acyl-CoA ox. Def. ($n=1$)	0.070	
Urine (mmol/mol creatinine)	*Age < 6 months*	*Age > 6 months*
Controls ($n=50$)	0.55–24.1 ($n=35$)	0.01–1.54 ($n=15$)
Per. Def. Dis. ($n=14$)	20.2–270 ($n=8$)	0.29–209 ($n=6$)
RCDP ($n=1$)		1.69
X-ALD ($n=1$)		0.02
Acyl-CoA ox. Def. ($n=2$)	7.28	0.49
Bifunct/Thiolase Def. ($n=2$)		0.87, 1.27
Amniotic fluid (μmol/L)	*Gestational age: 16–18 weeks*	
Controls ($n=27$)	1.50–8.40	
Per. Def. Dis. ($n=3$)	3.10–7.26	
Acyl-CoA ox. Def. ($n=1$)	1.67	

Per.Def.Dis. = peroxisome deficiency disorders; Acyl-CoA ox.Def. = peroxisomal acyl-CoA oxidase deficiency; Bifunct/Thiolase Def. = bifunctional protein or peroxisomal 3-oxoacyl-CoA thiolase deficiency; n.d. = not detectable

the ratios $C_{26:0}/C_{22:0}$, $C_{26:1}/C_{22:0}$, THCA/CA, DHCA/CDCA and the pristanic/phytanic acid ratio must be calculated. These parameters are of great value in differential diagnosis of the distinct peroxisomal disorders. If no plasma is available, blood spots collected in neonatal screening programmes can be used diagnostically for peroxisome deficiency disorders in which VLCFA ratios and often phytanic acid and pristanic acid are elevated, and for RCDP in which the pristanic/phytanic acid ratio is strikingly lower than in controls. For X-ALD, blood spots do not seem to be of diagnostic value. It must be stated that plasma is by far the preferred medium for VLCFA and phytanic and pristanic acid quantification, and blood spots should only be used when no plasma is available. In urine, only analysis of pipecolic acid and bile acids is feasible, the other metabolites being of no diagnostic value. In CSF, only pipecolic acid seems to have diagnostic importance. Although, in some cases, levels of metabolites in CSF from patients with a peroxisomal disorder exceed the control range, measurement of VLCFA, bile acids, pristanic acid and phytanic acid does not seem to offer a diagnostic perspective, since this can be done more conveniently in plasma. Furthermore, it seems improbable that the metabolites investigated in CSF play a direct role in the pathogenesis of neurological symptoms in peroxisomal disorders. For prenatal diagnosis, $C_{26:0}/C_{22:0}$, $C_{26:1}/C_{22:0}$ ratios and more reliably THCA/CA and DHCA/CDCA ratios can be quantified in amniotic fluid from the 16th to 18th weeks of pregnancy. Neither pipecolic acid nor phytanic acid in amniotic fluid are informative.

ACKNOWLEDGEMENTS

The authors gratefully thank Dr R.J.A. Wanders for enzyme analysis, and Professor C.R. Roe and Dr R.J.A. Wanders for their helpful comments on the manuscript.

REFERENCES

Eyssen H, Eggermont E, Eldere J van, Jaeken J, Parmentier G, Janssen G (1985) Bile acid abnormalities and the diagnosis of cerebro-hepato-renal syndrome (Zellweger syndrome). *Acta Paediatr Scand* 74: 539–544.

Fournier B, Smeitink JAM, Dorland L, Berger R, Saudubray J-M, Poll-The BT (1994) Peroxisomal disorders: a review. *J Inher Metab Dis* 17: 470–486.

Gustafsson J, Sisfontes L, Bjorkhem I (1987) Diagnosis of Zellweger syndrome by analysis of bile acids and plasmalogens in stored blood collected at neonatal screening. *J Pediatr* 111: 264–267.

Harrison AG (1992) *Chemical Ionization Mass Spectrometry*, 2nd edn. Boca Raton: CRC Press.

Heymans HSA, Wanders RJA, Schutgens RBH (1990) Peroxisomal disorders. In Fernandez J, Saudubray J-M, Tada K, eds. *Inborn Errors of Metabolism*. Heidelberg: Springer Verlag,421–437.

Jakobs C (1994) Antenatal diagnosis of inherited metabolic disorders using gas chromatography– mass spectrometry. In Farriaux JP, Dhondt JL, eds. *New Horizons in Neonatal Screening*. Amsterdam: Excerpta Medica, 357–364.

Jakobs C, Heuvel CMM vd, Stellaard F, Largillière C, Skovby F, Christensen E (1993) Diagnosis of Zellweger syndrome by analysis of very long chain fatty acids in stored blood spots collected at neonatal screening. *J Inher Metab Dis* 16: 63–66.

Kok RM, Kaster L, Jong APJM de, Poll-The BT, Saudubray J-M, Jakobs C (1987) Stable isotope dilution analysis of pipecolic acid in CSF, plasma, urine and amniotic fluid using electron capture negative ion mass fragmentography. *Clin Chim Acta* 168: 143–152.

McCloskey JA, ed. (1990) *Methods in Enzymology*, vol. 193: *Mass Spectrometry*. London: Academic Press, 3–106.

Poggi-Travert F, Fournier B, Poll-The BT, Saudubray J-M (1995) Clinical approach to inherited peroxisomal disorders. *J Inher Metab Dis* 18(Suppl. 1): 1–18.

Singh I, Pahan K, Singh AK, Barbosa E (1993) Refsum disease: a defect in the α-oxidation of phytanic acid in peroxisomes. *J Lipid Res* 34: 1755–1764.

Steinberg D (1989) Refsum disease. In Scriver CR, Beaudet AL, Sly WS, Valle D, eds. *The Metabolic Basis of Inherited Disease*. New York: McGraw-Hill, 1533–1550.

Stellaard F, Langelaar SA, Kok RM, Jakobs C (1989) Determination of plasma bile acids by capillary gas liquid chromatography–electron capture negative chemical ionization mass fragmentography. *J Lipid Res* 30: 1647–1652.

Stellaard F, ten Brink HJ, Kok RM, Heuvel L vd, Jakobs C (1990) Stable isotope dilution analysis of very long chain fatty acids in plasma, urine, and amniotic fluid by electron capture negative ion mass fragmentography. *Clin Chim Acta* 192: 133–144.

Stellaard F, Kleijer WJ, Wanders RJA, Schutgens RBH, Jakobs C (1991) Bile acid in amniotic fluid: promising metabolites for the prenatal diagnosis of peroxisomal disorders. *J Inher Metab Dis* 14: 353–356.

ten Brink HJ, Jakobs C, Baan JL vd, Bickelhaupt F (1989a) Synthesis of deuterium labelled analogues of pristanic acid and phytanic acid for use as internal standards in stable isotope dilution analysis. In Baillie TA, Jones JR, eds. *Synthesis and Applications of Isotopically Labelled Compounds*. Amsterdam: Elsevier, 717–722.

ten Brink HJ, Jakobs C, Bickelhaupt F (1989b) Synthesis of deuterium labelled analogues of very long chain fatty acids C22:0, C24:0 and C26:0 for use as internal standards in stable isotope dilution analysis. In Baillie TA, Jones JR, eds. *Synthesis and Applications of Isotopically Labelled Compounds*. Amsterdam: Elsevier, 723–726.

ten Brink HJ, Stellaard F, Heuvel CMM vd, et al (1992) Pristanic acid and phytanic acid in plasma from patients with peroxisomal disorders: stable isotope dilution analysis with electron capture negative ion mass fragmentography. *J Lipid Res* **33**: 31–37.

ten Brink HJ, Heuvel CMM vd, Christensen E, Largillière C, Jakobs C (1993a) Diagnosis of peroxisomal disorders by analysis of phytanic and pristanic acids in stored blood spots collected at neonatal screening. *Clin Chem* **39**(9): 1904–1906.

ten Brink HJ, Heuvel CMM vd, Poll-The BT, Wanders RJA, Jakobs C (1993b) Peroxisomal disorders: concentrations of metabolites in cerebrospinal fluid compared with plasma. *J Inher Metab Dis* **16**: 587–590.

Wanders RJA, Heymans HSA, Schutgens RBH, Barth PG, Bosch H vd, Täger JM (1988) Peroxisomal disorders in neurology. *J Neurol Sci* **88**: 1–39.

Zee T, Stellaard F, Jakobs C (1992) Analysis of pipecolic acid in biological fluids using capillary gas chromatography with electron capture detection and [^{2}H$_{11}$]pipecolic acid as internal standard. *J Chromatography* **574**: 335–339.

J. Inher. Metab. Dis. 18 Suppl. 1 (1995) 61–75
© SSIEM and Kluwer Academic Publishers.

Polyunsaturated fatty acids in the developing human brain, erythrocytes and plasma in peroxisomal disease: therapeutic implications

M. MARTINEZ

Biomedical Research Unit, Autonomous University of Barcelona, Maternity-Children Hospital, Paseo Valle de Hebron, 119–129, 08035 Barcelona, Spain

Summary: Patients with Zellweger syndrome and related peroxisomal disorders have profound changes in the polyunsaturated fatty acid (PUFA) patterns in brain and other tissues, with a constant decrease in docosahexaenoic acid (DHA, $22:6\omega3$) concentration. Arachidonic acid (AA, $20:4\omega6$) concentration is normal or increased and linoleic acid (LA, $18:2\omega6$) is increased in the brain of Zellweger patients. In the retina of these patients, the levels of DHA are extremely low. Since these alterations are reflected elsewhere, they can be detected *in vivo* in patients with generalized peroxisomal disorders by measuring the PUFA content of plasma and erythrocytes, which show very low concentrations of DHA. The concentration of AA is low in plasma in generalized peroxisomal patients, although it is within normal limits in erythrocytes. Patients with X-linked adrenoleukodystrophy (X-ALD) or adreno-myeloneuropathy (AMN) have a normal DHA and AA content in both plasma and erythrocytes, unless they receive extremely low-PUFA diets.

Given the probable role of DHA deficiency in the pathogenesis of Zellweger syndrome (ZS), it is important to normalize concentrations of DHA, at least in blood, in an attempt to correct the DHA deficiency in brain. DHA ethyl ester was given orally to two infants with a peroxisome deficiency disorder for a year, and some favourable biochemical changes were produced in erythrocytes and plasma. Normalization of the DHA concentrations in erythrocytes was obtained in about 2 months, and the ratios $26:0/22:0$ and $26:1/22:0$ decreased markedly in plasma in the two patients. The plasmalogen ratio $18:0$ dimethyl acetal/$18:0$ in erythrocytes increased to virtually normal values in both patients. There was a clear clinical improvement in the two patients, which paralleled the increase in blood DHA. The concentrations of AA and other PUFAs were closely monitored and, when necessary, AA was added to the diet. Such a DHA therapy, given under close biochemical and clinical control, and accompanied by a diet rich in other long-chain PUFA, is strongly recommended in all patients with peroxisomal disorders in whom a DHA deficiency is detected in blood.

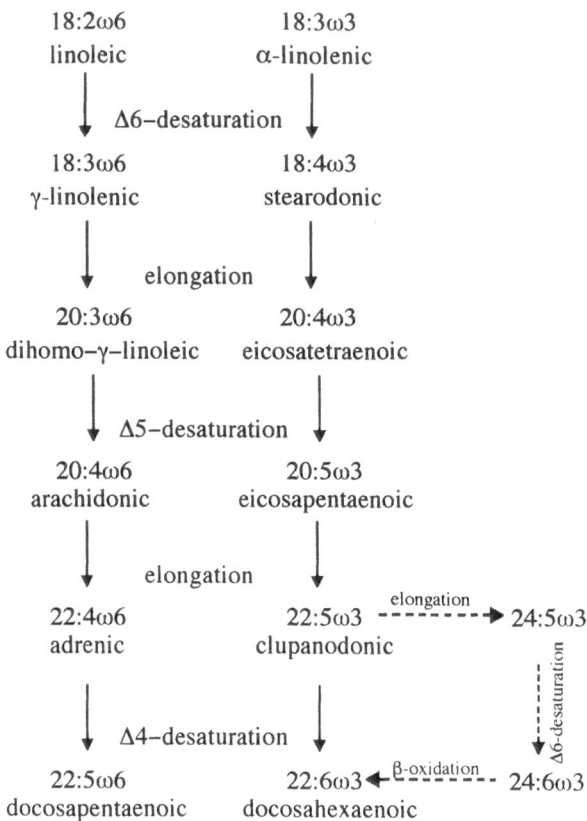

Figure 1   Metabolic pathways in the synthesis of long-chain PUFA. The classic route is indicated by solid arrows; the new route (Voss et al 1991), by dashed arrows

POLYUNSATURATED FATTY ACIDS DURING NORMAL HUMAN DEVELOPMENT

Phospholipids are ubiquitous constituents of cell membranes. In the brain and photo-receptor cells of the retina, phosphoglycerolipids esterified to polyunsaturated fatty acids (PUFA) with 20–22 carbon atoms are abundant. Docosahexaenoic acid (DHA, 22:6ω3) constitutes about 30% of brain ethanolamine phosphoglycerides (EPG), and even more of retinal EPG. DHA is also believed to play an important role in membrane physical properties and function (King et al 1977). In the brain, the particular abundance of DHA in synaptic plasma membranes suggests a role of this PUFA in neurotransmission (Breckenridge et al 1972). In the retina, DHA is especially enriched in the external segments of the rods, and its influence in the movements of rhodopsin seems to be of importance in the photoreceptive process (Anderson et al 1974). Significantly, monkeys deprived of ω3 fatty acids show visual abnormalities (Neuringer et al 1986). Arachidonic acid (AA, 20:4ω6) is a general constituent of cell membranes and the main prostaglandin precursor. Another ω6 long-chain PUFA, adrenic acid (22:4ω6) is particularly enriched

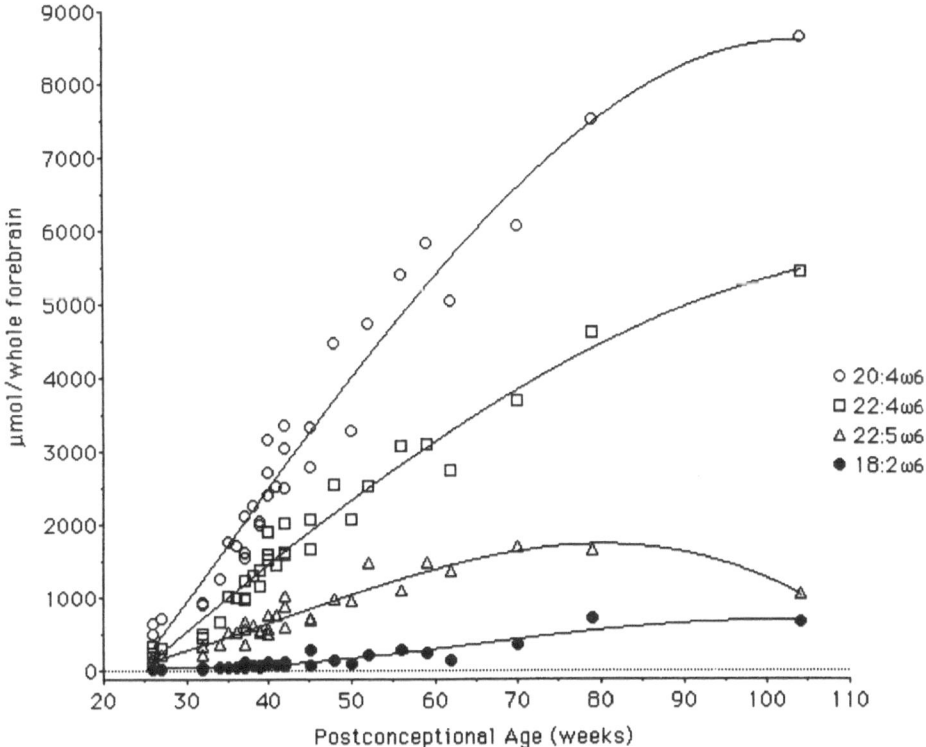

Figure 2 Total amount of the main $\omega 6$ PUFA in the normal human developing cerebrum versus postconceptional age in weeks

in myelin (O'Brien and Sampson 1965). These long-chain PUFA of the $\omega 3$ and $\omega 6$ families are derived from the two essential fatty acids, α-linolenic ($18:3\omega 3$) and linoleic ($18:2\omega 6$) acids, respectively, by successive desaturation and elongation steps (Figure 1).

When the accretion of long-chain PUFA is studied in the human brain during normal development, important differences can be noticed between the $\omega 3$ and $\omega 6$ families (Martinez 1992a). It can be seen that among the latter (Figure 2) arachidonic acid is the most important $\omega 6$ PUFA throughout development, followed closely by adrenic acid ($22:4\omega 6$), whose levels increase steadily during myelination. In the $\omega 3$ family, however, DHA is the only PUFA with levels that increase dramatically during brain development (Figure 3). This, again, indicates that DHA must play an important role during neurogenesis and myelinogenesis.

POLYUNSATURATED FATTY ACIDS IN THE BRAIN, ERYTHROCYTES AND PLASMA IN PATIENTS WITH PEROXISOMAL DISORDERS

Recently, it has been found that patients with disorders of peroxisome biogenesis have a consistent decrease of DHA concentration in the brain and other tissues (Martinez 1989, 1990, 1992b), which could explain many of the symptoms in this group of diseases. The

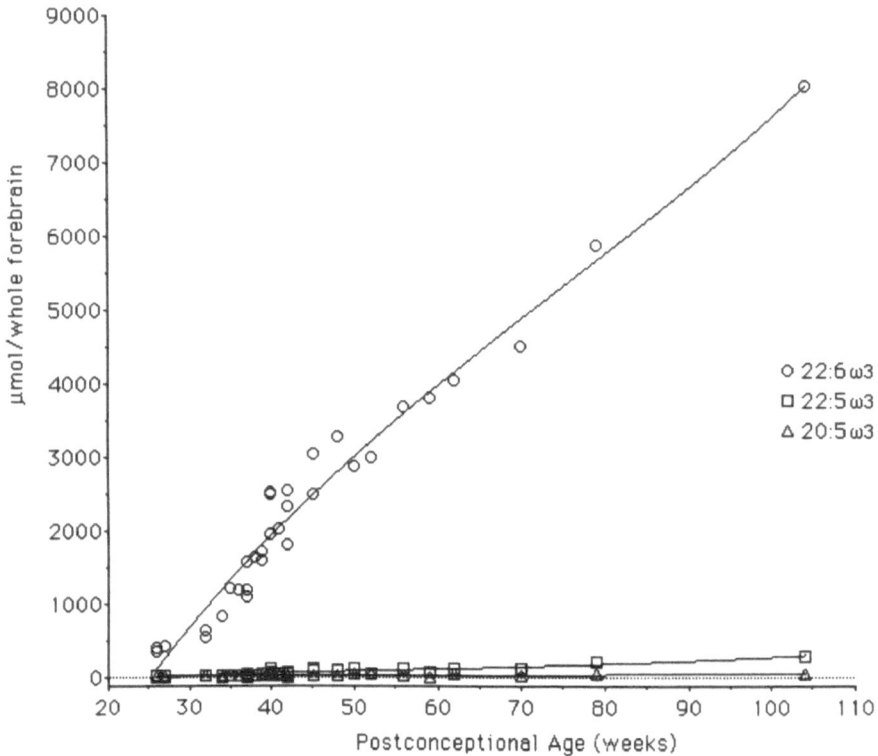

Figure 3   Total amount of ω3 PUFA in the normal human developing forebrain. In contrast to ω6 fatty acids, it can be seen that DHA is the only ω3 PUFA in the brain whose accumulation proceeds rapidly, prenatally as well as postnatally

DHA concentrations in the brains of these patients may be very low, as low as 20% of normal values or even less (Figure 4) in some cases. In the patient's retina, the concentration of DHA may be almost negligible (Martinez 1992b). The PUFA abnormality in neural tissues seems to be quite specific to DHA, since other long-chain polyenoic fatty acids, such as AA, are within normal limits, or even higher, in the patient's brain (Figure 5).

The DHA deficiency can be detected *in vivo* by measuring the concentrations of $22:6\omega3$ in plasma and erythrocytes. By analysing blood specimens of 53 patients with different peroxisomal disorders (Martinez et al 1994), it could be demonstrated that only those with defective peroxisome biogenesis have a DHA deficiency detectable in erythrocytes (Figure 6). On the other hand, patients with late-onset disorders, such as X-linked adrenoleuko-dystrophy (X-ALD) or adrenomyeloneuropathy (AMN), had normal blood concentrations unless there were very poor PUFA intakes. In contrast to DHA, arachidonic acid concentration was within normal limits in erythrocytes from all patients, although the mean values were lower than in normal controls (Figure 6). An unexpected finding in plasma was that the arachidonic acid concentrations were consistently decreased in all patients with generalized peroxisomal disorders (Figure 7).

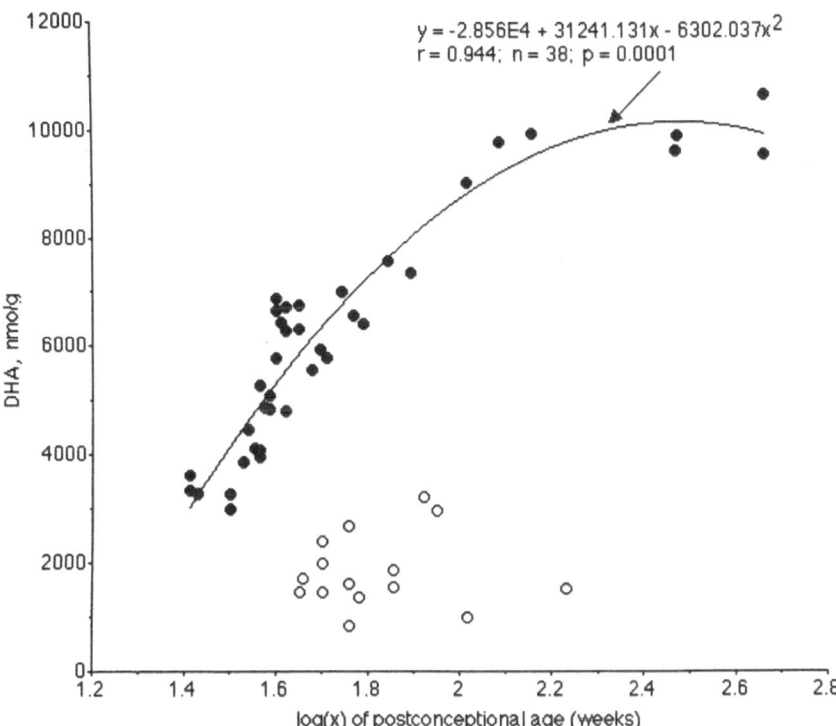

Figure 4 Concentration of DHA in the human cerebrum (nmol per g wet weight) throughout development versus the logarithm of postconceptional age (weeks). The equation corresponds to the normal forebrain (•, $n=38$). It can be seen that all the patients with disorders of peroxisome biogenesis (○, $n=15$) have extremely low DHA concentrations

These differences in the PUFA patterns in blood from patients with early-onset and late-onset peroxisomal disorders suggest that the pathogenesis of both groups of diseases is quite different. The site and mechanism by which DHA is generated remains unclear. For many years, it has been believed that DHA is synthesized in the endoplasmic reticulum by $\Delta 4$-desaturation of $22:5\omega 3$. It has recently been proposed (Voss et al 1991) that DHA is synthesized in the peroxisome by β-oxidation of the very long-chain precursor $24:6\omega 3$ (Figure 1). If this new route is the predominant one for DHA synthesis, then all patients with β-oxidation defects, and not only those with generalized peroxisomal disorders, should have a DHA deficiency (Roels et al 1993). As we have seen, however, this is not the case, at least if PUFA changes in blood are taken as reliable indicators of PUFA compositional changes in the brain.

Whatever the synthetic pathway for DHA in health and disease may be, it seems clear that peroxisomal disorder patients cannot synthesize all the DHA they need. Such a profound DHA deficiency may be pathogenetic in patients with generalized peroxisomal disorders. It was, therefore, of interest to test the effects of oral administration of pre-formed DHA in these patients. The first results with DHA therapy were quite encouraging (Martinez 1992c; Martinez et al 1993), although treatment was started at a very late stage of the disease. This paper shows the effects of a year's treatment with DHA ethyl ester in

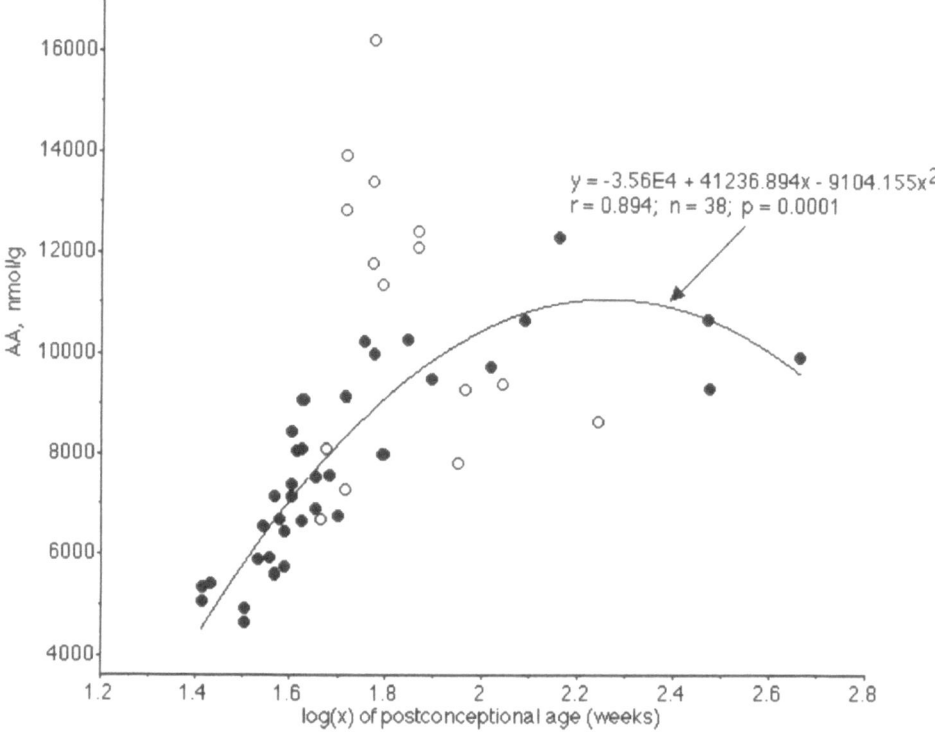

y = -3.56E4 + 41236.894x - 9104.155x²
r = 0.894; n = 38; p = 0.0001

Figure 5   Concentration of arachidonic acid (AA) in the developing human forebrain (nmol per g wet weight) versus the logarithm of postconceptional age (weeks). As in Figure 4, the equation corresponds to normal development (•). In contrast to DHA, the concentrations of AA in the peroxisomal brain (○) are either normal or increased; only two cases have somewhat low AA values

two patients diagnosed with Zellweger syndrome (McKusick 214100), who started the therapy during the first year of life.

THERAPEUTIC EFFECTS OF DOCOSAHEXAENOIC ACID ETHYL ESTER IN TWO YOUNG PATIENTS WITH GENERALIZED PEROXISOMAL DISORDERS

Patient 1 was a girl, the second child of a non-consanguineous couple, who presented at 4 months of age with failure to thrive, hepatosplenomegaly and gastrointestinal bleeding. In cultured skin fibroblasts, the β-oxidation of very long-chain fatty acids (VLCFA) and *de novo* plasmalogen synthesis (R.B.H. Schutgens and R.J.A. Wanders, Amsterdam) were markedly altered, in the range corresponding to Zellweger syndrome. There was dicarboxylic aciduria, and there were abnormally high concentrations of DHCA ($3\alpha,7\alpha$-dihydroxy-5β-cholestane-26-oic acid), THCA ($3\alpha,7\alpha,12\alpha$-trihydroxy-5β-cholestane-26-oic acid) and C_{29} ($3\alpha,7\alpha,12\alpha$-trihydroxy-$27\alpha,27\beta$–dihomo-5β-cholestane-$26,27\beta$-dioic acid). Before starting the treatment, at 9 months of age, the child was profoundly

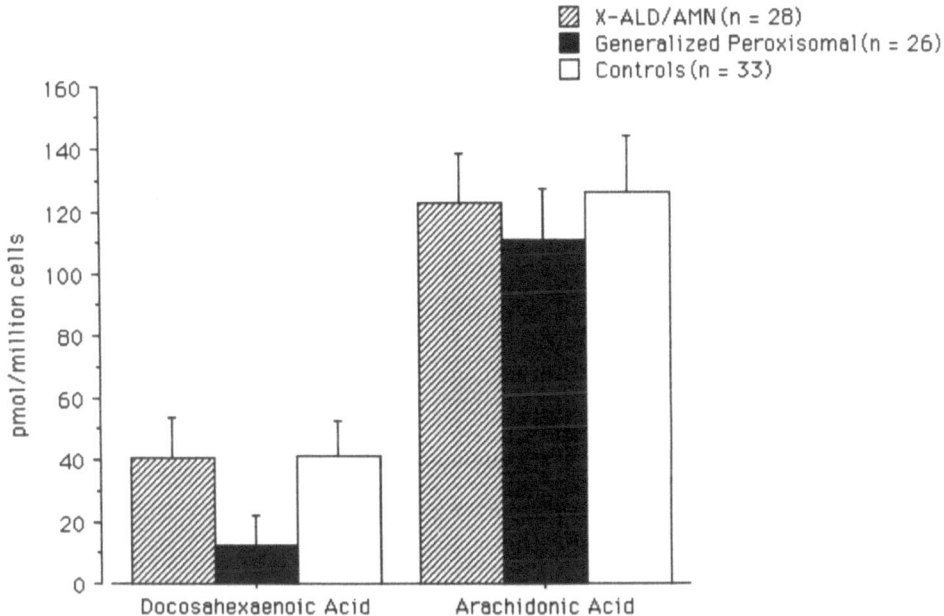

Figure 6 Concentration of DHA and AA in erythrocytes of peroxisomal disorder patients and controls. The bars are means + 1 SD. Notice that DHA is very significantly decreased only in patients with disorders of peroxisome biogenesis ($p < 0.0001$). AA is virtually within normal limits in erythrocytes

dystrophic, with a body weight of barely 4 kg. Fontanelles and cranial sutures were wide and there was generalized osteoporosis. Pendular nystagmus, strabismus and a pseudo-sunset phenomenon (Samson et al 1992) were seen. The patient was neurologically very depressed, with marked axial hypotonia and no social, visual or auditory contact.

Patient 2 was a boy, followed by C.A. Peltier and H. Souayah at the Saint Pierre University Hospital (Brussels). He was diagnosed at 2 months of age with Zellweger syndrome by the clinical picture and the demonstration of a very severe, generalized, defect involving all peroxisomal functions. The absence of peroxisomes was confirmed by liver biopsy (F. Roels, University of Gent). When this child started the DHA therapy, at 3 months of age, dysmyelination images, typical of Zellweger syndrome, were found by cranial MRI.

The first patient started the treatment with a daily dose of 100 mg of DHA ethyl ester (DHA-EE) for about 3 weeks, followed by 200 mg/day for 8 months, and then 300 mg/day. Particular attention was paid to the diet, which was as complete as possible for the age, and especially rich in liposoluble vitamins, meat and fish. The only exclusions were green leaf vegetables, the depot fat present in meat, and milk cream. After 1 month of treatment, when the blood concentrations of AA showed a tendency to decrease, the diet was enriched in arachidonate by including lamb's heart twice a week (equivalent to about 300 mg of AA per week). After 6 months, an $\omega 6$ mixture with about 45% of arachidonic acid, at a dosage of 400 mg of AA per day, was added to the treatment.

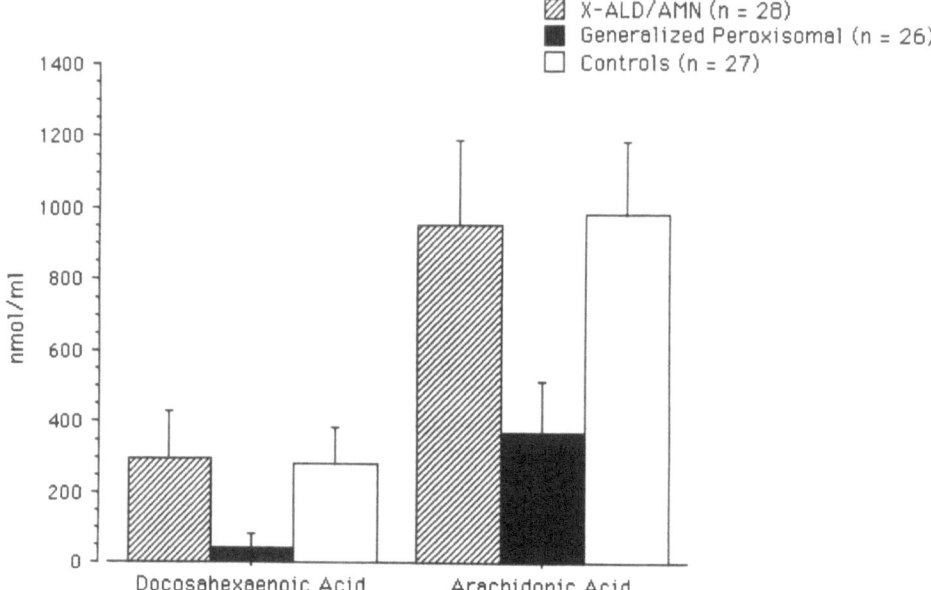

Figure 7   Plasma DHA and AA in peroxisomal disorder patients and controls. As in erythrocytes, DHA is markedly decreased in patients with generalized peroxisomal disorders ($p < 0.0001$). In contrast to erythrocyte AA, however, plasma arachidonic acid is also significantly decreased in these patients ($p < 0.0001$)

The diet of the second patient was also as complete as possible for age. A milk formula enriched in long-chain PUFA in a proportion similar to that of maternal milk (Milupa AG, Germany) was given to this child and meat was introduced early, but no food or special preparation rich in AA was given at any time. DHA therapy was started at 3 months of age. After 6 weeks with 50 mg/day of DHA-EE, the dosage was increased to 100 mg/day for 8 months, and then the daily dose was again increased to 200 mg.

Biochemical effects of DHA therapy

As expected, DHA increased to normal levels in erythrocytes and plasma in the two patients. A favourable effect of DHA therapy was that, in parallel to DHA normalization, the two diagnostic ratios $26:0/22:0$ (Figure 8) and $26:1/22:0$ (Figure 9) decreased abruptly in plasma in both patients, especially in the boy, in whom the $26:1/22:0$ ratio decreased to virtually normal values. The most favourable change in erythrocytes was the normalization of the plasmalogen ratio $18:0$ dimethyl acetal/$18:0$ after reaching a concentration of DHA of about 30 pmol/10^6 cells (Figure 10); this happened after 2 months of treatment in patient 1 and after 5 months of treatment in patient 2. The $16:0$ dimethyl acetal/$16:0$ and $18:1$ dimethyl acetal/$18:1$ ratios also increased, without reaching normal values.

The values for phytanic and pristanic acid concentrations increased at the beginning of the treatment but decreased later. A relatively constant biochemical change, in these and other patients treated with DHA ethyl ester, was a decrease, often to normal values, of

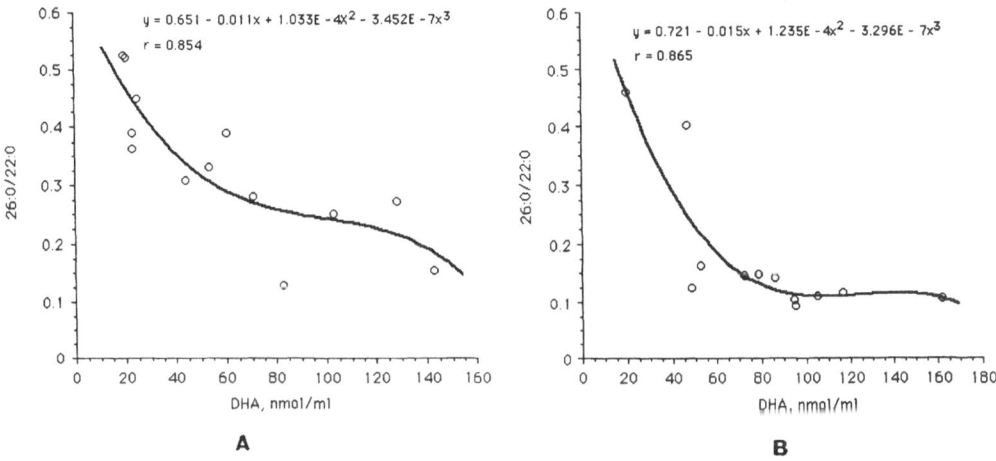

Figure 8 Plasma 26:0/22:0 ratio in patient 1 (A) and patient 2 (B), plotted against DHA concentration

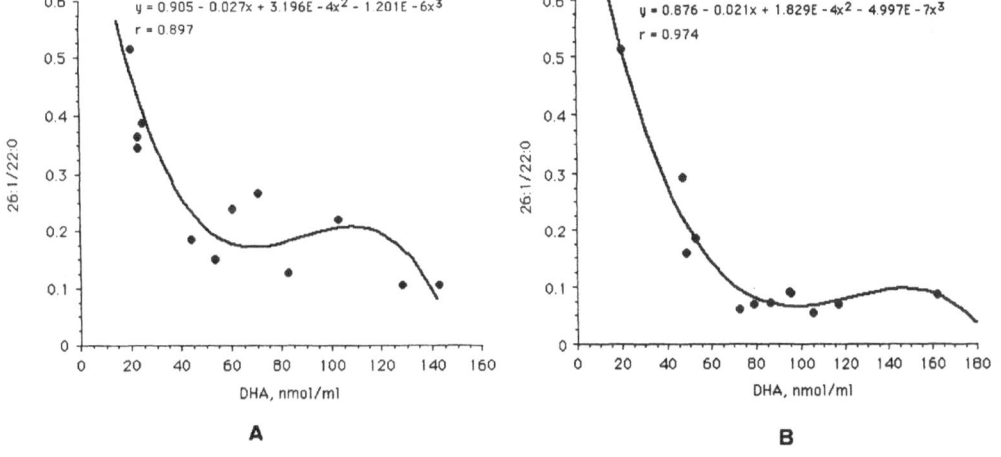

Figure 9 Plasma 26:1/22:0 ratio in patient 1 (A) and patient 2 (B), plotted against DHA concentration

plasma aspartate aminotransferase (AST, EC 2.6.1.1), alanine aminotransferase (ALT, EC 2.6.1.2) and γ-glutamyl transferase (γGT, EC 2.3.2.1) activities. In patients 1 and 2, all these activities were normal after, respectively, 1 year and 4 months of treatment. Other consistent changes during therapy were increases in the ratios $20:4\omega6/20:3\omega6$ and $20:4\omega6/18:2\omega6$, even when no arachidonic acid was provided.

Clinical effects of DHA therapy

Clinical improvement was noticed in the 9-month-old girl after only 2–3 weeks on 100 mg/day of DHA-EE. Although the DHA concentrations remained unchanged in

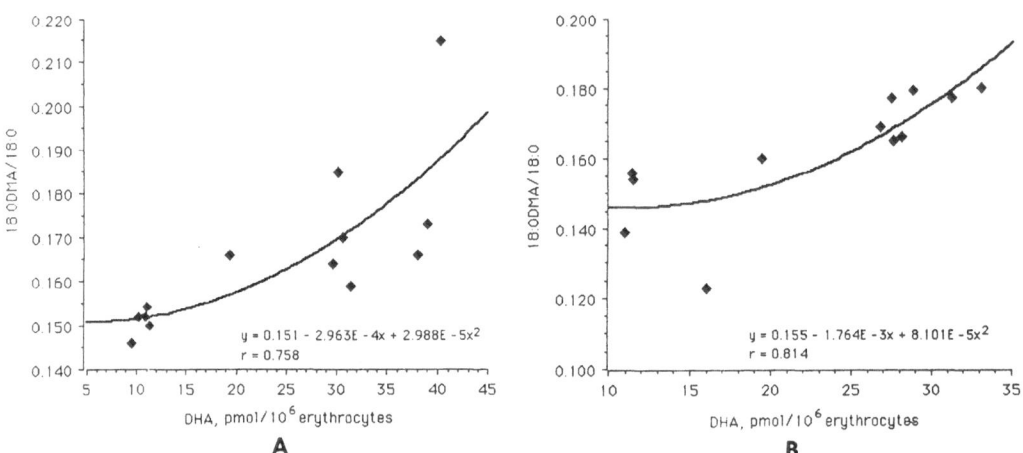

Figure 10 Plasmalogen ratio 18:0 dimethyl acetal/18:0 versus DHA concentration in erythrocytes from patients 1 (A) and patient 2 (B)

plasma and erythrocytes, the child became more active, and visual contact clearly started to improve. After 2 months with DHA therapy, the patient could follow objects and looked at people (Figure 11a,b), although she was still very dystrophic and hypotonic (Figure 11b,c). After 5 months with DHA-EE, growth and psychomotor development had improved. She had acquired head control, had good visual contact, smiled, played (Figure 11d–f) and uttered three monosyllabic and two bisyllabic words. After 1 year with DHA-EE, supplemented with AA for the last 7 months, the patient could sit unsupported, called her parents, played alone and with other people, manipulated small objects, clapped hands, sent kisses, waved goodbye, had an excellent visual contact and clearly responded to sounds (Figure 11g–l).

According to the medical team in Brussels, the other patient acquired good neurological development, and the MRI images normalized completely. Clinical details of this patient will appear in a collaborative paper with C.A. Peltier and H. Souayah.

CONCLUSIONS

Disorders of peroxisome biogenesis are extremely serious diseases which invariably lead to severe neurological and retinal involvement. Left untreated, patients with these disorders usually have a poor prognosis in the short term, although some patients have been reported to survive into their second decade. DHA deficiency is a consistent finding in peroxisomal disorder patients. DHA is a polyunsaturated fatty acid that apparently plays an important role in the plasma membranes of neurons and photoreceptor cells. This suggests a cause–effect relationship between the neurological and visual involvement in peroxisomal patients and their DHA deficiency. It was, therefore, worthwhile testing the possible beneficial effects of normalizing the DHA levels in these patients, especially after the encouraging preliminary results obtained with such therapy (Martinez et al 1993).

Among the biochemical changes found during the DHA treatment, the apparently unrelated improvement of the two diagnostic ratios 26:0/22:0 and 26:1/22:0 is especially

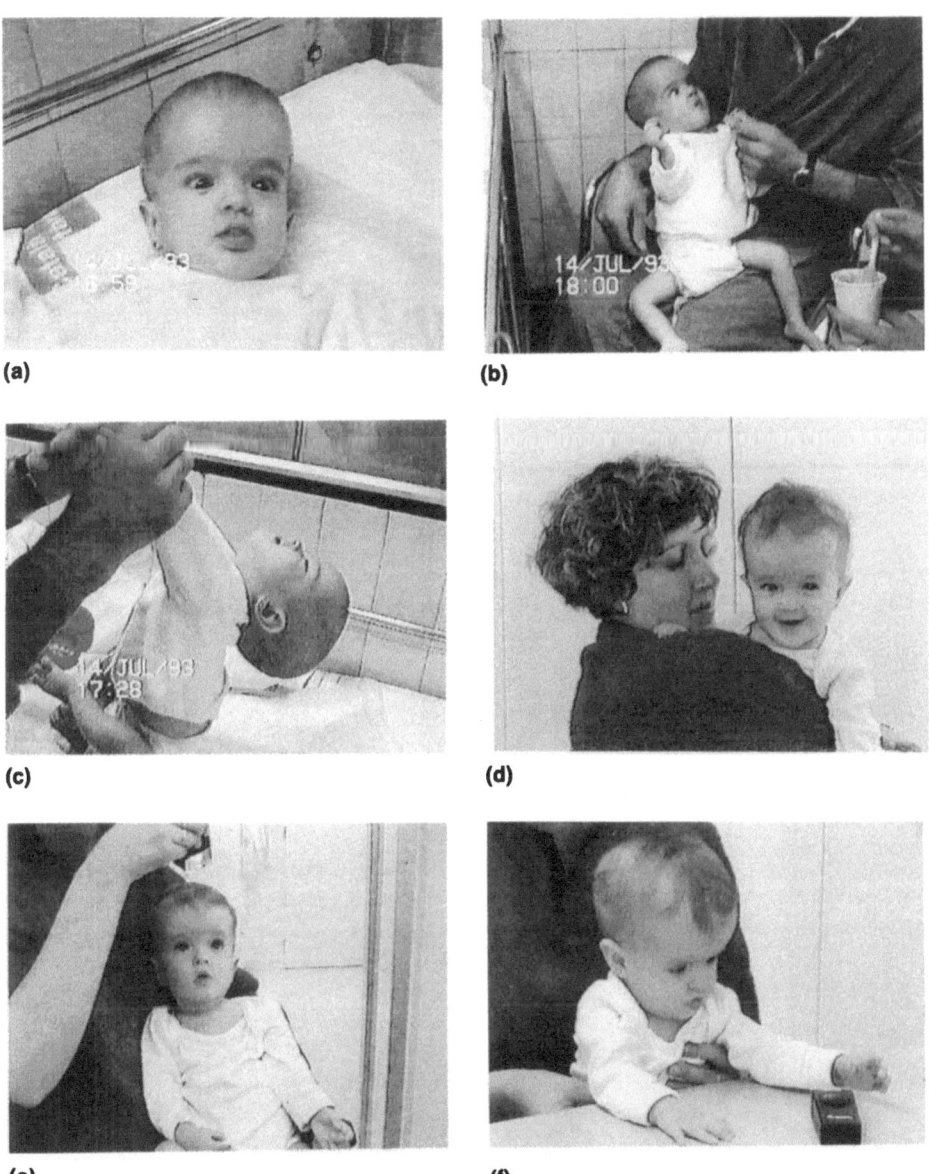

Figure 11 Video recordings of patient 1, taken after 2 months (a–c), 5 months (d–f) and 1 year (g–l) of treatment with DHA ethyl ester. The patient had an enlarged cranium with widely open fontanelles and sutures, nystagmus, strabismus and a pseudo-sunset phenomenon (a). She was still severely emaciated (b) and hypotonic (c) after 2 months of treatment, although she started to follow objects and look at her parents (b). Three months later, she smiled responsively and had good head control (d). She clearly followed a light source (e) and played with objects (f).

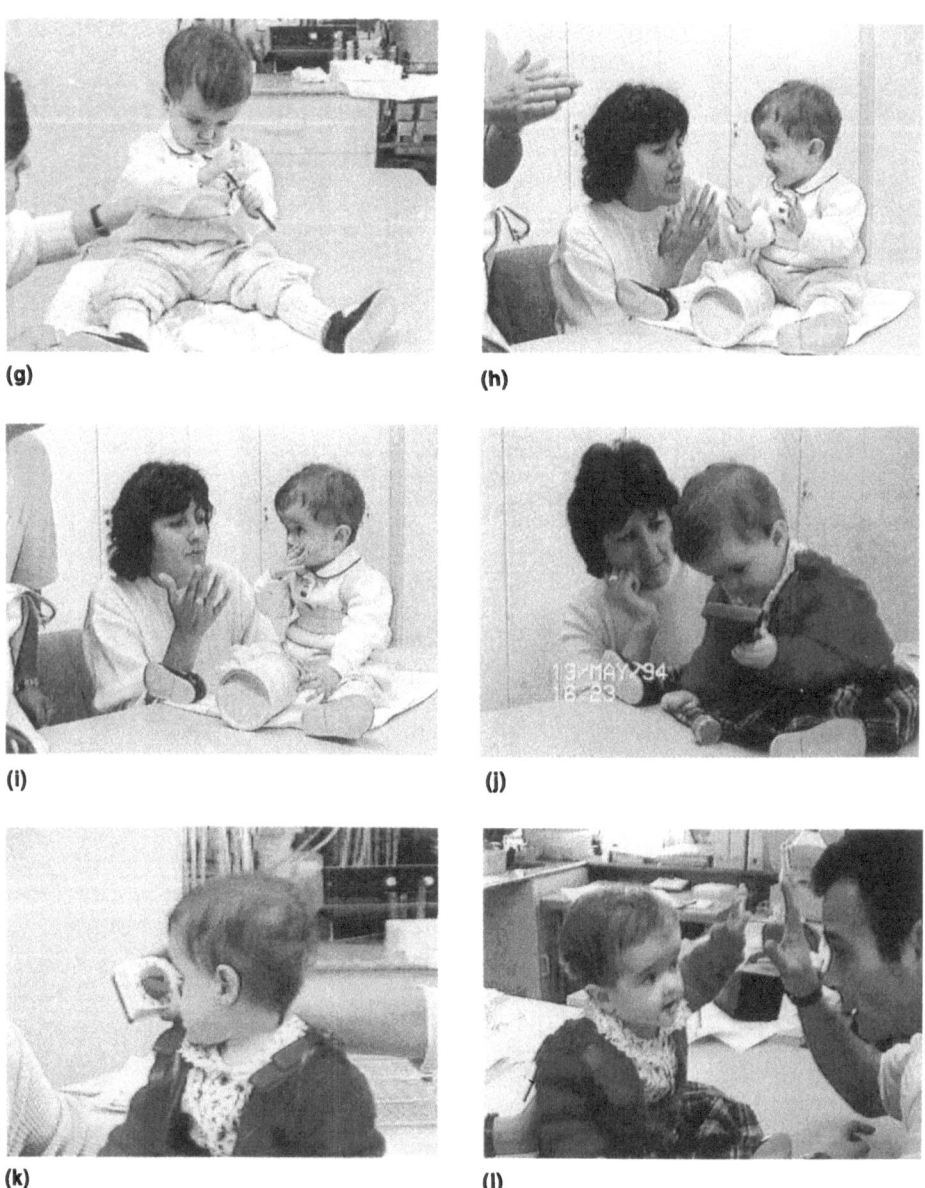

(g)   **(h)**

(i)   **(j)**

(k)   **(l)**

Figure 11   (*continued*)   After 1 year of DHA therapy, the patient could sit well without support, grasped and examined objects (g) and transferred them from hand to hand. She waved goodbye and imitated her parents in clapping hands (h) and sending kisses (i). The child could pull herself to sit again after reaching out for objects (j), she turned her head clearly in the direction of sound (k) and played with people (l)

interesting. Decreasing these ratios by other means such as reducing VLCFA intake or by inhibiting synthesis, or both has been attempted in X-ALD (Moser et al 1987; Rizzo et al 1989). However, in the two patients studied, DHA treatment was accompanied by an increase rather than a reduction in the supply of VLCFA. Therefore, it seems that the decrease in the VLCFA in the patients reported must be specifically due to some direct or indirect action of DHA on VLCFA metabolism.

A similar situation may apply to the increase in the plasmalogen levels, especially of the $18:0$ and $18:1$ molecular species, and in the corresponding ratios to their fatty acid counterparts. These changes are, again, consistent with an action of DHA on plasmalogen synthesis, also suggesting a possible role of DHA deficiency in some of the signs and symptoms found in generalized peroxisomal disorders.

Although we know that, in peroxisomal disorder patients, a spontaneous clinical amelioration cannot be ruled out, the coincidence between the biochemical and neurological improvements in the two patients is noteworthy. Also in favour of the beneficial influence of DHA in these patients is the fact that one was deteriorating rapidly until DHA therapy, when she started to improve in only a few weeks. The milestones of development were severely delayed in this 9-month-old girl, who was very inactive, totally emaciated, and practically blind and deaf before receiving DHA. At 14 months of age, the patient was well nourished and she could see, hear, smile, play and say some words. It seems improbable that DHA had nothing to do with these changes, especially since they were accompanied by a significant biochemical improvement.

As for arachidonic acid, the reason for its deficiency in plasma remains unclear, especially if one considers its concentrations found in the brain of patients with Zellweger syndrome (Martinez 1989, 1992b). However, the possible significance of the low values of the two ratios $20:4\omega6/20:3\omega6$ and $20:4\omega6/18:2\omega6$ in Zellweger patients (Martinez 1989) should not be overlooked. In any case, it seems reasonable not to allow the diminished arachidonic acid levels in plasma to decrease even further during DHA therapy, and therefore this was added to the diet.

It is important to emphasize that DHA therapy is a nutritional treatment and, as such, aims to normalize DHA values while maintaining a safe equilibrium with the other long-chain PUFA. It would be dangerous to consider DHA as a conventional drug, taken in isolation, disregarding the diet and the PUFA changes produced. We do not know how much of the DHA provided passes the blood–brain barrier or is incorporated into the brain. In the rat, we know that only $0.11-0.28\%$ of injected DHA is incorporated into the brain phospholipid pool (Sarda et al 1991). Recently, high purity ^{11}C-labelled DHA has been synthesized (Nariai et al 1993) and in the near future we hope to monitor such parameters *in vivo* by positron emission tomography (PET) or some other spectroscopic technique. In the meantime, we have to extrapolate from the PUFA values obtained in erythrocytes. It seems logical, therefore, to attempt to optimize these values as much as possible in an attempt to normalize concentrations in other tissues, especially in patients' brain.

It can be concluded that the DHA deficiency must be corrected in all patients in whom the DHA concentrations are low in blood, particularly in erythrocytes. At present, this therapy is being tested in seven other patients with disorders of peroxisome biogenesis. Normalization of the DHA concentrations in erythrocytes has been achieved in a short time, together with a parallel increase in plasmalogen values. Decreases in the concentrations of plasma VLCFA, particularly 26:1, and transaminase activities are common findings during DHA therapy. Although only fragmentary data have been obtained to date, clinical improvement, especially visual, has been noticed in most cases. These studies indicate that a close control of blood arachidonic acid concentrations is very important in order to avoid a possible deficiency and that, if found, such a deficiency should be corrected with an exogenous arachidonate supply.

ACKNOWLEDGEMENTS

DHA ethyl ester with a high degree of purity was obtained thanks to a grant of the National Institutes of Health (Fish Oil Test Materials Program, NIH/ADAMHA/DOC), with the cooperation of Harima Chemicals. Martek Biosciences Corporation provided the $\omega6$ mixture rich in arachidonic acid used in one of the patients. Dr C.A. Peltier and Dr H. Souayah (Hôpital Universitaire Saint Pierre, Brussels) collaborated with the study and treatment of one of the patients reported. Dr A. Verdú (Hospital Virgen de la Salud, Toledo) was the neurologist of the other patient, and Dr G. Lorenzo (Hospital Ramón y Cajal, Madrid) and Dr M. Girós (Instituto de Bioquímica Clínica, Barcelona) contributed to her diagnosis. Dr R.B.H. Schutgens and Dr R.J.A. Wanders (University Hospital, Amsterdam) studied *de novo* plasmalogen synthesis, β-oxidation of VLCFA and bile acids. The special contribution of Dr F. Roels (Universiteit Gent), who coordinates the European project on peroxisomal diseases, is gratefully acknowledged.

REFERENCES

Anderson E, Benolken RM, Dudley PA, Landis DJ, Wheeler TG (1974) Polyunsaturated fatty acids of photoreceptor membranes. *Exp Eye Res* **18**: 205–213.
Breckenridge WC, Gombos G, Morgan IG (1972) The lipid composition of adult rat brain synaptosomal plasma membranes. *Biochim Biophys Acta* **266**: 695–707.
King ME, Stevens BW, Spector AA (1977) Diet-induced changes in plasma membrane fatty acid composition affect physical properties detected with a spin-label probe. *Biochemistry* **16**: 5280–5285.
Martinez M (1989) Polyunsaturated fatty acid changes suggesting a new enzymatic defect in Zellweger syndrome. *Lipids* **24**: 261–265.
Martinez M (1990) Severe deficiency of docosahexaenoic acid in peroxisomal disorders: A defect of Δ4-desaturation? *Neurology* **40**: 1292–1298.
Martinez M (1992a) Tissue levels of polyunsaturated fatty acids during early human development. *J Pediatr* **120**: S129–138.
Martinez M (1992b) Abnormal profiles of polyunsaturated fatty acids in the brain, liver, kidney and retina of patients with peroxisomal disorders. *Brain Res* **583**: 171–182.
Martinez M (1992c) Treatment with docosahexaenoic acid favorably modifies the fatty acid composition of erythrocytes in peroxisomal patients. In Coates PM and Tanaka K, eds. *New Developments in Fatty Acid Oxidation*. New York: Wiley-Liss, 389–397.
Martinez M, Pineda M, Vidal R, Martin B (1993) Docosahexaenoic acid — a new therapeutic approach to peroxisomal-disorder patients: Experience with two cases. *Neurology* **43**: 1389–1397.
Martinez M, Mougan I, Roig M, Ballabriga A (1994) Blood polyunsaturated fatty acids in patients with peroxisomal disorders. A multicenter study. *Lipids* **29**: 273–280.
Moser AB, Borel JS, Odone A, et al (1987) A new dietary therapy for adrenoleukodystrophy: biochemical and preliminary clinical result in 36 patients. *Ann Neurol* **21**: 240–249.

Nariai T, Hirakawa K, Imahori Y, et al (1993) Synthesis of ¹⁴C labeled docosahexaenoic acid for imaging of neuronal structure and function. *J Cereb Blood Flow Metab* **13**: S704.

Neuringer M, Connor WE, Lin SD, Barstad L, Luck S (1986) Biochemical and functional effects of prenatal and postnatal 3 fatty acid deficiency on retina and brain in rhesus monkeys. *Proc Natl Acad Sci USA* **83**: 4021–4025.

O'Brien JS, Sampson EL (1965) Fatty acid and fatty aldehyde composition of the major brain lipids in normal human gray matter, white matter and myelin. *J Lipid Res* **6**: 545–551.

Rizzo WB, Leshner RT, Odone A, et al (1989) Dietary erucic acid therapy for X-linked adrenoleukodystrophy. *Neurology* **39**: 1415–1422.

Roels F, Fischer S, Kissling W (1993) Polyunsaturated fatty acids in peroxisomal disorders: a hypothesis and a proposal for treatment. *J Neurol Psychiatr* **56**: 937.

Samson JF, Jakobs C, van de Klei-van Moorsel J, Smit LME, Schutgens RBH, Wanders RJA (1992) Zellweger syndrome in a preterm small for gestational age infant. *J Inher Metab Dis* **15**: 75–83.

Sarda N, Gharib A, Moliere P, Grange E, Bobillier P, Lagarde M (1991) Docosahexaenoic acid (cervonic acid) incorporation into different brain regions in the awake rat. *Neurosci Lett* **123**: 57–60.

Voss A, Reinhart M, Sankarappa S, Sprecher H (1991) The metabolism of 7,10,13,16,19-docosapentaenoic acid to 4,7,10,13,16,19-docosahexaenoic acid in rat liver is independent of a 4-desaturase. *J Biol Chem* **266**: 19995–20000.

J. Inher. Metab. Dis. 18 Suppl. 1(1995) 76–83
© SSIEM and Kluwer Academic Publishers.

Measurement of very long-chain fatty acids, phytanic and pristanic acid in plasma and cultured fibroblasts by gas chromatography

G. DACREMONT*, G. COCQUYT and G. VINCENT
Department of Pediatrics, University Hospital Gent, Belgium

*Correspondence: Universitair Ziekenhuis, Kliniek voor Kinderziekten 'C. Hooft', De Pintelaan 185, 9000-Gent, Belgium

Summary: Two methods are described, both currently used in our laboratory, for the quantitative analysis of very long-chain fatty acids, phytanic acid and pristanic acid in plasma and cultured fibroblasts by gas–liquid chromatography. The first method is based on the procedure developed by Moser and Moser (1991) and the second is based on the method of Onkenhout and colleagues (1989), which is an application of the original method of Lepage and Roy for plasma and fibroblasts. A survey is given of the concentrations of very long-chain fatty acids, pristanic and phytanic acid in plasma and fibroblasts from control subjects and all patients investigated so far in our laboratory.

Measurement of very long straight-chain fatty acids (VLCFA) and of pristanic and phytanic acid concentrations in plasma or serum is an established technique for the screening of peroxisomal disorders.

Initially VLCFA assays were only used for the diagnosis of X-linked adrenoleukodystrophy (McKusick 300100). However, the discovery by Singh et al (1984) that VLCFA β-oxidation is mainly a peroxisomal process led to the concept that VLCFA accumulation can be used as a marker for other peroxisomal disorders in which there is a disturbance in the β-oxidation of these acids. Indeed eight other peroxisomal disorders have been identified so far in which there is an impairment of the β-oxidation of VLCFA, resulting in their accumulation in body fluids and tissues. These disorders include:

(i) The generalized peroxisomal disorders or peroxisome assembly disorders: classical Zellweger syndrome (McKusick 214100), neonatal adrenoleukodystrophy (NALD; McKusick 20370), infantile Refsum disease (IRD; McKusick 266500) and pseudo-infantile Refsum disease.

(ii) The disorders caused by a deficiency of a single peroxisomal enzyme in the β-oxidation pathway: acyl-CoA oxidase deficiency (pseudo-NALD; McKusick 264470), bi- (tri-)functional enzyme deficiency and acyl-CoA thiolase deficiency (pseudo-Zellweger syndrome; McKusick 261510).

(iii) Zellweger-like syndrome, a disease with multiple enzyme deficiencies.

In man, very long straight-chain fatty acids and 2-methyl branched-chain fatty acids such as pristanic acid and the bile acid precursors di- and trihydroxycoprostanoic acid (2-methyl-substituted in their side-chains) are initially oxidized by two different acyl-CoA oxidases (Vanhove et al 1993). Hence, pristanic acid accumulation is a marker for a disturbance in the β-oxidation of branched-chain fatty acids. Pristanic acid accumulates in the peroxisome assembly disorders, in bifunctional enzyme deficiency and in acyl-CoA thiolase deficiency, two disorders that also have a single defect in the β-oxidation of VLCFA. In trihydroxycholestanoic acidaemia, only the branched-chain fatty acids accumulate, not VLCFA.

Phytanic acid, the precursor of pristanic acid, accumulates in the peroxisome deficiency disorders, in rhizomelic chondrodysplasia punctata and in Refsum disease. Additionally, it is also elevated in the two above-mentioned disorders with isolated β-oxidation defects where pristanic acid also accumulates. In these latter diseases the phytanic acid accumulation is probably due to the inhibition of phytanic acid α-oxidation by accumulating pristanic acid, while in the other diseases it is possibly due to a separate peroxisomal enzyme defect in the conversion of phytanic to pristanic acid. It is still not clear which step in the oxidation of phytanic acid is a peroxisomal process.

A variety of methods have been described for the measurement of VLCFA. Most laboratories use the original procedure developed by Moser and Moser (1991), which involves preparation of a total lipid extract, treatment of this extract with methanolic hydrochloric acid, which yields the methylesters, purification of the methylesters by thin-layer chromatography (TLC) and quantification by capillary gas–liquid chromatography (GLC). This procedure was in current use in our laboratory for 6 years and has been applied to about 4000 samples.

Two years ago we introduced the simple one-step procedure described by Onkenhout et al (1989). In this method total fatty acids are directly transesterified with acetyl chloride in the presence of methanol and benzene and the fatty acid methyl esters subsequently extracted and purified by TLC.

A detailed description of the two procedures is given below.

EQUIPMENT, CHEMICALS, MATERIALS AND SAMPLES

Equipment

- Gas chromatograph for capillary columns equipped with an oven with reproducible temperature control, a flame ionization detector, a splitless or on-column injection system and an accurate and sensitive integrator
- Capillary column (Ultra II Hewlett Packard) for VLCFA analysis
- Capillary column (OV 1701 Betron Scientific) for pristanic and phytanic acid analysis
- Vortex mixer
- Thermostated heating block or oven
- Laboratory centrifuge (preferably cooled)

Chemicals

- Solvents: chloroform, methanol, n-hexane, toluene and diethylether for *procedure 1*; benzene, methanol, n-hexane, toluene and diethylether for *procedure 2*. All solvents must be of the highest purity with the lowest amount of residue
- Fatty acid standards: pentadecanoic acid ($C_{15:0}$ FA) for pristanic and phytanic acid measurements; heptacosanoic acid ($C_{27:0}$ FA) for C_{22}, C_{24} and C_{26} measurements
- 2 mol/L Methanolic HCl (*procedure 1*)
- Iodine crystals
- Potassium carbonate (*procedure 2*)
- Potassium chloride (*procedure 1*)
- Acetyl chloride (*procedure 2*)

Solutions

- Pentadecanoic acid solution, 100 μg/ml in chloroform – methanol 2 : 1
- Heptacosanoic acid solution, 40 μg/ml in chloroform – methanol 2 : 1
- Mixture of methanol – benzene 4 : 1 (v/v) (*procedure 2*)
- 6% Potassium carbonate (*procedure 2*)
- Mixture of chloroform – methanol 1 : 1 (v/v) (*procedure 1*)
- 0.1 mol/L Potassium chloride (*procedure 1*)

Materials

- Precoated thin-layer chromatography (TLC) plates, silica 60 (20×20 cm, 0.25 mm thickness (e.g. Merck 5721))
- Borosilicate glass tubes (18×100 and 13×100 mm) with Teflon-lined screwcaps
- Thin-layer chromatography tanks
- Organic solvent-resistant dispensers
- Teflon-lined screwcapped vials, 2 ml

Biological samples

- *Plasma or serum:* Venous blood (with or without anticoagulant) 2–5 ml, is centrifuged for 10 min at 800g, preferably within 1 h of collection to avoid haemolysis, and the plasma or serum is separated. Normally the sample is stored at −4°C or lower temperature until analysis can be performed. However, since concentrations of saturated VLCFA, pristanic acid and phytanic acid remain unchanged by storage at room temperature for at least 10 days, samples can be mailed at ambient temperature.
- *Cultured fibroblasts:* Fibroblasts grown to confluency are harvested either by trypsinization or by 'scraping'. The culture medium is removed after centrifugation for 15 min at 800g and the fibroblast pellet is washed three times with buffered saline. After the last wash, the liquid is drained and the pellet is stored at −20°C or lower until analysis.
- For VLCFA analysis, at least 300–500 μg of fibroblast protein is required. In our experience, two culture flasks with a growth surface of 75 cm² are sufficient.

PROCEDURES

Preparation of fatty acid methylesters (FAME)

Procedure 1 (see Moser and Moser, 1991): preparation of the total lipid extract; acid methanolysis of the total lipid extract; purification of the fatty acid methylesters by TLC.

Plasma or serum

- Pipette $250 \mu l$ of plasma or serum into a 18×100 mm tube with Teflon-lined screwcap; add $50 \mu l$ ($= 5 \mu g$) of $C_{15:0}$ and $50 \mu l$ ($= 2 \mu g$) of $C_{27:0}$ solution as internal standard, and then 3.1 ml of chloroform–methanol $1:2$ (v/v); cap and vortex vigorously for 5 min.
- Centrifuge the extracted plasma at $1000g$ for 15 min; transfer the supernatant to another screwcapped tube, taking care that the protein pellet is left. Add to the supernatant 1.55 ml of chloroform and 1 ml of 0.1 mol/L KCl; cap and vortex for 1 min.
- Centrifuge the vortexed sample for 5 min at $500g$; remove the upper phase with a Pasteur pipette and dry the lower phase under N_2.
- To the dried lipid extract, add 2 ml of 2 mol/L methanolic HCl; cap the tube tightly with a Teflon-lined screwcap and heat the tube for at least 4 h at 80°C in a metal block thermostat or oven.
- After cooling to room temperature, add 2 ml of water and 4 ml of n-hexane; vortex for 3 min and centrifuge for 5 min. Transfer the upper n-hexane phase into another tube and repeat the extraction of the lower phase twice with 2 ml of hexane.
- Concentrate the combined hexane extracts to a smaller volume (~ 1 ml) under N_2; transfer the concentrated hexane extract to a 2 ml vial and evaporate further until dry.
- Dissolve the residue in $200 \mu l$ of n-hexane and apply $100 \mu l$ of the solution in a 1.5 cm band on a precoated TLC plate; a reference mixture and five samples can be run simultaneously. The plate is developed in 1 h in toluene–diethylether $97:3$ (v/v).
- After air drying for 30 min, the five sample lanes are covered with a glass plate and the first lane containing the reference mixture is rendered visible with iodine vapour in a chromatography tank. Using the reference mixture, the positions of the FAME in the other lanes are outlined with a needle.
- The marked areas are scraped on to weighing paper, transferred to a centrifuge tube, and extracted three times with 1.5 ml of n-hexane. For each cycle, the tube is vortexed, sonicated and centrifuged for 3 min. The combined extracts are then evaporated in a 2 ml vial under N_2 until dry.
- The dried residue is finally taken up in $50 \mu l$ of n-hexane and approx. $1 \mu l$ is subjected to GLC for the analysis of VLCFA.
- For the measurement of phytanic and pristanic acids, the non-purified FAME solution is usually used. The $100 \mu l$ solution is concentrated to approx. $50 \mu l$ and $1 \mu l$ is subjected to GLC.

Cultured fibroblasts

- For VLCFA analysis of fibroblasts, suspend the fibroblast pellet (containing at least $500 \mu g$ of protein) in $400 \mu l$ of water and disrupt the cells by sonication to form a homogeneous suspension.

- Take two aliquots (10 and $20 \mu l$) of this suspension for duplicate protein analysis and transfer $200 \mu l$ of the remaining suspension to a glass tube (18×100 mm) with Teflon-lined screwcap.
- Add $25 \mu l$ ($= 1 \mu g$) of the heptacosanoic acid solution and proceed as described above for plasma.

Procedure 2

Based on the method described by Lepage and Roy (1986) and applied to serum by Onkenhout et al (1989), procedure 2 involves direct transesterification of the lipid in the plasma matrix and purification of the fatty acid methylesters by TLC.

Plasma or serum

- Into a 13×100 mm glass tube with Teflon-lined screwcap, pipette $250 \mu l$ of serum or plasma; add $50 \mu l$ of the $C_{15:0}$ and $50 \mu l$ of the $C_{27:0}$ solution and then 2 ml of methanol –benzene 4:1 (v/v) and mix.
- Add slowly, while stirring in an ice bath, $200 \mu l$ of acetyl chloride.
- Close the tube tightly and heat for 4h at 85°C in a heating block.
- Cool to room temperature and add, while cooling in an ice bath, 5 ml of 6% aqueous potassium carbonate.
- Vortex briefly, close the tube and centrifuge for 10 min at 750g at 4°C.
- With a Pasteur pipette, transfer the upper benzene layer into a 2 ml vial.
- Evaporate under N_2 and dissolve the residue in $200 \mu l$ of hexane.
- Apply $100 \mu l$ of this solution to a precoated TLC plate and proceed with the purification of the FAME as described in *procedure 1*.

Fibroblasts

- Transfer $250 \mu l$ of homogeneous suspension (see *procedure 1*) to a 13×100 mm screw-capped glass tube.
- Add $25 \mu l$ (1 μg) of the $C_{27:0}$ solution and 2 ml of methanol–benzene 4:1 (v/v) and proceed as described for *plasma*.

Alternative to procedure 2

The direct-one step transesterification can also be performed by using methanolic HCl instead of methanol–benzene and acetyl chloride.

- To $250 \mu l$ of plasma or fibroblast suspension and internal standard is added 2 ml of 2mol/L methanolic HCl.
- The tubes are heated for 4h at 85°C and the FAME is subsequently isolated and purified as described in *procedure 1*.

Capillary gas chromatography

For the assay of VLCFA in our laboratory we use a 25 m$\times 0.32$ mm ID cross-linked 5% phenylmethylsilicone (0.17μm) capillary column (Hewlett Packard Ultra II). Helium is used as the carrier gas, at a flow rate of 1 ml/min. The detector temperature is 320°C and the on-column injector temperature 60°C. The oven temperature is programmed as follows:

- The initial temperature of 60°C is immediately increased to 220°C at 50°C/min and held there for 15 min.
- Subsequently the temperature is increased to 250°C at 4°C/min and maintained there for 3 min.
- The final temperature of 285°C is reached by a temperature rise of 8°C/min and maintained for 15 min.

For the analysis of pristanic and phytanic acid an OV 1701 (14% cyanopropylphenyl) column (Betron) is used. The oven temperature is programmed as follows:

- The initial temperature of 60°C is maintained for 2 min and then increased to 200°C at a rate of 16°C/min and subsequently to 240°C at a rate of 4°C/min.
- The temperature of 240°C is maintained for 20 min and then increased to the final temperature of 280°C, which is maintained for 10 min.

RESULTS AND DISCUSSION

Table 1 summarizes the concentrations of VLCFA, phytanic acid and pristanic acid, in plasma and fibroblasts of controls and patients with different peroxisomal disorders as measured in our laboratory.

Note that our normal values are higher than those reported by Moser and Moser (1991) and by other laboratories that use essentially the same method (Schutgens et al 1993). The difference is most striking in fibroblasts. There is no clear explanation for this discrepancy unless our modes of extraction are more extensive, resulting in a higher recovery of the fatty acid methyl esters. Our control values are more in agreement with those reported by Onkenhout et al (1989), which were obtained by using *procedure 2*.

In our hands *procedure 1* and *procedure 2*, when they were compared in a number of duplicate plasma and fibroblast samples, gave similar results.

Our results obtained with X-linked adrenoleukodystrophy show that in virtually all affected patients the three parameters of $C_{26:0}$ concentration as well as $C_{26:0}/C_{22:0}$ and $C_{24:0}/C_{22:0}$ ratios are significantly elevated in both plasma and fibroblasts, although there is great variability among individuals.

In plasma of presumed heterozygotes we found abnormal values of the three parameters in only 60% of the cases. In the remaining cases sometimes only the $C_{26:0}$ or $C_{24:0}/C_{22:0}$ ratio was elevated. In 4 cases normal values of all three parameters were found. However, fibroblast analysis in these ambiguous cases revealed significantly elevated values for the three parameters, except in one case where the plasma values were normal. These findings are in agreement with the observation of Moser and Moser (1991) that 10–15% of het-erozygotes were not detectable by VLCFA analysis in either plasma or cultured fibroblasts. The X-ALD heterozygote identification can be made more secure by DNA linkage analysis (Willems et al 1990).

Table 1 Concentrations of VLCFA, pristanic acid and phytanic acid in plasma and fibroblasts of controls and patients affected by different peroxisomal diseases

Sample	$C_{26:0}$	$C_{26:0}/C_{22:0}$ ratio	$C_{24:0}/C_{22:0}$ ratio	Pristanic acid	Phytanic acid
Plasma/serum (μg/ml)					
Controls ($n=50$)	0.30–0.80	0.016–0.035	0.65–1.00	n.d.–0.2	0.4–4.0
	(0.48)	(0.025)	(0.83)	(0.1)	(0.8)
X-ALD-AMN ($n=35$)	0.70 –2.80	0.056–0.130	0.96–1.89	n.a.	n.a.
	(1.42)	(0.072)	(1.39)		
X-ALD heterozygotes ($n=40$)	0.55–1.75	0.018–0.108	0.68–1.44	n.a.	n.a.
	(1.10)	(0.055)	(1.10)		
Peroxisome deficiency disorders ($n=9$)	1.35–6.20	0.113–0.815	1.35–2.75	0.6–7.0	10.0–45.0
	(3.30)	(0.345)	(1.90)	(3.0)	(25.0)
β-Oxidation deficiencies ($n=4$)	3.5–6.10	0.360–0.525	1.70–2.90	12.0–50.0	8.0–21.0
	(4.40)	(0.410)	(2.20)	(31.0)	(12.0)
Refsum disease ($n=3$)	0.35–0.50	0.018–0.28	0.77–0.89	n.d.–0.20	800–1250
	(0.45)	(0.022)	(0.83)	(0.1)	(925)
Fibroblasts (μg/mg protein)					
Controls ($n=20$)	0.15–0.35	0.08–0.20	1.70–2.30	n.a.	n.a.
	(0.23)	(0.15)	(2.05)		
XALD-AMN ($n=12$)	0.55–1.00	0.40–0.75	2.55–3.20	n.a.	n.a.
	(0.65)	(0.51)	(2.70)		
XALD heterozygotes ($n=15$)	0.45–0.95	0.30–0.65	2.45–2.95	n.a.	n.a.
	(0.56)	(0.39)	(2.60)		
Peroxisome deficiency disorders ($n=6$)	0.95–1.70	0.60–1.10	2.90–4.55	n.a.	n.a.
	(1.45)	(0.85)	(3.50)		
β-Oxidation deficiencies ($n=3$)	0.85–2.10	0.50–1.30	3.10–4.60	n.a.	n.a.
	(1.70)	(0.75)	(3.70)		

n.a. = not analysed; n.d. = not detectable; mean values in parentheses

In newborns, the diagnosis of ALD by measurement of VLCFA in plasma is hampered by the fact that sometimes (we estimate in about 20% of cases) elevated values of some or all of the three parameters are found that appear not to be related to a peroxisomal defect but are probably due to an immaturity of the peroxisomal system. These abnormal levels normalize after some weeks or even months. False positive results may also be observed in children who are receiving a ketogenic diet (Moser and Moser 1991) and in cases of severe liver malfunction (own observations).

False negative results in plasma from XALD patients appear also to be possible, as reported by Wanders et al (1993). In two cases of clinically suspected patients the plasma VLCFA levels were found normal, whereas they were clearly elevated in cultured fibroblasts.

Whenever there is clinical evidence for XALD and plasma VLCFA concentrations are difficult to interpret or normal, analysis of fibroblasts is recommended.

In the 7 patients diagnosed with a generalized peroxisomal disorder, all three parameters for VLCFA were strongly elevated in both plasma and fibroblasts. Plasma phytanic acid in blood samples taken at least 6 weeks after birth was also elevated. Except in one case, pristanic acid concentration was also elevated.

The VLCFA concentrations were also very high in plasma and fibroblasts of 4 patients in whom plasmalogen biosynthesis was normal and hence who were classified as disorders with an isolated β-oxidation defect. In all these patients, plasma pristanic acid concentration was very high and phytanic acid was also elevated, but to a lesser extent.

As already stated by ten Brink et al (1992), the pristanic acid/phytanic acid ratio is high in isolated β-oxidation defects while it is normal in disorders of peroxisome assembly. One of these patients was identified by immunoblotting and complementation analysis as a functional defect of the bifunctional enzyme. The defect in the other cases is still unidentified. They all have normal acyl-CoA oxidase activities (of both palmitoyl-CoA and pristanoyl-CoA oxidase) and the three β-oxidation enzymes are immunologically present. Significantly there is no accumulation of bile acid precursors in two cases.

REFERENCES

Lepage G, Roy CC (1986) Direct transesterification of all classes of lipids in a one-step reaction. *J Lipid Res* **27**: 114–120.

Moser HW, Moser AB (1991) Measurement of saturated very long chain fatty acid in plasma. In Hommes FA, ed. *Techniques in Diagnostic Human Biochemical Genetics*. New York: Wiley-Liss, 177–191.

Onkenhout W, Pael PFH vd, Heuvel MPH vd (1989) Improved determination of very long-chain fatty acids in plasma and cultured skin fibroblasts: application to the diagnosis of peroxisomal disorders. *J Chromatography* **494**: 31–41.

Schutgens RBH, Bouman IW, Nijenhuis AA, Wanders RJA, Frumau MEJ (1993) Profiles of very long chain fatty acids in plasma, fibroblasts and blood cells in Zellweger syndrome, X-linked adrenoleukodystrophy and rhizomelic chondrodysplasia punctata. *Clin Chem* **39**: 1632–1637.

Singh I, Moser AB, Goldfisher S, Moser HW (1984) Lignoceric acid is oxidized in the peroxisome: implications for the Zellweger cerebro-hepato-renal syndrome and adrenoleukodystrophy. *Proc Natl Acad Sci USA* **81**, 4203–4207.

ten Brink HJ, Stellaard F, Heuvel CMM vd et al (1992) Pristanic and phytanic acid in plasma from patients with peroxisomal disorders: stable isotope dilution analysis with electron capture negative ion mass fragmentography. *J Lipid Res* **33**: 31–37.

Vanhove GF, Veldhoven PP v, Fransen M et al (1993) The CoA esters of 2-methyl branched chain fatty acids and of the bile acid intermediates di- and trihydroxycoprostanic acids are oxidized by one single peroxisomal branched chain acyl CoA oxidase in human liver and kidney. *J Biol Chem* **286**: 10335–10344.

Wanders RJA, Schutgens RBH, Barth PG, Tager JM, Bosch II vd (1993) Postnatal diagnosis of peroxisomal disorders: a biochemical approach. *Biochemie* **75**: 269–279.

Willems P, Vits L, Wanders RJA et al (1990) Linkage of DNA markers at Xq 28 to adrenoleukodystrophy and adrenomyeloneuropathy present within the same family. *Arch Neurol* **47**: 665–669.

J. Inher. Metab. Dis. 18 Suppl. 1 (1995) 84–89
© SSIEM and Kluwer Academic Publishers.

Assay of plasmalogens and polyunsaturated fatty acids (PUFA) in erythrocytes and fibroblasts

G. DACREMONT* and G. VINCENT
Department of Pediatrics, University Hospital Gent, Belgium

Correspondence: Universitair Ziekenhuis, Kliniek voor Kinderziekten 'C. Hooft', De Pintelaan 185, 9000 Gent, Belgium

Summary: The direct transesterification method of Lepage and Roy is described as used in our laboratory for the analysis of plasmalogens and polyunsaturated fatty acids in erythrocytes and cultured fibroblasts by gas chromatography. An overview is given of the plasmalogen ratios and docosahexaenoic acid concentrations from controls and patients with different peroxisomal disorders investigated in our laboratory.

Plasmalogens, or 1-0-alk-1'-enyl-2-acylphosphoglycerides, are the main end products of ether-phospholipid biosynthesis in mammals. Two enzymes, dihydroxyacetone-phosphate acyltransferase (DHAPAT) and alkyl dihydroxyacetone-phosphate synthetase (DHAPS) responsible for the introduction of the characteristic ether linkage, are localized in peroxisomes.

In disorders of peroxisome biogenesis (Zellweger syndrome, McKusick 214100; neonatal adrenoleukodystrophy, McKusick 202370; infantile Refsum disease, McKusick 266500; and pseudo infantile Refsum disease) these two enzymes are absent, resulting in reduced biosynthesis of plasmalogens and very low levels of these ether-phospholipids in cells. Both enzymes are also deficient in rhizomelic chondrodysplasia punctata (McKusick 215100), a disorder in which phytanic acid oxidase is deficient and peroxisomal thiolase occurs in a precursor form. Recently two variants of rhizomelic chondrodysplasia punctata were described in which there were isolated deficiencies of dihydroxyacetone-phosphate acyltransferase (Wanders et al 1992) or alkyldihydroxyacetone-phosphate synthetase (Wanders et al 1994) without any other enzymatic abnormality.

In these disorders, as well as in classical rhizomelic chondrodysplasia and disorders of peroxisome assembly, the plasmalogen concentrations are greatly decreased.

Estimation of plasmalogens in erythrocytes or fibroblasts, in combination with an assay of VLCFA, provides important information about the nature of the peroxisomal disorder and helps to classify it either as a generalized disorder or as a disorder with isolated defects.

The best procedure for measurement of erythrocyte or fibroblast plasmalogens is described by Björkheim et al (1986). The ether-linked chains are released from lipids as dimethylacetal derivatives following acidic methanolysis. The relative amount of plasmalogens is thus reflected in the ratio of $C_{16:0}$ dimethylacetal and $C_{18:0}$ dimethylacetal to their corresponding fatty acid methyl esters.

Polyunsaturated fatty acids (PUFA) are also involved in some peroxisomal disorders. Martinez (1989, 1990, 1992a,b, 1994a,b, 1995) has described changes in PUFA patterns in erythrocytes and tissues of patients with disorders of peroxisome biogenesis.

The most constant and severe abnormality is a drastic decrease in the total amount of docosahexaenoic acid ($C_{22:6}$ ω-3; DHA). The enzymatic defect responsible for this decrease of $C_{22:6}$ ω-3 is not known. Since $C_{22:6}$ ω-3 is mainly enriched in synaptic membranes and the retina, a drastic decrease might contribute to the mental deterioration and visual impairment. Treatment of patients by supplementation of $C_{22:6}$ ω-3 has been tested and in some cases resulted in restoration of normal levels and in some neurological improvement (Martinez 1992b). Consequently, estimation of PUFA levels, especially of $C_{22:6}$ ω-3 in erythrocytes and plasma, is important for nutritional assessment.

In our laboratory both plasmalogens and polyunsaturated fatty acids in erythrocytes and fibroblasts are measured by the direct transesterification method according to Lepage and Roy (1986) without any further purification.

A detailed description of the method is given below.

EQUIPMENT, CHEMICALS, SOLUTIONS AND SAMPLES

Equipment

- Gas chromatograph for capillary column equipped with an oven for reproducible temperature control, a flame ionization detector, a splitless or on-column injection system, and an accurate recorder
- Vortex mixer
- Thermostated heating block
- Laboratory centrifuge (preferably cooled)

Chemicals

- Solvents: n-hexane, benzene and methanol of highest purity and with the lowest amount of residue
- Potassium carbonate (analytical grade)
- Acetyl chloride 99% pure)
- Butylated hydroxytoluene (BHT) (analytical grade)
- Heptacosanoic acid ($C_{27:0}$ FA) as internal standard for PUFA
- $C_{16:0}$ and $C_{18:0}$: dimethylacetals for localization of plasmalogens

Solutions

- Heptacosanoic acid, 250 nmol/ml in chloroform – methanol 2:1 (v/v)
- Mixture of methanol – benzene 4:1 (v/v) containing 0.5 g BHT/L
- 6% Potassium carbonate (aqueous)

Materials

- Borosilicate glass tubes 15×100 mm (10 ml) with Teflon-lined screwcaps
- Adjustable automatic pipettes (glass)
- Organic solvent-resistant dispensers
- Teflon-lined screwcapped 2 ml vials
- Capillary column suitable for separation of polyunsaturated fatty acids.

Biological samples

- *Erythrocytes:* Venous blood (2–5 ml), taken into EDTA, is centrifuged for 10 min at 500*g*, preferably within 1 h of collection. The plasma is removed and the erythrocytes are washed three times with 2 ml of isotonic saline, each time with gentle mixing, centrifuging for 5 min at 500*g* and removing the supernatant.

  The packed cells are then diluted with an equal volume of saline and mixed thoroughly. Two aliquots of this suspension are taken for duplicate cell counting.

  If the test is not performed immediately, three or four 200 μl aliquots of the suspension are stored at $-30°C$ in 15×100 mm Teflon screwcapped tubes, preferably under N_2.

  For measurement of polyunsaturated fatty acids, it is preferable to collect the blood sample after an overnight fast, because the profile is strongly influenced by food intake.

  Samples can be mailed as whole EDTA blood. They should be sent cooled (on ice blocks) but not frozen, and must be delivered to the laboratory within 36 h.

- *Fibroblasts:* Fibroblasts grown to confluency are harvested either by trypsinization or by 'scraping'. Culture medium is removed after centrifugation for 15 min at 800*g* and the fibroblast pellet is washed three times. After the last wash, all liquid is removed as completely as possible and the pellet is stored at $-20°C$ or lower, until analysis.

For simultaneous analysis of plasmalogens and PUFA, 300–500 μg of fibroblast protein is required.

METHODS

Preparation of plasmalogen-dimethylacetals and PUFA-methyl esters

Erythrocytes

- Into a 15×100 mm glass tube with Teflon-lined screwcap, pipette 100 μl of internal standard solution (= 25 nmol of $C_{27:0}$) and 200 μl of erythrocyte suspension (see above).
- Add 2 ml of methanol–benzene 4:1 (v/v) containing 0.5 g/L of BHT.
- Add 200 μl of acetyl chloride slowly, drop by drop, while stirring in an ice bath.
- Close the tube tightly and heat for 4 h at 80°C in a heating block.
- When the tube is cool, carefully add 5 ml of aqueous 6% potassium carbonate.
- Vortex briefly, close the tube, and centrifuge for 10 min at 750*g*.
- Transfer the upper benzene layer to a 2 ml vial.
- Evaporate the benzene under nitrogen and take up the dry residue in approx. 100 μl of hexane and inject 0.25 μl.

Fibroblasts

- Suspend the fibroblast pellet (500 – 800 μg of protein) obtained from two 75 cm² culture flasks in 400 μl of water.
- Disrupt the cells by sonication to form a homogeneous suspension.
- Take two aliquots (e.g. 10 and 20 μl) of this suspension for duplicate protein analysis.
- Transfer 200 μl to a 10×15 mm Teflon-lined screwcapped tube and add 40 μl (10 nmol) of the $C_{27:0}$ internal standard solution.
- Proceed as for *erythrocytes*.

Capillary gas chromatography

In our laboratory a 30 m×0.32 mm ID biscyanopropyl 60% vinylmethylsilicone (0.20 μm) capillary column (SP2330 Supelco) is used for the determination of both plasmalogens and polyunsaturated fatty acids. Helium is used as the carrier gas at a flow rate of 1 ml/min. The detector temperature is 320°C and the on-column injector temperature is 60°C. The oven temperature is programmed as follows:

- The initial temperature of 60°C is immediately increased to 180°C at 2°C/min and held there for 1 min.
- Subsequently the temperature is increased to 250°C at 1.5°C/min and maintained there for 30 min.

RESULTS AND DISCUSSION

The plasmalogen concentrations in normal control erythrocytes and cultured fibroblasts, and values found in disorders of peroxisome biogenesis and rhizomelic chondrodysplasia, are given in Table 1. As can be seen, markedly reduced levels were found in both erythrocytes and fibroblasts from classical Zellweger syndrome (severe form) and rhizomelic chondrodysplasia punctata (RCDP). However, in the patients with neonatal adrenoleukodystrophy (NALD) and infantile Refsum disease (IRD), who were all older than 10 years of age, plasmalogen concentrations in erythrocytes were virtually normal, while in fibroblasts, where DHAPAT activity was clearly reduced, plasmalogen levels were also decreased in the one NALD case studied.

As described earlier by Wanders et al (1986), plasmalogen content in erythrocytes of patients with diseases of peroxisome biogenesis is age-related and normalizes by about 6 weeks after birth. This indicates that the diagnostic value of erythrocyte plasmalogen estimation is clearly limited. However, in rhizomelic chondrodysplasia punctata, both in its classical form as well as in forms with isolated deficiencies of dihydroxyacetone-phosphate acyltransferase or alkyldihydroxyacetone-phosphate synthetase, erythrocyte plasmalogen content appears to remain constantly deficient (Wanders et al 1992, 1994).

Concentrations of $C_{22:6}$ ω-3 in normal erythrocytes and fibroblasts were measured in our laboratory by the above described method (see Table 2). The values are in good agreement with those reported by Martinez (1989, 1990, 1992a,b). Our experience with polyunsaturated fatty acids in peroxisomal disorders is limited. Until now we have only investigated four cases: a 10-year-old patient with presumed NALD, two cases of IRD, respectively 15 and 17 years old, and recently a newborn with presumed isolated defect of β-oxidation

Table 1 Plasmalogen levels in erythrocytes and fibroblasts and of patients with disorders of peroxisome biogenesis and with rhizomelic chondrodysplasia punctata

	Normal controls			
	Erythrocytes ($n=25$)		*Fibroblasts* ($n=10$)	
	(Mean)	(Range)	(Mean)	(Range)
$C_{16}DMA/C_{16:0}$ ratio	0.085	0.065–0.110	0.145	0.140–0.160
$C_{18}DMA/C_{18:0}$ ratio	0.180	0.160–0.220	0.105	0.090–0.120
	Classical Zellweger			
	Erythrocytes ($n=4$)		*Fibroblasts* ($n=3$)	
$C_{16}DMA/C_{16:0}$ ratio	0.010	0.005–0.025	0.010	0.005–0.020
$C_{18}DMA/C_{18:0}$ ratio	0.035	0.010–0.050	0.008	0.004–0.015
	NALD			
	Erythrocytes ($n=2$)		*Fibroblasts* ($n=1$)	
$C_{16}DMA/C_{16:0}$ ratio	0.075	0.068, 0.082	0.050	DHAPAT ↓
$C_{18}DMA/C_{18:0}$ ratio	0.170	0.165, 0.175	0.030	
	IRD			
	Erythrocytes ($n=2$)		*Fibroblasts* ($n=1$)	
$C_{16}DMA/C_{16:0}$ ratio	0.080	0.075, 0.085	n.a.	DHAPAT ↓
$C_{18}DMA/C_{18:0}$ ratio	0.165	0.160, 0.170	n.a.	
	RCDP			
	Erythrocytes ($n=1$)		*Fibroblasts* ($n=4$)	
$C_{16}DMA/C_{16:0}$ ratio	0.005		0.005	0.–0.015
$C_{18}DMA/C_{18:0}$ ratio	0.010		0.005	0.–0.015

n.a. = not analysed

Table 2 Control values of $C_{22:6}$ ω-3 in erythrocytes and fibroblasts

	Erythrocytes ($n=10$)		*Fibroblasts*	
	(Mean)	(Range)	(Mean)	(Range)
$C_{22:6}$ ω-3	31.5	24.0–39.0	19.5	16.0–26.0

Age of controls 3–8 months

(abnormal VLCFA and normal plasmalogen concentration). Decreased levels of erythrocyte $C_{22:6}$ ω-3 were found only in the case of NALD and in the newborn.

REFERENCES

Bjorkheim I, Sisfontes L, Boström B, Kase BF, Blomstrand R (1986) Simple diagnosis of the Zellweger syndrome by gas–liquid chromatography. *J Lipid Res* **27**: 786–791.

Lepage G, Roy CC (1986) Direct transesterification of all classes of lipids in a one-step reaction. *J Lipid Res* **27**: 114–120.

Martinez M (1989) Polyunsaturated fatty acid changes suggesting a new enzymatic defect in Zellweger syndrome. *Lipids* **24**: 261–265.

Martinez M (1990) Severe deficiency of docosahexaenoic acid in peroxisomal diseases: a defect of Δ-4 desaturation? *Neurology* **40**: 1292–1298.

Martinez M (1992a) Abnormal profiles of polyunsaturated fatty acids in the brain, liver, kidney and retina of patients with peroxisomal disorders. *Brain Res* **583**: 171–182.

Martinez M (1992b) Treatment with docosahexaenoic acid favorably modifies the fatty acid composition of erythrocytes in peroxisomal patients. In Coates PM, Tanaka K, eds. *New Developments in Fatty Acid Oxidation.* New York: Wiley-Liss, 389–397.

Martinez M (1994a) Polyunsaturated fatty acids in the developing human brain, red cells and plasma: influence of nutrition and peroxisomal disease. In Galli C, Simopoulos AC, Tremoli E, eds. *Fatty Acids and Lipids: Biological Aspects. (World Rev Nutr Diet.* **75**) Karger: Basel, 70–78.

Martinez M (1995) Polyunsaturated fatty acids in the developing human brain, erythrocytes and plasma in peroxisomal disease: therapeutic implications. *J Inher Metab Dis* **18 (Suppl. 1)**: 61–75.

Martinez M, Mougan I, Ballabriga A (1994b) Blood polyunsaturated fatty acids in patients with peroxisomal disorders: a multicenter study. *Lipids* **29**: 273–280.

Wanders RJA, Purvis YR, Heymans HSA, et al. (1986) Age-related differences in plasmalogen content of erythrocytes from patients with the cerebro-hepato-renal (Zellweger) syndrome: implications for postnatal detection of the disease. *J Inher Metab Dis* **9**: 335–342.

Wanders RJA, Schumacher H, Heikoop J, Schutgens RBH, Tager JM (1992) Human dihydroxy-acetonephosphate acyltransferase deficiency: A new peroxisomal disorder. *J Inher Metab Dis* **15**: 389–399.

Wanders RJA, Schutgens RBH, Barth PG, Tager JM, Bosch H vd (1993) Postnatal diagnosis of peroxisomal disorders: a biochemical approach. *Biochemie* **75**: 269–279.

Wanders RJA, Dekker C, Horvath VAP et al (1994) Human alkyldihydroxyacetonephosphate synthase deficiency: a new peroxisomal disorder. *J Inher Metab Dis* **17**: 315–318.

J. Inher. Metab. Dis. 18 Suppl. 1 (1995) 90–100

Measurement of dihydroxyacetone-phosphate acyltransferase (DHAPAT) in chorionic villous samples, blood cells and cultured cells

R. J. A. Wanders[1]*, R. Ofman[1], G. J. Romeijn[1], R. B. H. Schutgens[1], P. A. W. Mooijer[1], C. Dekker[1] and H. van den Bosch[2]

[1]*University Hospital Amsterdam, Academic Medical Centre, Departments of Pediatrics and Clinical Chemistry, Laboratory of Pediatric Clinical Chemistry, Section for Clinical Enzymology;* [2]*Centre for Biomembranes and Lipid Enzymology, University of Utrecht, Utrecht, The Netherlands*

*Correspondence: University Hospital Amsterdam, AMC, Meibergdreef 9, Room F0-224, 1105 AZ Amsterdam, The Netherlands

Summary: Dihydroxyacetone-phosphate acyltransferase (DHAPAT) is a peroxisomal enzyme catalysing the first step in ether-phospholipid biosynthesis. DHAPAT is deficient in cells from patients suffering from a variety of peroxisomal disorders. Accurate measurement of the activity of this enzyme is of great importance, especially since it is a central parameter in the prenatal diagnosis of the disorders of peroxisome biogenesis, rhizomelic chondrodysplasia punctata and DHAPAT-deficiency. We describe a straightforward and accurate assay allowing the activity of DHAPAT to be measured reliably in chorionic villus samples, blood cells, cultured skin fibroblasts, cultured chorionic villus fibroblasts and cultured amniocytes.

Ether-phospholipids are phospholipids which differ from the generally known diacylglycerophospholipids in one major aspect, which is the occurrence of an ether bond rather than an ester bond at the *sn*-1 position of the glycerol backbone. In mammals the main end products of ether-phospholipid biosynthesis are the plasmalogens (1-0-alk-1'-enyl-2-acylphosphoglycerides) with a double bond between the two carbon atoms adjacent to the ether bond. Plasmalogens have long been known to be widely distributed in mammalian membranes, making up 5–20% of total phospholipids. They are particularly abundant in electrically active tissues such as brain. The physiological function of plasmalogens and of ether-phospholipids in general has remained an enigma except for that of platelet activating factor (PAF), which has been implicated in a range of (patho-) physiological processes (Chung 1992).

As first discovered by Hajra and co-workers (see Hajra and Bishop (1982) for review) peroxisomes play an essential role in ether-phospholipid biosynthesis. Indeed, the enzymes responsible for the introduction of the typical ether bond in ether-phospholipids, i.e. dihydroxyacetone-phosphate acyltransferase (DHAPAT) (EC 2.3.1.42) and alkyldihydroxyacetone-phosphate synthase (alkyl-DHAP synthase) (EC 2.5.1.26) are localized predominantly in peroxisomes (see Hajra and Bishop (1982) for review). DHAPAT has been found to be a membrane-bound protein with its active site facing the interior of the peroxisome (Jones and Hajra 1980; De Clerq et al 1984; Hardeman and Van den Bosch 1988). Purification of the enzyme protein has recently been achieved from guinea-pig liver (Webber and Hajra 1993) and human placenta (Ofman and Wanders 1994).

Measurement of DHAPAT activity is important for the correct identification of patients suspected to suffer from a peroxisomal disorder (Wanders et al 1988, 1993; Schutgens et al 1989). DHAPAT is deficient in patients suffering from disorders of peroxisome biogenesis that include Zellweger syndrome (ZS, McKusick 214100), neonatal adrenoleukodystrophy (NALD, McKusick 202370) and infantile Refsum disease (IRD, McKusick 266510). DHAPAT is also deficient in patients with the rhizomelic form of chondrodysplasia punctata (RCDP) (McKusick 215100), although residual activity is much higher in RCDP fibroblasts than in Zellweger fibroblasts (see Table 1). It should be noted that there are also other biochemical abnormalities in RCDP patients (Hoefler et al 1988; Van den Bosch et al 1992; Wanders et al 1993). Finally, DHAPAT is deficient in patients with a variant form of rhizomelic chondrodysplasia punctata characterized by an isolated deficiency of DHAPAT (Wanders et al 1992a; Barr et al 1993; Clayton et al 1994). Importantly, DHAPAT is also expressed in chorionic villus cells, either cultured or not (see Table 1), and in cultured amniocytes and is often used for prenatal diagnosis (Schutgens et al 1989; see also Table 1).

PRINCIPLE OF THE ASSAY METHOD USED TO MEASURE DHAPAT ACTIVITY

Acyl-CoA:dihydroxyacetone-phosphate acyltransferase (DHAPAT, EC 2.3.1.42) catalyses the reaction

dihydroxyacetone-phosphate + acyl-CoA → acyldihydroxyacetone-phosphate + CoA-SH

Activity is relatively low in tissues and fibroblasts and hence the use of a radioactively labelled substrate is mandatory. Since radiolabelled dihydroxyacetone phosphate (DHAP) is not available commercially, a method was developed for preparation of [^{14}C]DHAP from commercially available [^{14}C]glycerol-3-phosphate (Davis and Hajra 1979; Schutgens et al 1984, 1986; Wanders et al 1985; Ofman and Wanders 1994). This is done as shown schematically in Figure 1.

Owing to the unfavourable equilibrium constant of the *sn*-glycerol-3-phosphate dehydrogenase reaction (*sn*-glycerol-3-phosphate:NAD$^+$ 2-oxidoreductase, EC 1.1.1.8), pyruvate and lactate dehydrogenase (LDH, L-lactate:NAD$^+$ oxidoreductase, EC 1.1.1.27) have to be added to obtain full conversion of glycerol-3-phosphate to dihydroxyacetone-phosphate.

The [^{14}C]DHAP synthesized in a preincubation is then used for DHAPAT activity measurements, which are done at slightly acidic pH with palmitoyl-CoA as second substrate. Finally, the product acyl-[^{14}C]dihydroxyacetone-phosphate is separated from the substrate [^{14}C]DHAP and quantified.

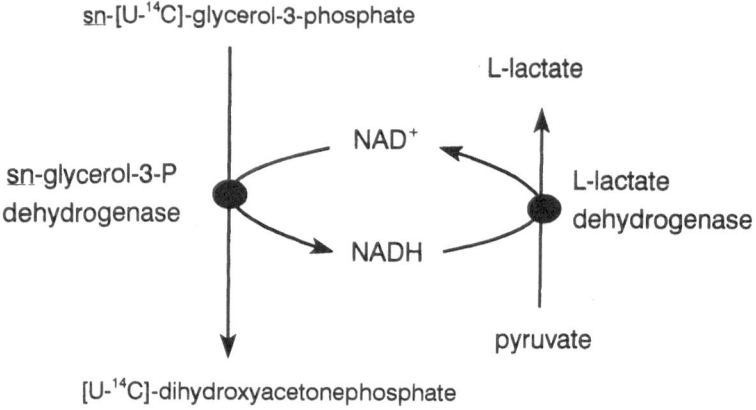

sn-[U-¹⁴C]-glycerol-3-phosphate

Figure 1 Schematic representation of the method used to prepare radiolabelled dihydroxyacetone-phosphate (DHAP) from glycerol-3-phosphate (G3P)

EQUIPMENT, CHEMICALS AND SOLUTIONS

1. Equipment

Liquid scintillation counter; vortex mixer; low-speed centrifuge; thermostated water bath; glass test tubes ($11/12 \times 65$ mm approx.) with caps; scintillation vials; adjustable automatic pipettes and tips; a pipette or syringe with a Teflon plunger for 100 and 600μl; gloves; Pasteur pipettes and bulbs; timer; container with ice; rocking table; needles (1.1×40 mm); cellulose filter papers (2.1 cm diameter); sonicator with a small tip (tip diameter of 0.25 cm).

2. Chemicals

- Lactate dehydrogenase (LDH, L-lactate : NAD oxidoreductase, EC 1.1.1.27, from hog muscle, specific activity ~550 U/mg (25°C); Boehringer Mannheim, cat. no. 127221)
- Glycerol-3-phosphate dehydrogenase (G3PDH, *sn*-glycerol-3-phosphate : NAD⁺ 2-oxidoreductase, EC 1.1.1.8, from rabbit muscle, specific activity ~170 U/mg (25°C); Boehringer Mannheim, cat. no. 127124)
- Triethanolamine hydrochloride (2,2′,2″-nitrilotriethanol hydrochloride; Sigma Chemicals, St Louis , MO, USA, cat. no. T 1502)
- NAD (β-nicotinamide-adenine dinucleotide, free acid; Boehringer Mannheim, cat. no. 127302)
- Magnesium chloride ($MgCl_2 \cdot 6H_2O$; Merck, Darmstadt, Germany, art 5833)
- Pyruvate (pyruvate, monosodium salt; Boehringer Mannheim, cat. no. 128147)
- Palmitoyl-CoA (*S*-palmitoyl-coenzyme A, tetrapotassium salt; Boehringer Mannheim, cat. no. 663387)
- Glycerol-3-phosphate (L-glycerol-3-phosphate, dicyclohexylammonium salt dihydrate; Boehringer Mannheim, cat. no. 105961)

- Radiolabelled glycerol-3-phosphate (L-[^{14}C]glycerol-3-phosphate, ammonium salt, aqueous solution containing 2% (v/v) ethanol, >100 mCi/mmol; Amersham International, UK, cat. no. CFB 171)
- Sodium fluoride (NaF, purity ~99% (pfs); Sigma Chemicals, St Louis, MO, USA, cat. no. 51504)
- Bovine serum albumin (BSA, fatty acid free; Boehringer Mannheim, cat. no. 775827)
- Trichloroacetic acid (TCA, purity >99%; Sigma Chemicals, St Louis, MO, USA, cat. no. T 6399)
- Phosphoric acid (H_3PO_4; Sigma Chemicals, St Louis, MO, USA, cat. no. P 6560)
- Trizma hydrochloride (tris(hydroxymethyl)aminomethane hydrochloride; Sigma Chemicals, St Louis, MO, USA, cat. no. T 3253)
- KCl (potassium chloride; Sigma Chemicals; St Louis, MO, USA, cat no. P 4504)
- NaCl (sodium chloride; Sigma Chemicals, St Louis, MO, USA, cat no. S 7653)
- Chloroform and methanol, glass-redistilled, of highest purity
- Buffered saline: 154 mmol/L sodium chloride plus 10.5 mmol/L phosphate (pH 7.4)

3. Solutions necessary for the DHAPAT assay

A. Stock solutions required for the preincubation (synthesis of [^{14}C]DHAP from [^{14}C]G3P)

- 0.35 mol/L triethanolamine-HCl (pH 7.6). Dissolve 55.16 g in water, adjust pH to 7.6 and make up to 1 litre.
- 100 mmol/L sodium pyruvate. Dissolve 110 mg sodium pyruvate in 10 ml water. Prepare fresh each time.
- 20 mmol/L NAD. Dissolve 66.34 mg in 5 ml water. Store in 100 μl aliquots at −80°C.
- 12 mmol/L glycerol-3-phosphate. Dissolve 48.78 mg dicyclohexylammonium glycerol-3-phosphate dihydrate in 10 ml water. Store in aliquots at −80°C.

B. Stock solutions required for the actual DHAPAT assay

- 0.5 mol/L sodium acetate (pH 5.4). Dissolve 41.02 g sodium acetate in water, adjust pH and make up to 1 litre. Store at 4°C.
- 0.5 mol/L magnesium chloride. Dissolve 101.65 g magnesium chloride hexahydrate in 1 litre water. Store at 4°C.
- 0.5 mol/L sodium fluoride. Dissolve 1.05 g sodium fluoride in 50 ml water. Store at 4°C.
- 66.7 mg/ml BSA. Dissolve 66.7 mg in 1 ml water. Store deep-frozen (−20°C or −80°C).
- 1.0 mmol/L palmitoyl-CoA in 10 mmol/L sodium acetate buffer (pH 5.4). To 11.62 mg of palmitoyl-CoA add 10 ml of 10 mmol/L sodium acetate buffer prepared by diluting 50-fold the stock solution of 0.5 mol/L (see above). Store in portions of 1 ml at −80°C.
- 1:1 (v/v) chloroform−methanol
- 2 mol/L KCl plus 0.2 mol/L H_3PO_4. For 1 litre, weigh 149.1 g potassium chloride, add 11.63 ml H_3PO_4, and make up to 1 litre.

- TCA solutions. For 10%, 5% and 1% (w/v) solutions, weigh 100, 50 and 10 g TCA, respectively, and make up to 1 litre.
- 50 mmol/L sodium chloride plus 5 mmol/L Tris-HCl (pH 7.4). For 1 litre, weigh 2.92 g sodium chloride, add 1 litre 5 mmol/L Tris-HCl (pH 7.4) prepared by dilution from the 0.5 mol/L stock solution (see above).

EXPERIMENTAL PROCEDURE

1. Synthesis of [^{14}C]DHAP from [^{14}C]G3P (*preincubation step*)

(i)   Add to a 5 ml glass tube: 145 μl 0.35 mol/L triethanolamine-HCl (pH 7.6), 50 μl 100 mmol/L Na-pyruvate, 50 μl 20 mmol/L NAD$^+$, 15 μl LDH (10 mg/ml), 100 μl *sn*-[^{14}C]G3P, 580 μl water, 15 μl glycerol-3-phosphate dehydrogenase, plus 45 μl 12 mmol/L G3P (total volume 1000 μl).

(ii) Close tube, mix gently and incubate at 25°C for 1 h. During this period, prepare the cell homogenates (see below) and mark tubes for the second step.

(iii) Stop the preincubation reaction by adding an equal volume of chloroform and mix thoroughly. Centrifuge at 2000 rpm for 5 min and transfer upper layer containing [^{14}C]DHAP to a clean tube.

Note: This quantity is enough for at least 40 incubations.

2. Preparation of chorionic villus biopsy specimens, blood platelets, cultured skin fibroblasts, cultured amniocytes and cultured chorionic villous fibroblasts

Chorionic villus biopsy specimens are usually split into two portions so that one can be used for direct analysis and the other cultured for future biochemical analysis if necessary. For direct analysis, the specimen is washed in physiological saline (see above) three times. After the final wash, the material is centrifuged (12000g_{av}, 5 min, 4°C) and stored at −80°C.

Blood platelets are prepared from venous blood samples using EDTA as anticoagulant (Wanders et al 1985). The final pellet is stored at −80°C.

Skin fibroblasts are cultured according to standard procedures and collected by trypsin treatment after the cells have reached confluency (see Wanders et al (1987) for full details). Cells are washed several times in physiological saline and the final pellet is stored at −80°C containing 1−2 mg protein.

Amniotic fluid cells and chorionic villus fibroblasts are cultured essentially as described for cultured skin fibroblasts.

3. Preparation of homogenates

On the day of the experiment the different cell pellets, prepared as described above and stored at −80°C, are taken followed by addition of 0.5 ml of a solution containing 50 mmol/l sodium chloride plus 5 mmol/l Tris-HCl (pH 7.4) and ultrasonic disruption by sonication for 3 periods of 15 s at 70−80 W. Homogenates should be kept in ice-water during sonication and there should be an interval of at least 45 s between two rounds of sonication.

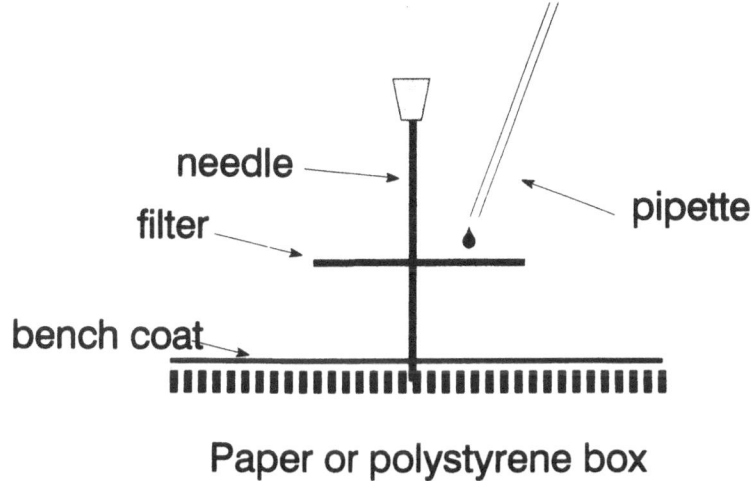

Paper or polystyrene box

Figure 2 Schematic representation of the system used to prepare the filter papers

Protein concentrations can be measured using a variety of different methods using albumin as a standard. Protein content should be between 1 and 5 mg/ml.

4. DHAPAT assay

In our experience it is best first to prepare a reaction medium containing all necessary components at double their final concentrations with the exception of palmitoyl-CoA and radiolabelled [^{14}C]DHAP, which are added separately.

For preparation of the reaction medium, add in a 5 ml glass tube (total incubation volume 2.5 ml): 750 μl sodium acetate buffer (pH 5.4), 80 μl 0.5 mol/L MgCl$_2$, 80 μl 0.5 mol/L sodium fluoride, and 250 μl 66.7 mg/ml BSA. Make up to 2500 μl. Check pH. This quantity is enough for 40 assays.

Once the incubation medium has been prepared, the test tubes are prepared by adding the following constituents: 60 μl incubation medium prepared as described above, 20 μl 1 mmol/L palmitoyl-CoA, and 20 μl [^{14}C]DHAP as prepared enzymatically (upper phase obtained after addition of chloroform as described above). Reactions are then initiated by adding a 20 μl aliquot of cell homogenate (20–100 μg protein) or 20 μl 50 mmol/L sodium chloride plus 5 mmol/L Tris-HCl (blank). For all cell types analysed, enough acyl-[^{14}C]DHAP is formed under such conditions to ensure reliable results.

Reactions are allowed to proceed for 2 h except for platelets, in which case a 30 min incubation period has been selected. Reactions are terminated by adding 600 μl chloroform –methanol (1 : 1, v/v) plus 150 μl KCl/H$_3$PO$_4$ to each tube followed by thorough mixing.

At the end of the experiment, tubes are centrifuged (500g$_{av}$, 5 min, 4°C) and the upper phase is discarded using a Pasteur pipette. Subsequently, 200 μl of the lower chloroform layer is pipetted on to a filter paper by adding two 100 μl portions with a syringe or pipette with Teflon plunger. After applying the first aliquot, wait until the filter is dry before adding the second aliquot (see Figure 2).

The dried filters, which have been marked previously with a pencil, are then added to ice-cold 10% (w/v) TCA in a large beaker (0.5–1.0 litre). After 15 min the TCA is removed and fresh 10% (w/v) TCA is added. After an additional 15 min the filters are washed using 5% (w/v) TCA and 1% (w/v) TCA. Finally, the filters are collected and transferred to scintillation vials followed by addition of 10 ml scintillation cocktail. After thorough mixing, radioactivity is counted overnight.

5. Calculation of results

From the counts (in dpm), determine the specific activity of DHAPAT in the samples (nmoles of acyl-[^{14}C]DHAP formed/2 h per mg protein; for platelets, nmoles of acyl-[^{14}C]DHAP formed/30 min per mg protein).

$$\text{Activity} = (\text{counts} - \text{blank}) \times 12 \times 1.5 \times 1000 \times 1/\mu g \text{ protein}$$

RESULTS AND DISCUSSION

Table 1 summarizes the results of DHAPAT activity measurements in a variety of different cell types including chorionic villus biopsy specimens, blood platelets, cultured skin fibroblasts, cultured chorionic villus fibroblasts and cultured amniocytes.

If the results in cultured skin fibroblasts are considered first, it is clear that the deficiency of DHAPAT is most pronounced in cells from patients suffering from a variant form of rhizomelic chondrodysplasia punctata due to isolated DHAPAT deficiency (compare Wanders et al 1992a; Barr et al 1993; Clayton et al 1994). In cells from patients suffering from classical Zellweger syndrome (ZS) DHAPAT is strongly deficient, although not completely deficient as in DHAPAT deficiency. In patients suffering from milder types of peroxisomal deficiency disorder including infantile Refsum disease (Poll-Thé et al 1987), neonatal adrenoleukodystrophy (Kelley et al 1986) and other phenotypes not easily assigned to either of these two entities (Bleeker-Wagemaker et al 1986; Barth et al 1987) residual DHAPAT activity is higher both in cultured skin fibroblasts and blood platelets. In some patients DHAPAT activity may even be near-normal, which can lead to erroneous conclusions in the absence of additional peroxisomal investigations (compare patient VII in Table IV of Wanders et al. (1993)).

As shown in Table 1, DHAPAT is also deficient in cells from patients suffering from the classical form of chondrodysplasia punctata. On the average, residual DHAPAT activity is higher in fibroblasts (and platelets) from RCDP patients than in Zellweger patients. It should be noted that biochemical abnormalities in RCDP are not restricted to the partial deficiency of DHAPAT (Heÿmans et al 1985; Hoefler et al 1988; Schutgens et al 1988). Indeed, alkyldihydroxyacetone-phosphate synthase and phytanic acid α-oxidation are also deficient. In addition, peroxisomal thiolase occurs in its precursor (44 kDa) rather than in its mature (41 kDa) form (Hoefler et al 1988).

Table 1 further shows normal activity of DHAPAT in fibroblasts from patients with defects in peroxisomal β-oxidation. This includes X-linked adrenoleukodystrophy, acyl-CoA oxidase deficiency (Poll-Thé et al 1988; Wanders et al 1990a; Suzuki et al 1994), bifunctional protein deficiency (Watkins et al 1989; Wanders et al 1992b, 1993; Suzuki et al 1994), peroxisomal thiolase deficiency (Goldfischer et al 1986; Schram et al 1987) and

Table 1 Dihydroxyacetone-phosphate acyltransferase activity measurements in blood platelets, cultured skin fibroblasts, chorionic villus biopsy specimens, cultured chorionic villus fibroblasts and amniotic fluid cells from peroxisomal disorder patients

	n	*DHAPAT-activity*[a]	*DHAPAT/GluDH ratio*[b]
Blood platelets			
Control	123	3.5±0.9 (2.2–4.9)	6.9±1.6 (4.5–9.1)
Classical ZS	11	0.24±0.29 (0–1.1)	0.57±0.60 (0–2.0)
IRD, NALD, variant ZS[c]	5	1.96±0.71 (1.3–3.3)	4.72±1.82 (2.6–8.0)
RCDP	6	0.65±0.37 (0.3–1.3)	1.42±0.88 (0.6–3.2)
X-ALD	5	3.7±0.7	7.2±0.8
DHAPAT deficiency	1	0.10	0.10
Cultured skin fibroblasts			
Control	78	8.1±2.5 (3.2–14.3)	6.8±3.4 (2.2–15.6)
Classical ZS	23	0.6±0.5 (0 1.9)	0.5±0.3 (0–1.2)
IRD, NALD, variant ZS[c]	12	2.4±1.5 (0.6–5.9)	1.9±1.7 (0.4–6.5)
RCDP	39	1.9±0.9 (0.4–4.4)	1.7±1.8 (0.2–3.3)
DHAPAT deficiency	6	0.0±0.0	0.0±0.0
X-ALD	5	8.7±1.3 (5.6–9.6)	7.0±0.9 (5.1–8.7)
AOX deficiency[d]	3	9.1±2.0 (8.1–10.1)	6.7±1.0 (6.0–7.8)
BP deficiency[e]	3	8.8±1.8 (7.4–11.1)	7.1±2.4 (5.9–8.2)
Chorionic villi			
Control	68	7.0±2.2 (3.0–10.8)	4.2±1.5 (1.8–9.4)
ZS	15	0.6±0.5 (0.0–1.7)	0.5±0.7 (0–0.8)
RCDP	2	0.2; 3.8	0.2; 1.1
Cultured chorionic villus fibroblasts			
Control	11	6.3±2.2 (3.7–10.3)	n.d.
ZS	8	0.1±0.2 (0.0–0.4)	n.d.
RCDP	2	0.2; 1.9	n.d.
Cultured amniotic fluid cells			
Control	21	6.5±2.5 (3.1–10.9)	n.d.
ZS	4	0.2±0.2 (0.1–0.5)	n.d.

Results are given as mean±SD with the range in parentheses. *n* denotes the number of different patients studied; n.d. = not determined
[a]DHAPAT-activity is expressed in nmol/2h per mg protein except for platelets in which case activity is in nmol/30 min per mg protein, or [b]relative to the activity of glutamate dehydrogenase (GluDH activity in nmol/min per mg protein) measured in the same sample
[c]This includes patients with infantile Refsum disease (IRD), neonatal adrenoleukodystrophy (NALD) and other patients with a deficiency of peroxisomes but with a phenotype different from ZS, IRD or NALD
[d]AOX deficiency = acyl-CoA oxidase deficiency
[e]BP deficiency = bifunctional protein deficiency

other disorders of peroxisomal β-oxidation of unknown aetiology (Clayton et al 1988; Naidu et al 1988; Barth et al 1990; Wanders et al 1990b; Espeel et al 1991; Mandel et al 1992; Van Maldergem et al 1992; Wanders et al 1992b; Santer et al 1993).

When the results obtained in blood platelets are considered, it is clear that these are in good agreement with the data on cultured skin fibroblasts. Finally, the data of Table 1 show that the activity of DHAPAT in chorionic villus biopsy specimens, cultured chorionic villus fibroblasts and cultured amniocytes compares well with the activity in cultured skin fibroblasts and that the activity of DHAPAT is strongly deficient in cases of Zellweger syndrome and rhizomelic chondrodysplasia punctata. Based on these latter results, it is clear that DHAPAT activity measurement is of central importance in the prenatal diagnosis of Zellweger syndrome, other disorders of peroxisome biogenesis (infantile Refsum disease, neonatal adrenoleukodystrophy and other variant forms) rhizomelic chondrodysplasia punctata, and DHAPAT deficiency.

ACKNOWLEDGEMENTS

The authors gratefully acknowledge The Princess Beatrix Fund, The Hague, The Netherlands for financial support through the years and Mrs Iet van der Gracht for preparation of the manuscript. Dr N. Leschot, Dr M. Verjaal and their co-workers, Department of Anthropogenetics, University of Amsterdam, are gratefully acknowledged for their generous help in culturing chorionic villus fibroblasts and amniocytes. Mrs E. de Jonge-Meijboom and P. Huizinga-Veltman are thanked for expert preparation of the cell cultures.

REFERENCES

Barr DGD, Kirk JM, Al Howasi M, Wanders RJA, Schutgens RBH (1993) Rhizomelic chondrodysplasia punctata with isolated DHAP-AT deficiency. *Arch Dis Child* **68**: 415–417.

Barth PG, Schutgens RBH, Wanders RJA, et al (1987) A sibship with a mild variant of Zellweger syndrome. *J Inher Metab Dis* **10**: 253–259.

Barth PG, Wanders RJA, Schutgens RBH, Bleeker-Wagemakers EM, Van Heemstra D (1990) β-Oxidation defect with detectable peroxisomes: a case with neonatal onset and progressive course. *Eur J Pediatr* **149**: 722–726.

Bleeker-Wagemaker EM, Oorthuys JWE, Wanders RJA, Schutgens RBH (1986) Long term survival of a patient with the cerebro-hepato-renal (Zellweger) syndrome. *Clin Genet* **29**: 160–164.

Chung KF (1992) Platelet-activating factor in inflammation and pulmonary disorders. *Clin Sci* **83**: 127–138.

Clayton PT, Lake BD, Hjelm M, et al (1988) Bile acid analyses in pseudo-Zellweger syndrome: clues to the defect in peroxisomal β-oxidation. *J Inher Metab Dis* **11**: 165–168.

Clayton PT, Eckhardt S, Wilson J, et al (1994) Isolated dihydroxyacetone-phosphate acyltransferase deficiency presenting with developmental delay. *J Inher Metab Dis* **17**: 533–540.

Davis PA, Hajra AK (1979) Stereochemical specificity of the biosynthesis of the alkyl ether bond in ether lipids. *J Biol Chem* **254**: 4760–4763.

De Clerq PE, Haagsman HP, Van Veldhoven PP, Debeer LJ, Van Golde LM, Mannaerts GP (1984) Rat liver dihydroxyacetone-phosphate acyltransferase and their contribution to glycerolipid synthesis. *J Biol Chem* **259**: 9064–9075.

Espeel M, Roels F, Van Maldergem L, et al (1991) Peroxisomal localization of the immunoreactive β-oxidation enzymes in a neonate with a β-oxidation defect: pathological observations in liver, adrenal cortex and kidney. *Virchows Arch A (Pathol Anat)* **419**: 339–347.

Goldfischer S, Collins J, Rapin I, et al (1986) Pseudo-Zellweger syndrome: deficiencies in several peroxisomal oxidative activities. *J Pediatr* **108**: 25–32.

Hajra AK, Bishop JE (1982) Glycerolipid biosynthesis in peroxisomes via the acyldihydroxyacetone phosphate pathway. *Ann NY Acad Sci* **386**: 170-182.

Hardeman D, Van den Bosch H (1988) Rat liver dihydroxyacetone-phosphate acyltransferase: enzyme characteristics and localization studies. *Biochim Biophys Acta* **963**: 1–9.

Heymans HSA, Oorthuys JWE, Nelck G, Wanders RJA, Schutgens RBH (1985) Rhizomelic chondrodysplasia punctata: another peroxisomal disorder. *N Engl J Med* **313**: 187–188.

Hoefler G, Hoefler S, Watkins PA, et al (1988) Biochemical abnormalities in rhizomelic chondrodysplasia punctata. *J Pediatr* **112**: 726–733.

Jones CL, Hajra AK (1980) Properties of guinea pig liver peroxisomal dihydroxyacetone phosphate acyltransferase. *J Biol Chem* **255**: 8289–8295.

Kelley RI, Datta NS, Dobijns WS, et al (1986) Neonatal adrenoleukodystrophy: new cases, biochemical studies, and differentiation from Zellweger syndrome and related peroxisomal polydystrophy syndromes. *Am J Med Genet* **23**: 869–901.

Mandel H, Berant M, Aizin A, et al (1992) Zellweger-like phenotype in two siblings: a defect in peroxisomal β-oxidation with elevated very-long-chain fatty acids but normal bile acids. *J Inher Metab Dis* **15**: 381–384.

Naidu S, Hoefler G, Watkins PA, et al (1988) Neonatal seizures and retardation in a girl with biochemical features of X-linked adrenoleukodystrophy. *Neurology* **38**: 1100–1107.

Ofman R, Wanders RJA (1994) Purification of peroxisomal acyl-CoA: dihydroxyacetonephosphate acyltransferase from human placenta. *Biochim Biophys Acta* **1206**: 27–34.

Poll-Thé BT, Saudubray JM, Ogier HAM, et al (1987) Infantile Refsum disease: an inherited peroxisomal disorder and comparison with Zellweger syndrome and neonatal adrenoleukodystrophy. *Eur J Pediatr* **146**: 477–483.

Poll-Thé BT, Roels F, Ogier HAM, et al (1988) A new peroxisomal disorder with enlarged peroxisomes and a specific deficiency of acyl-CoA oxidase (pseudo neonatal adrenoleukodystrophy). *Am J Hum Genet* **42**: 422–434.

Santer R, Claviez A, Oldigs HD, Schaub J, Schutgens RBH, Wanders RJA (1993) Isolated defect of peroxisomal β-oxidation in a 16-year-old patient. *Eur J Pediatr* **152**: 339–342.

Schram AW, Goldfischer S, Van Roermund CWT, et al (1987) Human peroxisomal 3-oxoacyl-coenzyme A thiolase deficiency. *Proc Natl Acad Sci USA* **84**: 2494–2496.

Schutgens RBH, Romeijn GJ, Wanders RJA, et al (1984) Deficiency of acyl-CoA: dihydroxyacetonephosphate acyltransferase in patients with Zellweger (cerebro-hepato-renal) syndrome. *Biochem Biophys Res Commun* **120**: 179–184.

Schutgens RBH, Romeijn GJ, Ofman R, Van den Bosch H, Tager JM, Wanders RJA (1986) Acyl-CoA: dihydroxyacetonephosphate acyltransferase: study of its properties using a new assay method. *Biochim Biophys Acta* **879**: 286–291.

Schutgens RBH, Heijmans HSA, Wanders RJA, et al (1988) Multiple peroxisomal enzyme deficiencies in rhizomelic chondrodysplasia punctata: comparison with Zellweger syndrome, Conradi–Hunermann syndrome and X-linked dominant chondrodysplasia punctata. *Adv Clin Enzymol* **6**: 57–65.

Schutgens RBH, Schrakamp G, Wanders RJA, Heijmans HSA, Tager JM, Van den Bosch H (1989) Pre- and perinatal diagnosis of peroxisomal disorders. *J Inher Metab Dis* **12**: 118–134.

Suzuki Y, Shimozawa N, Yajima S, et al (1994) Novel subtype of peroxisomal acyl-CoA oxidase deficiency and bifunctional enzyme deficiency with detectable enzyme proteins: identification by means of complementation analysis. *Am J Hum Genet* **54**: 36–43.

Van den Bosch H, Schutgens RBH, Wanders RJA, Tager JM (1992) Biochemistry of peroxisomes. *Annu Rev Biochem* **61**: 157–197.

Van Maldergem L, Espeel M, Wanders RJA, et al (1992) Neonatal seizures and severe hypotonia in a male infant suffering from a defect in peroxisomal β-oxidation. *Neuromusc Disord* **2**: 217–224.

Wanders RJA, Van Weringh G, Schrakamp G, Tager JM, Van den Bosch H, Schutgens RBH (1985) Deficiency of acyl-CoA: dihydroxyacetonephosphate acyltransferase in thrombocytes of Zellweger patients: a simple postnatal test. *Clin Chim Acta* **151**: 217–221.

Wanders RJA, Van Roermund CWT, Van Wijland MJA, et al (1987) Peroxisomal fatty acid β-oxidation in relation to the accumulation of very-long-chain fatty acids in cultured skin fibroblasts from patients with Zellweger syndrome and other peroxisomal disorders. *J Clin Invest* **80**: 1778–1783.

Wanders RJA, Heijmans HSA, Schutgens RBH, Barth PG, Van den Bosch H, Tager JM (1988) Peroxisomal disorders in neurology. *J Neurol Sci* **88**: 1–39.

Wanders RJA, Schelen A, Feller N, et al (1990a) First prenatal diagnosis of acyl-CoA oxidase deficiency. *J Inher Metab Dis* **13**: 371–374.

Wanders RJA, Van Roermund CWT, Schelen A, et al (1990b) Bifunctional protein with deficient activity: identification of a new peroxisomal disorder. *J Inher Metab Dis* **13**: 375–379.

Wanders RJA, Schumacher H, Heikoop J, Schutgens RBH, Tager JM (1992a) Human dihydroxy-acetonephosphate acyltransferase deficiency: A new peroxisomal disorder. *J Inher Metab Dis* **15**: 389–391.

Wanders RJA, Van Roermund CWT, Brul S, Schutgens RBH, Tager JM (1992b) Bifunctional enzyme deficiency: identification of a new type of peroxisomal disorder in a patient with an impairment in peroxisomal β-oxidation of unknown aetiology by means of complementation analysis. *J Inher Metab Dis* **15**: 385–388.

Wanders RJA, Schutgens RBH, Barth PG, Tager JM, Van den Bosch H (1993) Postnatal diagnosis of peroxisomal disorders: a biochemical approach. *Biochimie* **75**: 269–279.

Watkins PA, Chen WW, Harris CJ, et al (1989) Peroxisomal bifunctional enzyme deficiency. *J Clin Invest* **83**: 771–777.

Webber KO, Hajra AK (1993) Purification of dihydroxyacetonephosphate acyltransferase from guinea pig liver peroxisomes. *Arch Biochem Biophys* **300**: 88–97.

J. Inher. Metab. Dis. 18 Suppl. 1 (1995) 101–112
© SSIEM and Kluwer Academic Publishers.

Immunoblot analysis of peroxisomal proteins in liver and fibroblasts from patients

R. J. A. WANDERS*, C. DEKKER, R. OFMAN, R. B. H. SCHUTGENS and P. MOOIJER
University Hospital Amsterdam, Academic Medical Centre, Departments of Pediatrics and Clinical Chemistry, Laboratory of Pediatric Clinical Chemistry, Section for Clinical Enzymology, Meibergdreef 9, PO Box 22700, 1100 DE, Amsterdam, The Netherlands

*Correspondence

Summary: Identification of a patient as suffering from a peroxisomal disorder usually starts by the finding of elevated very long-chain fatty acids in plasma and/or serum. This is followed by more detailed studies in blood, fibroblasts and tissues, including immunoblot analysis. Indeed, immunoblot analysis has become a valuable tool in the correct diagnosis and assignment of individual patients, except for X-linked adrenoleukodystrophy (X-ALD). We describe a simple immunoblotting procedure applicable to liver and fibroblast homo-genates using antibodies raised against catalase and the three β-oxidation enzyme proteins acyl-CoA oxidase I, bifunctional protein and peroxisomal thiolase. The same procedure can also be used for chorionic villus biopsy specimens and has now become the method of choice for the prenatal diagnosis of Zellweger syndrome (and other disorders of peroxisome biogenesis) and rhizomelic chondrodysplasia punctata.

Mixtures of proteins can be separated successfully on the basis of their rates of movement in an electric field. In most procedures the detergent sodium dodecyl sulphate (SDS) is used, which dissociates proteins into their subunits and unfolds each polypeptide chain to form a long rod-like SDS–polypeptide complex. In this complex the polypeptide chain is coated with a layer of SDS molecules in such a way that their hydrocarbon chains are in tight hydrophobic association with the polypeptide chain and the charged sulphate groups of the detergent are exposed to the aqueous medium. Such complexes contain a constant ratio of SDS to protein (about 1.4:1 by weight) and differ only in mass. Proteins are usually also treated with a reducing agent such as β-mercaptoethanol in order to break disulphide bonds. When an SDS-treated single chain protein is subjected to electro-phoresis in a molecular-sieve gel containing SDS, its rate of migration is determined primarily by the mass of the SDS–polypeptide particle, through the molecular exclusion principle using a polyacrylamide matrix. To calibrate a given gel system, marker proteins of known molecular weight are run for comparison. There is a linear relationship between the logarithm of the molecular weight of a polypeptide and its migration coefficient (R_f) which makes SDS–polyacrylamide gel electrophoresis (SDS-PAGE) a powerful technique for determining the molecular weight of a protein.

In principle there are several ways of staining a gel after electrophoresis. Coomassie brilliant blue is often used for staining, although staining with silver provides superior sensitivity and is used increasingly. A highly sensitive technique for characterizing the presence, quantity or molecular weight of a particular protein in a mixture of proteins as resolved by SDS-PAGE is the immunoblot procedure. In this procedure proteins are transferred from the gel to a membrane (usually nitrocellulose) that binds the proteins non-specifically. Transfer is usually achieved by means of an electric field. After transfer, non-specific binding sites on the membrane are usually blocked to reduce background signals. Blocking can be achieved using various proteins (bovine serum albumin, horse serum, defatted milk, among others) and/or detergent solutions (e.g. Tween-20). Subsequently, the sheets are incubated with a specific antibody at an appropriate dilution. For immunoblot analysis, polyclonal antibodies are used rather than monoclonal antibodies, since poly-clonal antibodies, provided they are specific, usually give higher signal intensities. After binding of the primary antibody, visualization of the bound immunoglobulins must occur. In principle, the primary antibody can be labelled directly, allowing straightforward identification of the antigen. In most procedures, however, a secondary antibody is used containing a particular label. There are a large number of methods for labelling the second antibody. Usually these secondary antibodies are labelled with enzymes such as peroxidase (EC 1.11.1.7), alkaline phosphatase (EC 3.1.3.1) or β-galactosidase (EC 3.2.1.23), which allow visualization of the bound antibodies using appropriate substrates for peroxidase, alkaline phosphatase and β-galactosidase, respectively. In the present study, use is made of anti-rabbit IgG labelled with alkaline phosphatase. Detection of bound IgG is then done using the substrate couple bromochloroindolyl phosphate – nitrobluetetrazolium (BCIP/NBT) which generates an intense black/purple precipitate at the site of binding.

Immunoblot analysis has become a valuable tool in the correct diagnosis of patients suffering from certain peroxisomal disorders. In principle, patients are usually identified by finding elevated very long-chain fatty acids in plasma and/or serum (Wanders et al 1993). This is followed by more detailed studies in blood (bile acid intermediates, phytanic acid, pristanic acid, pipecolic acid and polyunsaturated fatty acids in serum and/or plasma, plasmalogens in erythrocytes, dihydroxyacetonephosphate acyltransferase (DHAPAT) (EC 2.3.1.42) activity measurements in platelets and/or leukocytes). In addition, detailed studies are done on cultured skin fibroblasts, including enzyme activity measurements, immunoblot analysis and immunofluorescence studies. In each patient, a liver biopsy should be taken for detailed morphology (see Roels et al (1991) for review), including enzyme cytochemistry of catalase (EC 1.11.1.6), a peroxisomal marker enzyme (Roels and Goldfischer 1979 Roels et al 1983, 1993; Roels and Cornelis 1989), and immuno-localization studies at the light and electron microscopic level (Espeel et al 1990, 1991, 1993, 1995; Hughes et al 1990, 1992, 1993).

Liver biopsy studies are especially important because recent reports have described patients with variable levels of expression in liver and fibroblasts (Roels et al 1993; Mandel et al 1994; Schutgens et al 1994; Espeel et al 1995). Indeed, Mandel and co-workers (1994) described a patient with all the biochemical signs and symptoms of a peroxisomal disorder but no peroxisomal abnormalities in fibroblasts. The liver was found to display mosaicism; that is, parenchymal cells with peroxisomes were found adjacent to cells without peroxisomes (Mandel et al 1994; Espeel et al 1995).

As described above, immunoblot analysis is an indispensable tool in the correct diagnosis of peroxisomal patients (other than X-ALD, where no additional information is obtained). In our own laboratory we perform immunoblot analysis using antibodies raised against catalase (EC 1.11.1.6), acyl-CoA oxidase (EC 1.3.99.3), bi(tri)functional protein (EC 4.2.1.17; EC 1.1.1.35), peroxisomal thiolase (EC 2.3.1.16) and alanine:glyoxylate aminotransferase (AGT) (EC 2.6.1.44). The latter is used only in cases of hyperoxaluria type I (McKusick 259900) (Horvath and Wanders 1994). Catalase can be purified relatively easily from erythrocytes and used to generate a polyclonal antibody (Tager et al 1985). Alternatively, catalase can be bought commercially (Sigma Chemicals, St Louis, MO, USA), but needs some additional purification before it can be injected into rabbits. Polyclonal antibodies against catalase are also commercially available (see Espeel and Van Limbergen, 1995). Acyl-CoA oxidase (Osumi et al 1980), bi(tri)functional protein (Furuta et al 1980) and peroxisomal thiolase (Miyazawa et al 1980) can be purified from livers from rats treated with clofibrate or some other hypolipidaemic drugs. These drugs lead to a drastic increase in the relative number of peroxisomes in liver and induce the expression of acyl-CoA oxidase, bi(tri)functional protein and peroxisomal thiolase dramatically.

The antibodies generated by injection of these proteins into rabbits cross-react with the corresponding human enzyme proteins and can be used for immunoblot analysis (Tager et al 1985; Lazarow et al 1986; Suzuki et al 1988; Wiemer et al 1989), immunofluorescence (Wiemer et al 1989; Van Roermund et al 1991; Santos et al 1992; Suzuki et al 1992) and immunoelectronmicroscopy (Espeel et al 1991, 1993, 1995) in liver samples and cultured skin fibroblasts.

Recent studies have led to the identification of additional peroxisomal acyl-CoA oxidases as well as the one identified and purified by Osumi et al (1980). Indeed, rat liver peroxisomes appear to contain at least three acyl-CoA oxidases including the clofibrate-inducible acyl-CoA oxidase (Osumi et al 1980), a clofibrate non-inducible pristanoyl-CoA oxidase (Van Veldhoven et al 1991, 1992; Wanders et al 1992) and a clofibrate non-inducible trihydroxycholestanoyl-CoA oxidase expressed in liver only (Casteels et al 1990; Schepers et al 1990; Van Veldhoven et al 1992). Interestingly, human liver peroxisomes seem to contain two acyl-CoA oxidases only, the first being the human homologue of the clofibrate inducible acyl-CoA oxidase purified by Hashimoto and coworkers (Osumi et al 1980). The second acyl-CoA oxidase identified in human liver peroxisomes is a branched-chain acyl-CoA oxidase processing both pristanoyl-CoA and trihydroxycholestanoyl-CoA as a substrate. The protein is monomeric with a molecular weight of 70 kDa (Van Hove et al 1993).

Because a mixture of straight-chain and branched-chain oxidases (from rat liver) is used for immunization, our polyclonal antibodies react with both oxidases. The reaction with the 50 kDa and 20 kDa bands is likely to be specific for the straight-chain oxidase, however. These considerations are of importance for the interpretation of the immunoblots (see Figures 1 and 2). Van Hove et al (1993) used antibodies raised against the 52 kDa and 21 kDa subunits of rat straight-chain acyl-CoA oxidase; this explains the absence of cross-reaction with the branched-chain acyl-CoA oxidase reported in their paper.

The immunoblot procedure has been used successfully through the years and has led to the identification of patients with acyl-CoA oxidase deficiency (pseudo-neonatal adrenoleukodystrophy; McKusick 264470) (Poll-The et al 1988; Wanders et al 1990), bifunctional enzyme deficiency (Watkins et al 1989) and peroxisomal thiolase deficiency (pseudo-Zellweger syndrome; McKusick 261510) (Schram et al 1987). In this paper we describe a procedure for immunoblot analysis applicable to liver samples and cultured skin fibroblasts.

EQUIPMENT, CHEMICALS AND SOLUTIONS

Equipment

Access to: a vertical slab gel electrophoresis apparatus with a power supply, semi-dry electrotransfer apparatus (2117 Multiphor II Electrophoresis Unit, LKB), water bath, microcentrifuge for Eppendorf tubes, rocking table, heating plate.

On the bench: Pasteur pipettes, adjustable automatic pipettes with tips, Hamilton syringe (100 μl), Eppendorf tubes, timer, gloves, forceps, transfer membrane, nitrocellulose (Schleicher & Schuell, Dassel, Germany; BA 85 (0.45 μm); cat. no. 401196), filter papers (Whatman).

Chemicals

- Sodium dodecyl sulphate (SDS) (Bio-Rad Laboratories, UK: cat. no. 142.805).
- Glycine (Sigma Chemicals, St Louis, MO, USA; cat. no. G7403).
- Glycerol (Janssen Chemica, Belgium; cat. no. 15.892.81).
- β-Mercaptoethanol (Bio-Rad Laboratories, UK; cat. no. 161-0710).
- TEMED (*N,N,N',N'*-tetramethylethylenediamine-HCl) (Bio-Rad Laboratories, UK; cat. no. 161-0800).
- Morpholinopropanesulphonic acid (MOPS) (Sigma Chemicals, St Louis, MO, USA; cat. no. M5162).
- Phenylmethylsulphonyl fluoride (PMSF) (Boehringer Mannheim, Germany; cat. no. 837.091).
- Leupeptine (Boehringer Mannheim, Germany; cat. no. 1017.128).
- *N,N'*-dimethylformamide (Merck, Darmstadt, Germany; cat. no. 822275).
- Nitroblue tetrazolium (NBT) (Bio-Rad Laboratories, UK; cat. no. 170-6532).
- Pre-stained molecular weight (MW) markers (Gibco BRL, Paisley, UK; cat. no. 6041 LA).
- Goat (anti-rabbit) IgG-alkaline phosphatase (Bio-Rad Laboratories, UK; cat. no. 172-1016).
- Natural goat serum (Bio-trading, Wilnis, The Netherlands).
- Ammonium persulphate (Bio-Rad Laboratories, UK; cat. no. 161-0700).
- Bromophenol blue, sodium salt (Bio-Rad Laboratories, UK; cat. no. 161-0404).
- Tris (2-amino-2-(hydroxymethyl)-1,3-propanediol; Tris(hydroxymethyl)amino-methane (MW = 121.1)) (Sigma Chemicals, St Louis, MO, USA; cat. no. T 6791).
- Sodium chloride (NaCl; MW = 58.44) (Sigma Chemicals, St Louis, MO, USA; cat. no. S 7653).
- Magnesium chloride ($MgCl_2 \cdot 6H_2O$; MW = 203.3) (Sigma Chemicals, St Louis, MO, USA; cat. no. M 0250).

- Tween 20 (polyoxyethylene (20) sorbitan monolaurate) (Merck, Darmstadt, Germany; cat. no. 822184).
- 5-Bromo-4-chloro-3-indolyl phosphate (BCIP) (Bio-Rad Laboratories, UK; cat. no. 170-6532).
- Acrylamide/*N,N'*-methylene-bisacrylamide (Boehringer Mannheim, Germany). Available separately or premixed at several ratios. We use a premixture of 27.5:1 (w/w) (cat. no. 100153).
- Agarose (Sigma Chemicals, St Louis, MO, USA; cat. no. A6013).

Solutions required

SEPARATION OF PROTEINS ON SDS-PAGE (10%)

- 1% (w/v) agarose solution: Dissolve 1 g agarose in distilled H_2O and heat until complete dissolution.
- 10% (w/v) SDS: Dissolve 10 g SDS in H_2O (final volume 100 ml).
- Ammonium persulphate (APS) solution (0.75% w/v: Dissolve 7.5 mg APS in 1 ml H_2O).
- Bromophenol blue solution (0.05% (w/v)): Dissolve 50 mg in 100 ml ethanol.
- Buffer A (separating gel buffer): Weigh out 34.1 g Tris, add 7.5 ml 10% (w/v) SDS and make up to 250 ml with pH adjustment to 8.8 using concentrated hydrochloric acid (HCl).
- Buffer B (stacking gel buffer): Weigh out 7.6 g Tris, add 5.0 ml 10% (w/v) SDS and make up to 50 ml with pH adjustment to 6.8 using HCl.
- Acrylamide/bisacrylamide solution (30%/0.8% w/v): Dissolve 30 g acrylamide and 0.8 g bisacrylamide in H_2O (final volume 100 ml) or buy a 37.5:1 premixture.
- Loading buffer (2×) containing 4% (w/v) SDS, 20% (v/v) glycerol, 0.57 mol/L β-mercaptoethanol, 0.005% (w/v) bromophenol blue in 10% (v/v) buffer B: For 5 ml medium, mix 0.5 ml buffer B, 1.0 ml glycerol, 0.5 ml bromophenol blue solution, 2 ml 10% (w/v) SDS, 0.2 ml mercaptoethanol and 0.8 ml H_2O.
- Running buffer: Dissolve 14.4 g glycine, 3.03 g Tris and 1 g SDS in 1 litre H_2O.
- 0.5 mol/L MOPS buffer (pH 7.4): Dissolve 104.65 g morpholinopropanesulphonic acid (MOPS) in 1 litre H_2O with pH adjustment to 7.4 using sodium hydroxide.
- Lysis buffer: For 10 ml buffer combine 0.4 ml MOPS buffer (pH 7.4), 0.25 ml 10% (w/v) Triton X-100 in H_2O, 0.25 ml 1 mg/mL PMSF in H_2O and 0.25 ml 1 mg/ml leupeptine and make up to 10 ml.

ELECTROTRANSFER OF PROTEINS ON NITROCELLULOSE MEMBRANES USING SEMI-DRY BLOTTING

- Transfer buffer containing 39 mmol/L glycine, 48 mmol/L Tris, 0.0375% (w/v) SDS and 20% methanol: weigh out 14.4 g glycine and 3.03 g Tris, add 3.75 ml 10% (w/v) SDS and 200 ml methanol and make up to 1 litre.

BLOTTING OF MEMBRANE WITH SPECIFIC ANTIBODIES

- Phosphate-buffered saline (PBS) containing 140 mmol/L NaCl, 9.2 mmol/L Na_2HPO_4 and 1.2 mmol/L $NaH_2PO_4 \cdot 2H_2O$ (final pH 7.4): Dissolve 8.18 g/L NaCl, 1.306 g/L Na_2HPO_4 and 0.203 g/L $NaH_2PO_4 \cdot 2H_2O$ in 1 litre H_2O.

- Tween/PBS buffer (0.1% (v/v) Tween 20 in PBS): Dissolve $10\,\mu$l Tween 20 in 10 ml PBS.
- Blocking buffer containing 4% (v/v) normal goat serum (NGS) in Tween/PBS buffer: Dilute normal goat serum 25-fold in Tween/PBS buffer.
- Nitroblue tetrazolium solution (NBT solution): Weigh out 100 mg NBT and add $600\,\mu$l H_2O and $1400\,\mu$l N,N'-dimethylformamide.
- Bromochloroindolyl phosphate (BCIP) solution: Weigh out 100 mg BCIP and add 2 ml N,N'-dimethylformamide.
- Alkaline phosphatase (AP) buffer containing 100 mmol/L Tris, 100 mmol/L NaCl and 5 mmol/L $MgCl_2$ (pH 9.5): Weigh out 12.11 g Tris, 5.844 g NaCl and 1.016 g $MgCl_2 \cdot 6H_2O$ and fill up to 1 litre with pH adjustment to 9.5 using concentrated HCl.

EXPERIMENTAL PROCEDURE

Preparation of the gels

Note: The following volumes are given for 1 gel of 15×16 cm and 1.5 mm thickness.

For preparation of the separating gel (10%) add together 8.5 ml of buffer A, 8.3 ml of acrylamide/bisacrylamide solution (30%/0.8% w/v), $15\,\mu$l TEMED, 1.5 ml 0.75% APS and make up to 25 ml with distilled H_2O.

Mix the components well, pour the solution between the prepared plates with a Pasteur pipette and gently layer 5 mm of water-saturated butanol on top to form a flat surface while the acrylamide polymerizes. Wait for polymerization for about 60 min. After polymerization is complete, discard the butanol, rinse with H_2O several times and dry with a filter paper.

When this is done, the stacking gel (5%) is prepared by adding together the following components: 1.0 ml buffer B, 1.7 ml acrylamide/bisacrylamide (30%/0.8% w/v) solution, $6\,\mu$l TEMED, 0.6 ml 0.75% (w/v) APS and H_2O to a total volume of 10 ml.

Mix well, pour on top of the separating gel and immediately insert the comb (20 wells, maximum volume to load is $100\,\mu$l/well). Wait for polymerization for about 1 h. Remove the comb and wash the wells with H_2O (add and remove water to and from each well a few times). Fit the gel in an electrophoresis unit and add the running buffer.

The system is now ready to operate.

Preparation of the liver and fibroblasts homogenates

Human liver specimens are kept at $-80°C$ in small pieces and a small aliquot (0.5 mg protein (this corresponds to 3.5 mg wet weight); 6 mm of a needle biopsy core are sufficient to provide this quantity) is taken and added to $200-500\,\mu$l (depending on the quantity of liver material) of cold lysis buffer containing MOPS, Triton X-100, PMSF and leupeptine as described above. The mixture is gently homogenized by hand using a small glass homogenizer followed by ultrasonic disruption (3 periods of 15 s at $70-80$ W). This is followed by protein measurement. The protein concentration is subsequently adjusted to 2 mg per ml by appropriate dilution.

Human skin fibroblasts are kept at $-80°C$ as dry pellets in aliquots of $1-2$ mg per tube (see Wanders et al (1995) elsewhere in this handbook for details of the methods used to culture cells). Add $250\,\mu$l of lysis buffer, homogenize gently by hand, sonicate as described above and determine the protein concentration.

Preparation of the samples for electrophoresis: Prepare the different samples in Eppendorf tubes as follows. Dilute the liver and fibroblast homogenate samples prepared as described above in $2\times$ loading buffer (maximum dilution 1:1) to a protein concentration of $833\,\mu$g/l. Take care that one part is made up by loading buffer and the other part by the homogenate plus lysis buffer. Mix well, centrifuge the sample for a few seconds (to collect the full sample at the bottom of the tubes), and heat to 95°C for 3 min.

Electrophoresis: Run the gel at 30 mA for about 1 h to concentrate the proteins at the top of the separating gel. Run the gel at 45 mA for about 2.5 h to cause the proteins to migrate into the separating gel.

Electrotransfer of proteins using a semi-dry apparatus: Cut filter paper (Whatman) according to the size of the gel. Equilibrate the gel, the filter paper and the nitrocellulose membrane in transfer buffer for 1 min.

Assemble the sandwich as follows:

TOP (–)
3 sheets of Whatman filter paper
GEL (proteins are negatively charged)
NC membrane
3 sheets of Whatman filter paper
BOTTOM (+)

Keep the sandwich wet and avoid air bubbles. Connect the semi-dry apparatus to a power supply and run at $0.8\,$mA/cm^2 of the trans unit for about 1 h.

Blotting of the membrane with a specific antibody: All incubations are done at room temperature with gentle agitation in a plastic dish, which should be as small as possible to minimize the amount of antibody used.

Blocking step: Incubate the membrane with 100 ml blocking buffer for 30 min. This step reduces the amount of non-specific binding and gives cleaner blots.

Primary antibody binding step: Dilute the specific antibody in 50 ml blocking buffer and incubate for 2 h. The appropriate dilution should be determined for each antibody separately.

Washing steps: Wash the membrane three times (for 10 min each) with Tween/PBS buffer.

Second antibody–alkaline phosphatase conjugate binding step: Dilute the anti-rabbit IgG–alkaline phosphatase conjugate with blocking buffer to 1:3000 dilution. Incubate the membrane with the diluted conjugate for 1 h.

Washing steps: Wash the membrane three times (for 10 min each) with Tween/PBS.

IMMUNOBLOT-ANALYSIS IN LIVER HOMOGENATES

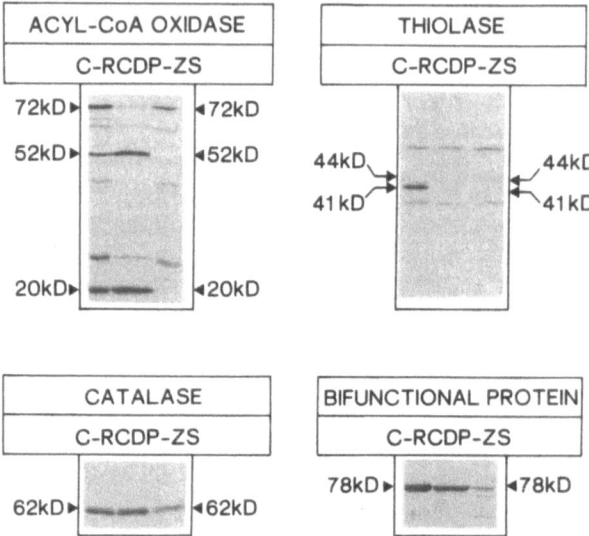

Figure 1   Immunoblot analysis of acyl-CoA oxidase, bi(tri)functional protein, peroxisomal thiolase and catalase in liver homogenates from a control subject (C), a Zellweger patient (ZS) and a patient with classical rhizomelic chondrodysplasia punctata (RCDP). To each lane 50 μg of protein was applied except for thiolase, in which case only 5 μg was applied

Alkaline phosphatase reaction step: Prepare the colour substrate solution just before use: Mix 264 μl NBT solution and 132 μl BCIP solution in 40 ml AP buffer containing Tris, $MgCl_2$ and NaCl. Wash the membrane once with PBS, incubate with colour substrate solution and time the incubation to observe the development (30 s to 5 min). Stop the reaction by washing the membrane with H_2O several times. The membrane can be stored dry.

RESULTS

Figure 1 shows the results of immunoblotting experiments performed with human liver and fibroblast homogenates. The antibodies used were raised against acyl-CoA oxidase (Osumi et al 1980), bifunctional protein (Furuta et al 1980), and peroxisomal thiolase (Miyazawa et al 1980) as purified from rat liver and catalase as purified from human erythrocytes (Tager et al 1985). Studies in rat liver have shown that acyl-CoA oxidase is synthesized as a 72 kDa (A) precursor which undergoes proteolytic processing after import into the peroxisome to a 52 kDa (B) and 20 kDa (C) protein. Figure 1 shows that the 72 kDa form of acyl-CoA oxidase is normally present in both ZS and RCDP liver, which is in accordance with the literature (Tager et al 1985; Hoefler et al 1988). Pulse-chase studies by Schram and co-workers (1986) have shown that acyl-CoA oxidase is, indeed, normally synthesized in Zellweger fibroblasts as a 72 kDa protein, although there is no processing to the 52 and 20 kDa forms.

Inspection of the blots reveals the presence of a variety of cross-reacting bands,

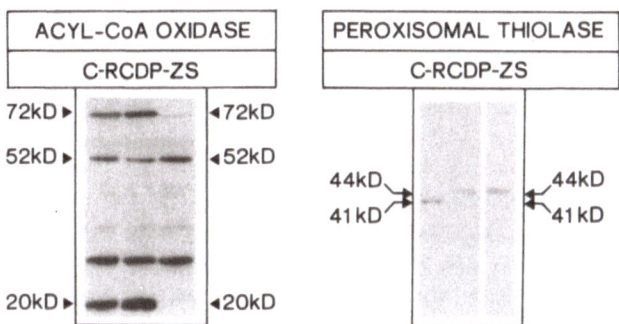

Figure 2 Immunoblot analysis of acyl-CoA oxidase and peroxisomal thiolase in fibroblast homogenates from a control subject (C), a Zellweger patient (ZS) and a patient with rhizomelic chondrodysplasia punctata (RCDP). To each lane $50\,\mu g$ of protein was applied

suggesting that the antibodies used are not monospecific. It should be noted that the reactivity of the antiserum with the 72kDa band may in part be due to reactivity with the 72kDa branched-chain acyl-CoA oxidase. The reactivity with the 52kDa and 20kDa bands, however, is likely to be specifically due to the 52kDa and 20kDa components of the acyl-CoA oxidase I, the human homologue of the acyl-CoA oxidase purified by Hashimoto and co-workers (Osumi et al 1980).

The second and third steps in peroxisomal β-oxidation are catalysed by a single protein called bifunctional protein, although recent studies have shown that the enzyme protein, a monomer of 78kD, also carries a third catalytic activity (Palosaari and Hiltunen 1990). Figure 1 shows the normal presence of bifunctional protein in RCDP liver and a strong deficiency of this protein in Zellweger liver.

The last step in peroxisomal β-oxidation is catalysed by a specific 3-oxoacyl-CoA thiolase which is a dimer of 89kD. Interestingly, peroxisomal thiolase is synthesized as a 44kDa precursor which undergoes proteolytic processing to a 41kDa form after import into the peroxisome. Figure 1 shows the absence of mature 41kDa precursor thiolase in both Zellweger and RCDP liver. Instead, low levels of 44kDa thiolase are present both in ZS and RCDP liver, which contrasts sharply with the findings in control liver. Finally, Figure 1 shows the normal presence of catalase in ZS and RCDP liver.

Figure 2 shows the results of similar experiments using fibroblast homogenates. Immunoblot analysis can be done reliably in the case of acyl-CoA oxidase, peroxisomal thiolase and catalase, whereas immunoblot analysis of bifunctional protein is impossible in fibroblast homogenates, probably owing to the presence of a major cross-reactive protein in fibroblasts. The results of Figure 2 show essentially similar results in fibroblast homogenates to those in liver (Figure 1) with one exception which involves the presence of a 52kDa band in fibroblast homogenates from ZS patients but its absence in liver homogenates from ZS patients using anti-(acyl-CoA) oxidase. The 52kDa band observed in fibroblast homogenates is probably unrelated to acyl-CoA oxidase. Accordingly, only the 20kDa band should be inspected.

In summary, we have described a simple, straightforward immunoblotting procedure applicable to liver (and other tissues) and fibroblasts. The method can also be applied to chorionic villus biopsy specimens; one frond of a villus provides sufficient material for the test. The results obtained in chorionic villus biopsies are very promising, especially since the level of expression of acyl-CoA oxidase and peroxisomal thiolase is much higher than in cultured cells.

Accordingly, immunoblot analysis of acyl-CoA oxidase and peroxisomal thiolase in chorionic villi has now become the method of choice for the prenatal diagnosis of Zellweger syndrome, other disorders of peroxisome biogenesis and rhizomelic chondro-dysplasia punctata, since only little protein is required for the analysis of acyl-CoA oxidase and peroxisomal 3-ketothiolase ($50\,\mu g$ protein per lane: compare Figures 1 and 2).

A problem with the method described here is that the antibodies against acyl-CoA oxidase, bi(tri)functional protein and peroxisomal thiolase are not commercially available and have to be generated by each laboratory or obtained as a gift. The antibodies used in this study were prepared by ourselves after purification of the enzyme proteins by the methods described by Hashimoto and co-workers (Osumi et al 1980; Furuta et al 1980; Miyazawa et al 1980).

We are currently investigating other possibilities for obtaining good antibodies. One way would be to try to prepare specific antibodies using small synthetic peptides selected from the known amino acid sequences of the different peroxisomal β-oxidation enzyme proteins. This would obviate the need to purify the different enzyme proteins from tissues.

ACKNOWLEDGEMENTS

The authors gratefully thank Mrs Iet van der Gracht for expert preparation of the manu-script and the Prinses Beatrix Fonds, Den Haag, The Netherlands for financial support.

REFERENCES

Casteels M, Schepers L, Van Veldhoven PP, Eyssen HJ, Mannaerts GP (1990) Separate peroxisomal oxidases for fatty acyl-CoAs and trihydroxycoprostanoyl-CoA in human liver. *J Lipid Res* **31**: 1865–1872.

Espeel M, Van Limbergen G (1995) Immunocytochemical localization of peroxisomal proteins in human liver and kidney. *J Inher Metab Dis* **18** (**Suppl. 1**): 135 –154.

Espeel M, Hashimoto T, De Craemer D, Roels F (1990) Immunocytochemical detection of peroxisomal β-oxidation enzymes in cryostat and paraffin sections of human post mortem liver. *Histochem J* **22**: 57–62.

Espeel M, Roels F, Van Malderghem L, et al (1991) Peroxisomal localization of the immunoreactive β-oxidation enzymes in a neonate with a β-oxidation defect. *Virchows Arch A Pathol Anat* **419**: 301–308.

Espeel M, Heikoop JC, Smeitink JAM, et al (1993) Cytoplasmic catalase and ghost-like peroxi-somes in the liver from a child with atypical chondrodysplasia punctata. *Ultrastruct Pathol* **17**: 625–637.

Espeel M, Mandel H, Poggi F, et al (1995) Peroxisome mosaicism in the livers of peroxisomal deficiency patients. *Hepatology*, **22**.

Furuta S, Miyazawa S, Osumi T, Hashimoto T, Ui N (1980) Properties of mitochondrial and peroxisomal enoyl-CoA hydratases from rat liver. *J Biochem* **88**: 1059–1070.

Hoefler G Hoefler S, Watkins P, et al (1988) Biochemical abnormalities in rhizomelic chondrodys-plasia punctata. *J Pediatr* **112**: 726–733.

Horvath VAP, Wanders RJA (1994) Re-evaluation of conditions required for measurement of true alanine/glyoxylate aminotransferase activity in human liver: implications for the diagnosis of hyperoxaluria type I. *Ann Clin Biochem* **31**: 361–366.

Hughes JL, Poulos A, Robertson E, et al (1990) Pathology of hepatic peroxisomes and mitochondria in patients with peroxisomal disorders. *Virchows Arch A Pathol Anat* **416**: 255–264.

Hughes JL, Poulos A, Crane De, Chow CW, Sheffield LJ, Silence D (1992) Ultrastructure and immunocytochemistry of hepatic peroxisomes in rhizomelic chondrodysplasia punctata. *Eur J Pediatr* **151**: 829–836.

Hughes JL, Bourne AJ, Poulos A (1993) Morphometry of peroxisomes and immunolocalisation of peroxisomal proteins in the liver of patients with generalised peroxisomal disorders. *Virchows Arch A Pathol Anat* **423**: 453–457.

Lazarow PB, Fujiki Y, Small GM, Watkins P, Moser HW (1986) Presence of the peroxisomal 22-kDa integral membrane protein in the liver of a person lacking recognizable peroxisomes (Zellweger syndrome). *Proc Natl Acad Sci USA* **83**: 9193–9196.

Mandel H, Espeel M, Roels F, et al (1994) A new type of peroxisomal disorder with variable expression in liver and fibroblasts. *J Pediatr* **125**: 549–555.

Miyazawa S, Osumi T, Hashimoto T (1980) The presence of a new 3-oxoacylCoA thiolase in rat liver peroxisomes. *Eur J Biochem* **103**: 589–596.

Novikow DK, Vanhove GK, Carchon H, et al (1994) Peroxisomal β–oxidation: purification of four novel 3-hydroxyacyl-CoA dehydrogenases from rat liver peroxisomes. *J Biol Chem* **269**: 27125–27135.

Osumi T, Hashimoto T, Ui N (1980) Purification and properties of acyl-CoA oxidase from rat liver. *J Biochem* **87**: 1735–1746.

Palosaari PM, Hiltunen JK (1990) Peroxisomal bifunctional protein from rat liver is a trifunctional enzyme possessing 2-enoyl-CoA hydratase, 3-hydroxyacyl-CoA dehydrogenase and $\Delta 3,\Delta 2$-enoyl-CoA isomerase activities. *J Biol Chem* **265**: 2446–2449.

Poll-The BT, Roels F, Ogier H, et al (1988) A new peroxisomal disorder with enlarged peroxisomes and a specific deficiency of acyl-CoA oxidase (pseudo-neonatal adrenoleukodystrophy). *Am J Hum Genet* **42**: 422–434.

Roels F, Cornelis A (1989) Heterogeneity of catalase staining in human hepatocellular peroxisomes. *J Histochem Cytochem* **37**: 331–337.

Roels F, Goldfischer S (1979) Cytochemistry of human catalase. The demonstration of hepatic and renal peroxisomes by a high temperature procedure. *J Histochem Cytochem* **27**: 1471–1477.

Roels F, Pauwels M, Cornelis A, et al (1983) Peroxisomes (microbodies) in human liver. Cytochemical and quantitative studies in 85 biopsies. *J Histochem Cytochem* **31**: 235–237.

Roels F, Espeel M, De Craemer D (1991) Liver pathology and immunocytochemistry in peroxisomal disorders: A review. *J Inher Metab Dis* **14**: 853–875.

Roels F, Espeel M, Poggi F, Mandel H, Van Maldergem L, Saudubray JM (1993) Human liver pathology in peroxisomal diseases: a review including novel data. *Biochimie* **75**: 281–292.

Santos MJ, Hoefler S, Moser AB, Moser HW, Lazarow PB (1992) Peroxisome assembly mutations in humans: structural heterogeneity in Zellweger syndrome. *J Cell Physiol* **151**: 103–112.

Schepers L, Van Veldhoven PP, Casteels M, Eyssen HJ, Mannaerts GP (1990) Presence of three acyl-CoA oxidases in rat liver peroxisomes. *J Biol Chem* **265**: 5242–5246.

Schram AW, Strijland A, Hashimoto T, et al (1986) Biosynthesis and maturation of peroxisomal β-oxidation enzymes in fibroblasts in relation to the Zellweger syndrome and infantile Refsum disease. *Proc Natl Acad Sci USA* **83**: 6156–6158.

Schram AW, Goldfischer S, Van Roermund CWT, et al (1987) Human peroxisomal 3-oxoacyl-CoA thiolase deficiency. *Proc Natl Acad Sci USA* **84**: 2494–2496.

Schutgens RBH, Wanders RJA, Jakobs C, et al (1994) A new variant of Zellweger syndrome with normal peroxisomal functions in cultured fibroblasts. *J Inher Metab Dis* **17**: 319–322.

Suzuki Y, Shimozowa N, Orii T, et al (1988) Zellweger-like syndrome with detectable hepatic peroxisomes: a variant peroxisomal disorder. *J Pediatr* **113**: 841–845.

Suzuki Y, Shimozowa N, Yajima S, et al (1992) Different intracellular localization of peroxisomal proteins in fibroblasts from patients with aberrant peroxisome assembly. *Cell Struct Funct* **17**: 1–8.

Tager JM, Ten Harmsen van de Beek WA, Wanders RJA, et al (1985) Peroxisomal β-oxidation enzyme proteins in the Zellweger syndrome. *Biochem Biophys Res Commun* **126**: 1269–1275.

Van Hove GF, Van Veldhoven PP, Fransen M, Denis S, Wanders RJA, Mannaerts GP (1993) The CoA-esters of 2-methyl-branched chain fatty acids and of the bile acid intermediates di- and trihydroxycoprostanoic acid are oxidized by one single peroxisomal branched chain acyl-CoA oxidase in human liver and kidney. *J Biol Chem* **268**: 10335–10344.

Van Veldhoven PP, Van Hove G, Van Houtte F, et al (1991) Identification and purification of a peroxisomal branched chain fatty acyl-CoA oxidase. *J Biol Chem* **266**: 24676–24683.

Van Veldhoven PP, Van Hove G, Asselberghs S, Eyssen HJ, Mannaerts GP (1992) Substrate specificities of rat liver peroxisomal acyl-CoA oxidases. *J Biol Chem* **267**: 20065–20074.

Van Roermund CWT, Brul S, Tager JM, Schutgens RBH, Wanders RJA (1991) Acyl-CoA oxidase, peroxisomal thiolase and dihydroxyacetone phosphate acyltransferase: aberrant subcellular localization in Zellweger syndrome. *J Inher Metab Dis* **14**: 152–164.

Wanders RJA, Schelen A, Feller N, et al (1990) First prenatal diagnosis of acyl-CoA oxidase deficiency. *J Inher Metab Dis* **13**: 371–374.

Wanders RJA, Denis S, Jakobs C, Ten Brink HJ (1992) Identification of pristanoyl-CoA oxidase as a distinct clofibrate non-inducible enzyme in rat liver peroxisomes. *Biochim Biophys Acta* **1124**: 199–202.

Wanders RJA, Schutgens RBH, Barth PG, Tager JM, Van den Bosch H (1993) Postnatal diagnosis of peroxisomal disorders: a biochemical approach. *Biochimie* **75**: 269–279.

Wanders RJA, Denis S, Ruiter JPN, et al (1995) Measurement of peroxisomal fatty acid β-oxidation in cultured human skin fibroblasts. *J Inher Metab Dis* **18** (Suppl. 1): 113–124.

Watkins PA, Chen WW, Harris CJ, et al (1989) Peroxisomal bifunctional enzyme deficiency. *J Clin Invest* **83**: 771–777.

Wiemer EC, Brul S, Just WW, et al (1989) Presence of peroxisomal membrane proteins in liver and fibroblasts from patients with the Zellweger syndrome and related disorders: evidence for the existence of peroxisomal ghosts. *Eur J Cell Biol* **50**: 407–417.

J. Inher. Metab. Dis. 18 Suppl. 1 (1995) 113–124
© SSIEM and Kluwer Academic Publishers.

Measurement of peroxisomal fatty acid β-oxidation in cultured human skin fibroblasts

R. J. A. Wanders*, S. Denis, J. P. N. Ruiter, R. B. H. Schutgens,
C. W. T. van Roermund and B. S. Jacobs
*University Hospital Amsterdam, Departments of Pediatrics and Clinical Chemistry,
Laboratory of Pediatric Clinical Biochemistry, Section for Clinical Enzymology,
Meibergdreef 9, PO Box 22700, 1100 DE Amsterdam, The Netherlands*

*Correspondence

Summary: One of the main functions of mammalian peroxisomes is the β-oxidation of a variety of fatty acids and fatty acid derivatives, including very long-chain fatty acids. Oxidation of these fatty acids is deficient in a number of different peroxisomal disorders, including the disorders of peroxisome biogenesis (Zellweger syndrome, neonatal adrenoleukodystrophy and infantile Refsum disease), X-linked adrenoleukodystrophy and a number of other disorders of peroxisomal β-oxidation of known and unknown aetiology. Accurate measurement of peroxisomal fatty acid oxidation is of utmost importance for correct postnatal and prenatal diagnosis of these disorders. In this paper we describe a straightforward and accurate assay method to measure the β-oxidation of palmitic acid ($C_{16:0}$), hexacosanoic acid ($C_{26:0}$) and pristanic acid in intact fibroblasts.

One of the main functions of peroxisomes in higher eukaryotes is the β-oxidation of a range of fatty acids and fatty acid derivatives. Lazarow and de Duve (1976) were the first to describe the presence of a fatty acid β-oxidation system in peroxisomes, although the functional significance of a second β-oxidation system in peroxisomes next to that in mitochondria was initially not very clear. Now, however, some 20 years after Lazarow and de Duve's first report, we know that peroxisomes catalyse the β-oxidation of a distinct group of fatty acid derivatives. These include very long-chain fatty acids (Singh et al 1984), di- and trihydroxycholestanoic acids (Kase et al 1983), pristanic acid (Singh et al 1990), prostaglandins (Diczfalusy and Alexson 1988, 1991; Schepers et al 1988), certain leukotrienes (Jedlitschky et al 1991), 12- and 15-hydroxyeicosatetraenoic acids (Gordon et al 1990), long-chain dicarboxylic acids (Suzuki et al 1994) and certain thromboxanes (Diczfalusy et al 1993; De Waart et al 1994).

Like mitochondria, peroxisomes contain the full enzymic machinery to catalyse β-oxidation. Fatty acids need to be activated to their coenzyme A esters before β-oxidation can occur. For some fatty acids, activation occurs at the peroxisomal membrane. This is true, for instance, for very long-chain fatty acids which can be activated by a specific acyl-CoA synthetase present in peroxisomes and the endoplasmic reticulum but not in mitochondria (Wanders et al 1987b; Singh et al 1987). There is evidence that very long-chain fatty acyl-CoA esters activated at the site of the endoplasmic reticulum are not available for β-oxidation in the peroxisome despite the fact that the catalytic site of microsomal very long-chain acyl-CoA synthetase faces the cytosol (Singh et al 1985). Apparently, there is a tight coupling between activation of very long-chain fatty acids at the site of the peroxisomal membrane and β-oxidation in the peroxisome (see Wanders et al (1992d) for discussion). There is controversy about the location of the synthetase at the inner or outer site of the peroxisomal membrane. According to Lageweg et al (1991) the catalytic site of the synthetase faces the cytosol, whereas according to Lazo et al (1990) the enzyme faces the interior of the peroxisome. This important point needs to be resolved in the near future.

The obligatory coupling between activation at the peroxisomal membrane and β-oxidation within the peroxisome is not observed with other fatty acids. Indeed, dicarboxylic acids (Vamecq et al 1985), di- and trihydroxycholestanoic acids (Prydz et al 1988; Schepers et al 1989) and prostaglandins (Schepers et al 1988) are activated by different synthetases located at the endoplasmic reticulum membrane. It is unclear how the acyl-CoA esters synthesized at the endoplasmic reticulum membrane traverse through the cytoplasm to the peroxisome and in what way the acyl-CoA esters are transported across the peroxisomal membrane. There is some evidence that ATP is involved (Wolvetang et al 1990).

Activation of pristanic acid can occur at three subcellular sites in mitochondria, peroxisomes and endoplasmic reticulum (Wanders et al 1992b), analogously to the activation of long-chain fatty acids (Miyazawa et al 1985). It is not known whether β-oxidation of pristanic acid in peroxisomes requires activation at the peroxisomal membrane or whether activation may also occur at the mitochondrial or endoplasmic reticulum membrane.

After entry into the peroxisomal interior, the different acyl-CoA esters are subjected to the actual β-oxidative process involving sequential steps of dehydrogenation, hydration, dehydrogenation again, and thiolytic cleavage. Dehydrogenation of acyl-CoA esters to their corresponding *trans*-2-enoyl-CoA esters is brought about by various acyl-CoA oxidases accepting molecular oxygen as hydrogen acceptor. In rat liver three distinct acyl-CoA oxidases have been identified (Schepers et al 1990). The first acyl-CoA oxidase was identified and characterized by Hashimoto and co-workers (Osumi et al 1980) and accepts a range of saturated acyl-CoA esters but not pristanoyl-CoA (2,6,10,14-tetramethyl-pentadecanoyl-CoA), a branched-chain fatty acyl-CoA ester. The activity of this oxidase is strongly stimulated in livers from rats fed a variety of hypolipidaemic drugs (e.g. clofibrate). The second oxidase, called pristanoyl-CoA oxidase, does accept branched-chain acyl-CoAs such as pristanoyl-CoA but also straight-chain acyl-CoAs (Van Veldhoven et al 1991; Wanders et al 1992a; Wanders et al 1993a). However, the branched-chain acyl-CoA ester trihydroxycholestanoyl-CoA, which also contains a methyl group at the 2-position, is not accepted by rat liver pristanoyl-CoA oxidase. Instead, a third oxidase expressed only in liver reacts with trihydroxycholestanoyl-CoA (Van Veldhoven et al 1991).

Remarkably, only two oxidases appear to be present in human liver (Vanhove et al 1993). The first closely resembles the acyl-CoA oxidase isolated from rat liver by Hashimoto and co-workers (Osumi et al 1980). The other oxidase accepts both pristanoyl-

CoA and trihydroxycholestanoyl-CoA as substrates (Vanhove et al 1993). Figure 1 shows a schematic illustration of the enzymes involved in the β-oxidation of very long-chain fatty acids, pristanic acid and trihydroxycholestanoic acid in rat and human liver peroxisomes.

Until recently it was assumed that peroxisomes contain a single bifunctional enzyme catalysing the subsequent hydration and 3-hydroxyacyl-CoA dehydrogenation of all the different enoyl-CoA esters produced in peroxisomes. Mannaerts and co-workers, however, recently identified several additional enoyl-CoA hydratases/3-hydroxyacyl-CoA dehydrogenases in peroxisomes (Novikov et al 1994). It is to be expected that there are also additional thiolases. We are currently exploring the latter possibility.

Peroxisomal fatty acid β-oxidation can be studied conveniently in human skin fibroblasts using radiolabelled substrates preferably containing the radioactive label in the 1-position. Tetracosanoic ($C_{24:0}$) and hexacosanoic ($C_{26:0}$) acid can best be used to study peroxisomal β-oxidation. Oxidation of [1-^{14}C]hexadecanoic acid (palmitic acid, $C_{16:0}$), which takes place predominantly in mitochondria, is studied also in order to ascertain the functional integrity of the fibroblasts analysed.

Oxidation of very long-chain fatty acids is deficient in a number of different peroxisomal disorders. This includes first Zellweger syndrome and other disorders of peroxisome biogenesis (neonatal adrenoleukodystrophy and infantile Refsum disease) and second those peroxisomal disorders in which peroxisomal β-oxidation is impaired owing to an isolated deficiency of one of the peroxisomal β-oxidation enzymes. The latter group of disorders includes:

(1)   X-linked adrenoleukodystrophy (peroxisomal very long-chain fatty acid activating enzyme deficiency) (Singh et al 1984; Lazo et al 1988; Wanders et al 1988).

(2)   Pseudo-Zellweger syndrome (peroxisomal thiolase deficiency) (Goldfischer et al 1986; Schram et al 1987).

(3)   Bifunctional enzyme deficiency (Watkins et al 1989; Wanders et al 1990b, 1992c; Suzuki et al 1994).

(4)   Pseudo-neonatal adrenoleukodystrophy (acyl-CoA oxidase deficiency) (Poll-Thé et al 1988; Wanders et al 1990a; Suzuki et al 1994). In addition, a number of patients have been described with a defect in peroxisomal β-oxidation of unknown aetiology (Clayton et al 1988; Naidu et al 1988; Barth et al 1990; Espeel et al 1991; Mandel et al 1992; Van Maldergem et al 1992; Santer et al 1993; Vanhole et al 1994).

PRINCIPLE OF THE METHOD USED TO STUDY FATTY ACID β-OXIDATION IN FIBROBLASTS

Peroxisomal fatty acid β-oxidation using 1-[^{14}C]-labelled $C_{26:0}$ as a substrate can best be studied in intact human skin fibroblasts. When the cells have reached confluency, they are harvested using trypsin–EDTA to dissociate the cells. Subsequently the cells are seeded in glass vials at a concentration of $100\,\mu g$ protein per vial and allowed to adhere and grow overnight in 5 ml culture medium. The following day the medium is removed carefully, followed by washing of the fibroblast monolayer and addition of F10 (Ham) medium containing radiolabelled fatty acid. Reactions are allowed to proceed for 2 h and stopped by adding perchloric acid. [^{14}C]CO_2 and [^{14}C]-labelled acid-soluble material (which

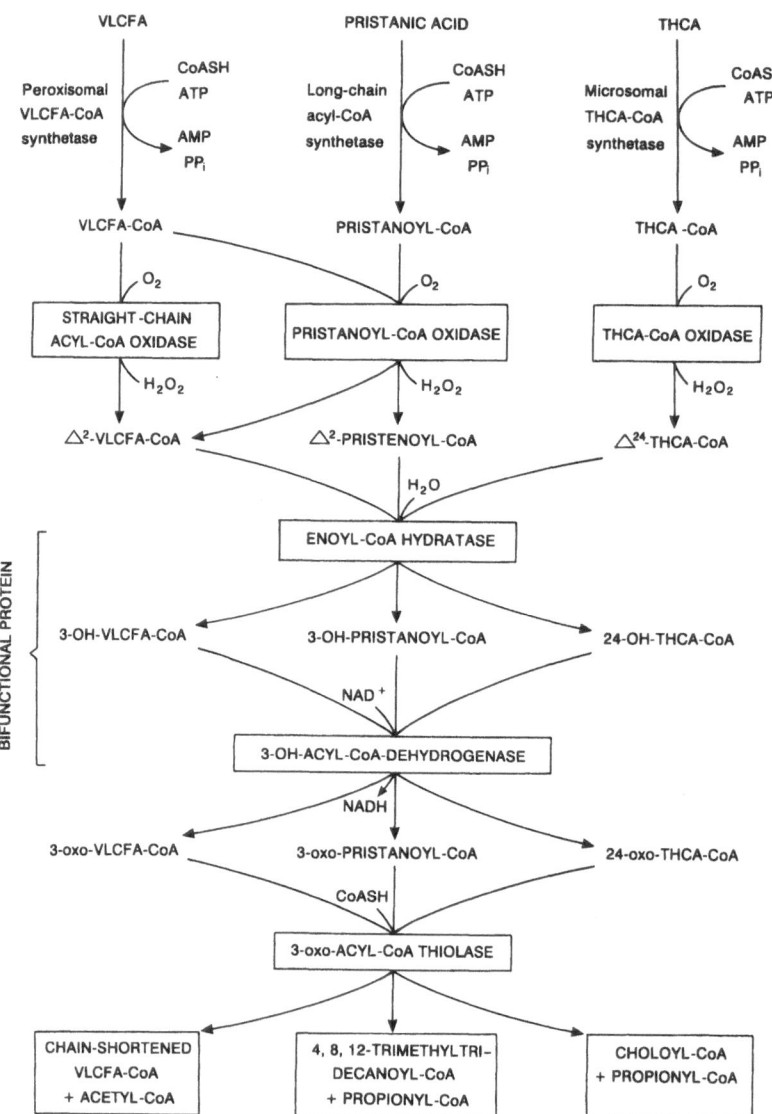

β-oxidation of very-long-chain fatty acids (VLCFA), pristanic acid and trihydroxycholestanoic acid (THCA) in rat peroxisomes

Figure 1 Schematic illustration of the enzymes involved in the activation and β-oxidation of very long-chain fatty acids, pristanic acid and trihydroxycholestanoic acid in rat (left) and human (right) liver

β-oxidation of very-long-chain fatty acids (VLCFA), pristanic acid and trihydroxycholestanoic acid (THCA) in human peroxisomes

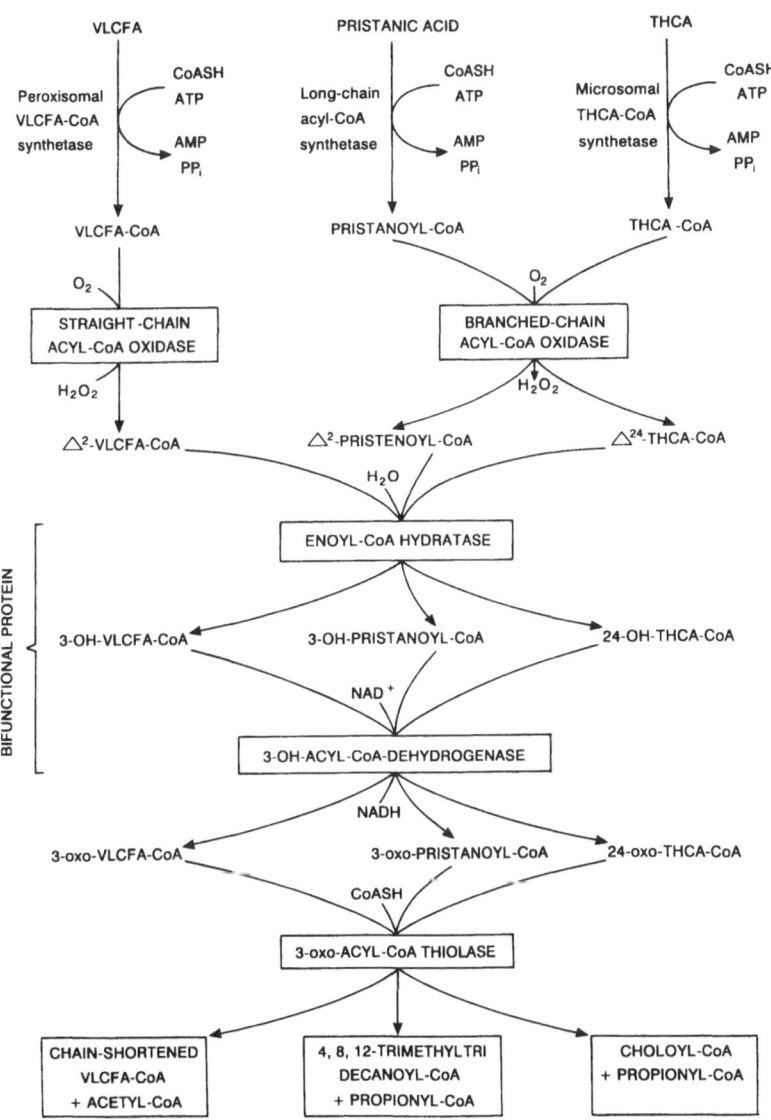

Figure 1 *Continued*

represents radiolabelled acetate, Krebs cycle intermediates, etc.) are then collected and results are calculated.

In principle, fatty acid oxidation can be studied not only in intact cells but also in permeabilized cells using digitonin, for example (Singh et al 1990), and/or in homogenates (Singh et al 1984; Wanders et al 1987a). Nevertheless, we prefer to study fatty acid oxidation in intact cells since this mimics the *in vivo* situation most closely.

EQUIPMENT, CHEMICALS AND SOLUTIONS

Equipment

- *Access to:* bench centrifuge, spectrophotometer, water bath, Eppendorf tubes (with and without caps), incubator, liquid scintillation counter.
- *On the bench:* adjustable automatic pipettes with tips, a Hamilton syringe ($100\,\mu$l), Eppendorf tubes, timer, gloves, sterile vials (2.6 cm diameter) with perforated caps and rubber seals, forceps, plastic cuvettes, scintillation vials.

Chemicals

- Sodium chloride (NaCl; Merck, Darmstadt, Germany; cat. no. 6400.1000E).
- Sodium dihydrogen phosphate dihydrate ($NaH_2PO_4\cdot 2H_2O$; Merck, Darmstadt, Germany; cat. no. 6345.1000).
- Disodium hydrogen phosphate (anhydrous) (Na_2HPO_4; Merck, Darmstadt, Germany; cat. no. 6586.0500).
- Trypsin–EDTA solution (Gibco BRL, Paisley, UK; cat. no. 35400.027).
- F10 (Ham) medium fortified with 10 mmol/L Hepes (Gibco BRL, Paisley, UK; cat. no. 22390.025).
- Penicillin (10 000 U/ml) plus streptomycin ($10\,000\,\mu$g/ml) solution (Imperial Chemicals, UK; cat. no. 4-804/07).
- Amphotericin B (Sigma Chemicals, St Louis, MO, USA; cat. no. A4888).
- Bovine calf serum (Hyclone Laboratories, Logan, UT, USA; cat. no. A2151/L).
- L-Carnitine (3-hydroxy-4-trimethylaminobutyric acid, Sigma Chemicals, MO, USA; cat. no. C0158).
- Tris (tris(hydroxymethyl)aminomethane) (Merck, Darmstadt, Germany; cat. no. 8382.0500).
- α-Cyclodextrin (Sigma Chemicals, St Louis, MO, USA; cat. no. C4642).
- Sodium hydroxide (NaOH, Sigma Chemicals, St Louis, MO, USA; cat. no. S-5881).
- Bovine serum albumin (essentially fatty acid free) (BSA; Sigma Chemicals, St Louis, MO, USA; cat. no. A6003).

Solutions required for growth of fibroblasts and fatty acid oxidation activity measurements

- Phosphate buffered saline (PBS) containing 140 mmol/L NaCl, 9.2 mmol/L Na_2HPO_4 and 1.2 mmol/L $NaH_2PO_4\cdot 2H_2O$ (final pH 7.4). Dissolve 8.18 g/L NaCl, 1.306 g/L Na_2HPO_4 and 0.203 g/L $NaH_2PO_4\cdot 2H_2O$ in 1 litre H_2O.

- Culture medium containing F10 (Ham), bovine calf serum, penicillin, streptomycin and amphotericin. For 1 litre, weigh out $250 \mu g$ amphotericin B, add 150 ml bovine calf serum, add 10 ml penicillin/streptomycin solution and make up with F10 (Ham) to 1 litre.
- Stock solutions of $100 \mu mol/L$ [1-^{14}C]palmitic acid (hexadecanoic acid $C_{16:0}$), [1-^{14}C]cerotic acid (hexadocosanoic acid, $C_{26:0}$) and [1-^{14}C]pristanic acid (2,6,10,14-tetramethylpentadecanoic acid). Labelled fatty acids (55 mCi/mmol) dissolved in benzene or toluene are pipetted into glass tubes followed by evaporation to dryness under a stream of nitrogen. Subsequently, a solution containing 100 mmol/L Tris-HCl plus 10 mg/l α-cyclodextrin (pH 8.0) is added followed by sonication in a water bath to obtain full solubilization. Preparation of the Tris/α-cyclodextrin solution is done as follows: first prepare a stock solution of 100 mmol/L Tris-HCl by dissolving 12.114 g/L Tris with pH adjustment to 8.0. Then add 10 ml of this solution to 100 mg α-cyclodextrin. Vortex rigorously. Keep frozen (−80°C) in small aliquots.
- 2 mol/L sodium hydroxide. Dissolve 80 g sodium hydroxide in 1 litre H_2O.
- 20% (w/v) bovine serum albumin (BSA) in PBS. Dissolve 20 g of albumin in 100 ml PBS.
- 0.2 mol/L L-carnitine. Dissolve 3.224 g in about 80 ml H_2O, neutralize with sodium hydroxide (2 mol/L) to pH 7.4 and make up to 100 ml. Store deep-frozen in small aliquots.

EXPERIMENTAL PROCEDURE

(1) Preparation of the fibroblasts

Fibroblasts are grown in a standard medium containing F10 (Ham), bovine calf serum penicillin, streptomycin and amphotericin B. At the time of confluency the fibroblasts (162 cm^2 flask) are washed twice with PBS, followed by addition of 3 ml trypsin−EDTA and incubation for about 5 min. Culture medium (7 ml) is subsequently added to inhibit further action of the trypsin. After gentle mixing, the cells are spun down ($400 g_{av}$, 5 min). The supernatant is removed and the cells are suspended in 2 ml F10 (Ham).

(2) Protein measurement (estimation)

Obtain a rough estimate of the amount of protein by measuring the OD at 600 nm. For this purpose dilute the cells 10 times in F10 (Ham). The correlation between protein and absorption should be checked in each individual laboratory. In our hands an absorbance of 0.7 corresponds to 0.5 mg protein per ml and hence the stock solution contains about 5 mg protein/ml. The exact protein concentration has to be determined later according to standard procedures.

(3) Allow the fibroblasts to adhere overnight in an incubator at 37°C

Fibroblast protein (100 μg) is added to sterile vials followed by addition of 2 ml of culture medium. The vials are gently shaken to suspend the cells in the medium and to allow them to adhere to the bottom by overnight incubation in an incubator at 37°C.

(4)   Incubation of cells with fatty acids

Remove the culture medium and wash the fibroblasts carefully with sterile PBS three times. After the final wash the remaining fluid should be removed as completely as possible. Then place small Eppendorf tubes containing $500 \mu l$ 2 mmol/L NaOH in the glass vials (2.6 cm diameter) and ensure that the Eppendorf tubes will stand upright during incubation. Start the oxidation assay by adding $500 \mu l$ of incubation medium made up of $445 \mu l$ F10 (Ham), $5 \mu l$ 200 mmol/L L-carnitine and $50 \mu l$ [1-^{14}C]-labelled fatty acid. Close each vial tightly with a perforated cap equipped with a rubber septum. Incubate for 2–4 h at 37°C in a slowly rocking water bath. Reactions are terminated by adding $100 \mu l$ 2 mol/L PCA using a Hamilton syringe.

(5)   Collection of [^{14}C]CO_2 and acid-soluble-products (ASP), the end-products of β-oxidation

After addition of PCA, vials are left for 3 h at room temperature to ensure complete trapping of [^{14}C]CO_2 as sodium carbonate in the NaOH present in the Eppendorf tube. After this period the vials are uncapped and the Eppendorf tubes are taken out, followed by quantification of the radioactivity in the sodium hydroxide. The acidified material $(600 \mu l)$ is pipetted out and transferred to an Eppendorf tube containing $60 \mu l$ of 20% (w/v) BSA. The tubes are vortexed vigorously and centrifuged to precipitate the denatured protein $(12000g_{av}, 2 \text{ min}, 4°C)$. The radioactivity present in the supernatant is quantified using a liquid scintillation counter.

Calculation of results

The sum of radioactive [^{14}C]CO_2 and [^{14}C]-labelled acid-soluble-products is taken as a measure of fatty acid oxidation. Rates are expressed as pmol/h per mg protein and can simply be calculated from the radioactivity recovered as [^{14}C]CO_2 and [^{14}C]-labelled acid-soluble products.

RESULTS AND DISCUSSION

Table 1 summarizes the results of fatty acid oxidation activity measurements in fibroblasts from patients affected with different peroxisomal disorders. The data show that $C_{26:0}$ β-oxidation is strongly deficient in Zellweger fibroblasts but also in X-linked adreno-leukodystrophy, although the extent of the deficiency is much less in X-linked ALD as compared to Zellweger cells. Deficient $C_{26:0}$ β-oxidation is also found in fibroblasts with a deficiency of acyl-CoA oxidase or bifunctional protein, whereas normal rates are found in rhizomelic chondrodysplasia punctata fibroblasts despite the fact that these cells lack mature peroxisomal thiolase (Hoefler et al 1988). In order to ascertain that the cells studied are functionally intact, we always include [1-^{14}C]palmitic acid as a substrate. The latter substrate is primarily (>95%) degraded in mitochondria, at least in fibroblasts (Jakobs and Wanders 1991).

The same system as described above for the oxidation of $C_{26:0}$ can also be used to study the β-oxidation of pristanic acid (2,6,10,14-tetramethylpentadecanoic acid), which is primarily oxidized in peroxisomes (Singh et al 1990; Jakobs et al 1994). However, the

Table 1 $C_{26:0}$, pristanic and palmitic acid oxidation in cultured skin fibroblasts from controls and patients affected by different peroxisomal disorders

Fibroblasts studied	Rate of fatty acid oxidation (pmol/h per mg protein)		
	$C_{26:0}$	Pristanic acid	$C_{16:0}$
Controls (>50)	1002±336	1147±325	4086±1524
Zellweger (12)	117±82	12±12	4417±1813
X-ALD (9)	309±151	1133±181	3803±644
RCDP (3)	751±369	919±28	5028±1930
DHAPAT-deficiency (3)	1051±303	1202±198	5140±1120
Bifunctional protein deficiency			
Patient 1	168	0	4763
Patient 2	137	15	4846

Results represent the mean±SD with the number of different patients studied between parentheses.
X-ALD=X-linked adrenoleukodystrophy; RCDP=rhizomelic chondrodysplasia punctata; DHAPAT= dihydroxyacetonephosphate acyltransferase

pathway of oxidation of pristanic acid differs from that of $C_{26:0}$ in several aspects (Figure 1). First, activation of $C_{26:0}$ and pristanic acid is brought about by two distinct synthetases located at the peroxisomal membrane (see Figure 1). Second, the respective CoA-esters are handled by the two different oxidases identified in human peroxisomes (see Mannaerts and Van Veldhoven (1993) for review). Table 1 shows that oxidation of pristanic acid is deficient in Zellweger fibroblasts but normal in X-ALD and RCDP fibroblasts, which is in line with literature data (Singh et al 1990). Furthermore, pristanic acid β-oxidation is deficient in bifunctional protein-deficient cells, suggesting that bifunctional protein (or rather trifunctional protein (Palosaari et al 1990)) is involved in $C_{26:0}$ and pristanic acid oxidation. The results of Table 1 further show normal rates of oxidation of pristanic acid and $C_{26:0}$ in DHAPAT-deficient cells.

REFERENCES

Barth PG, Wanders RJA, Schutgens RBH, Bleeker-Wagemakers EM, Van Heemstra D (1990) β-Oxidation defect with detectable peroxisomes: a case with neonatal onset and progressive course. *Eur J Pediatr* **149**: 722–726.

Clayton PT, Lake BD, Hjelm M, et al (1988) Bile acid analyses in pseudo-Zellweger syndrome: clues to the defect in peroxisomal β-oxidation. *J Inher Metab Dis* **11**: 165–168.

De Waart DR, Koomen GCM, Wanders RJA (1994) Studies on the urinary excretion of thromboxane B2 in Zellweger patients and control subjects: evidence for a major role for peroxisomes in the β-oxidative chain-shortening of thromboxane B2. *Biochim Biophys Acta* **1226**: 44–48.

Diczfalusy U, Alexson SEH (1988) Peroxisomal chain-shortening of prostaglandin F2α. *J Lipid Res* **29**: 1629–1636.

Diczfalusy U, Kase BF, Alexson SEH, Björkhem I (1991) Metabolism of prostaglandin F2α in Zellweger syndrome: peroxisomal β-oxidation is of major importance for *in vivo* degradation of prostaglandins in humans. *J Clin Invest* **88**: 978–984.

Diczfalusy U, Versterqvist O, Kase BF, Lund E, Alexson SEH (1993) Peroxisomal chain-shortening of thromboxane B2: evidence for impaired degradation of thromboxane B2 in Zellweger syndrome. *J Lipid Res* **34**: 1107–1112.

Espeels M, Roels F, Van Maldergem L, et al (1991) Peroxisomal localization of the immunoreactive β-oxidation enzymes in a neonate with a β-oxidation defect: pathological observations in liver, adrenal cortex and kidney. *Virchows Arch (Pathol Anat)* **419**: 339–347.

Goldfischer S, Collins J, Rapin I, et al (1986) Pseudo-Zellweger syndrome: deficiencies in several peroxisomal oxidative activities. *J Pediatr* **108**: 25–32.

Gordon JA, Figard PH, Spector AA (1990) Hydroxy-eicosatetraenoic acid metabolism in cultured human skin fibroblasts. *J Clin Invest* **85**: 1173–1181.

Hoefler G, Hoefler S, Watkins PA, et al (1988) Biochemical abnormalities in rhizomelic chondrodysplasia punctata. *J Pediatr* **112**: 726–733.

Jakobs BS, Wanders RJA (1991) Conclusive evidence that very long chain fatty acids are oxidized exclusively in peroxisomes in human skin fibroblasts. *Biochem Biophys Res Commun* **178**: 842–847.

Jakobs BS, van den Bogert C, Dacremont G, Wanders RJA (1994) β-Oxidation of fatty acids in cultured human skin fibroblasts devoid of the capacity for oxidative phosphorylation. *Biochim Biophys Acta* **1121**: 37–43.

Jedlischky G, Huber M, Völkl A, et al (1991) Peroxisomal degradation of leukotrienes by β-oxidation from the ω-end. *J Biol Chem* **266**: 24763–24772.

Kase BF, Björkhem I, Pedersen JI (1983) Formation of cholic acid from $3\alpha,7\alpha,12\alpha$-trihydroxy-5β-cholestanoic acid by rat liver peroxisomes. *J Lipid Res* **24**: 1560–1567.

Lageweg W, Tager JM, Wanders RJA (1991) Topography of very long-chain acyl-CoA synthetase in peroxisomes from rat liver. *Biochem J* **276**: 53–56.

Lazo O, Contreras M, Hashmi M, Stanley W, Irazu C, Singh I (1988) Peroxisomal lignoceroyl-CoA ligase deficiency in childhood adrenoleukodystrophy and adrenomyeloneuropathy. *Proc Natl Acad Sci USA* **85**: 7647–7651.

Lazo A, Contreras M, Singh I (1990) Topographical localization of peroxisomal acyl-CoA ligases: differential localization of palmitoyl-CoA and lignocenoyl-CoA ligases. *Biochemistry* **29**: 2981–2986.

Lazarow PB, de Duve C (1976) A fatty acyl-CoA oxidizing system in rat liver peroxisomes; enhancement by clofibrate, a hypolipidemic drug. *Proc Natl Acad Sci USA* **73**: 2043–2046.

Mandel H, Berant M, Aizin A, et al (1992) Zellweger-like phenotype in two siblings: a defect in peroxisomal β-oxidation with elevated very long-chain fatty acids but normal bile acids. *J Inher Metab Dis* **15**: 381–384.

Mannaerts GP, Van Veldhoven PP (1993) Metabolic pathways in peroxisomes. *Biochimie* **75**: 147–159.

Miyazawa S, Hashimoto T, Yokota S (1985) Identity of long chain acyl-coenzyme A synthetase of microsomes, mitochondria and peroxisomes in rat liver. *J Biochem* **98**: 723–733.

Naidu S, Hoefler G, Watkins PA, et al (1988) Neonatal seizures and retardation in a girl with biochemical features of X-linked adrenoleukodystrophy. *Neurology* **38**: 1100–1107.

Novikov DK, Vanhove GF, Carchon H, et al (1994) Peroxisomal β-oxidation: purification of four novel 3-hydroxyacyl-CoA dehydrogenases from rat liver peroxisomes. *J Biol Chem* **269**: 27125–27135.

Osumi T, Hashimoto T, Ui N (1980) Purification and properties of acyl-CoA oxidase from rat liver. *J Biochem (Tokyo)* **87**: 1735–1746.

Palosaari PM, Hiltunen JK (1990) Peroxisomal bifunctional protein from rat liver is a trifunctional enzyme possessing 2-enoyl-CoA hydratase, 3-hydroxyacyl-CoA dehydrogenase and Δ3,Δ2-enoyl-CoA isomerase activities. *J Biol Chem* **265**: 2446–2449.

Poll-Thé BT, Roels F, Ogier HAM, et al (1988) A new peroxisomal disorder with enlarged peroxisomes and a specific deficiency of acyl-CoA oxidase (pseudo neonatal adrenoleukodystrophy). *Am J Hum Genet* **42**: 422–434.

Prydz K, Kase BF, Björkhem I, Pedersen JI (1988) Subcellular localization of $3\alpha,7\alpha$-dihydroxy- and $3\alpha,7\alpha,12\alpha$-trihydroxy-5β-cholestanoyl-CoA ligase(s) in rat liver. *J Lipid Res* **29**: 997–1004.

Santer R, Claviez A, Oldigs HD, Schaub J, Schutgens RBH, Wanders RJA (1993) Isolated defect of peroxisomal β-oxidation in a 16-year-old patient. *Eur J Pediatr* **152**: 339–342.

Schepers L, Casteels M, Vamecq J, Parmentier G, Van Veldhoven PP, Mannaerts GP (1988) β-Oxidation of the carboxyl side-chain of prostaglandin E2 in rat liver peroxisomes and mitochondria. *J Biol Chem* **263**: 2724–2731.

Schepers L, Casteels M, Verheyden K, et al (1989) Subcellular distribution and characteristics of trihydroxycoprostanoyl-CoA synthetase. *Biochem J* **257**: 221–229.

Schepers L, Van Veldhoven PP, Casteels M, Eyssen HJ, Mannaerts GP (1990) Presence of three acyl-CoA oxidases in rat liver peroxisomes: an inducible fatty acyl-CoA oxidase, a non-inducible fatty acyl-CoA oxidase and a non-inducible trihydroxycoprostanoyl-CoA oxidase. *J Biol Chem* **265**: 5242–5246.

Schram AW, Goldfischer S, Van Roermund CWT, et al (1987) Human peroxisomal 3-oxoacyl-coenzyme A thiolase deficiency. *Proc Natl Acad Sci USA* **84**: 2494–2496.

Singh H, Derwas N, Poulos A (1987) Very long chain fatty acid β-oxidation by rat liver mitochondria and peroxisomes. *Arch Biochem Biophys* **359**: 382–390.

Singh H, Usher S, Johnson D, Poulos A (1990) A comparative study of straight chain and branched chain fatty acid oxidation in skin fibroblasts from patients with peroxisomal disorders. *J Lipid Res* **31**: 271–275.

Singh I, Moser AE, Goldfischer S, Moser HW (1984) Lignoceric acid is oxidized in the peroxisome: implications for the cerebro-hepato-renal (Zellweger) syndrome and adrenoleukodystrophy. *Proc Natl Acad Sci USA* **81**: 4203–4207.

Singh I, Sing RP, Bhusman A, Singh AK (1985) Lignoceroyl-CoA ligase activity in rat brain microsomal fraction: topographical localization and effect of detergents and α-cyclodextrin. *Arch Biochem Biophys* **236**: 418–426.

Suzuki H, Yamada J, Watanabe T, Suga T (1989) Compartmentation of dicarboxylic acid β-oxidation in rat liver: importance of peroxisomes in the metabolism of dicarboxylic acids. *Biochim Biophys Acta* **990**: 25–30.

Suzuki Y, Shimozawa N, Yajima S, et al (1994) Novel subtype of peroxisomal acyl-CoA oxidase deficiency and bifunctional enzyme deficiency with detectable enzyme proteins: identification by means of complementation analysis. *Am J Hum Genet* **54**: 36–43.

Vamecq J, de Hoffmann E, van Hoof F (1985) The microsomal dicarboxylyl-CoA synthetase. *Biochem J* **102**: 225–234.

Vanhole C, de Zegher F, Casaer P, Devlieger H, Wanders RJA, Jaeken J (1994) A new peroxisomal disorder with fetal and neonatal adrenal insufficiency. *Arch Dis Child* **71**: 55–57.

Vanhove GF, Van Veldhoven PP, Fransen M, Denis S, Wanders RJA, Mannaerts GP (1993) The CoA-esters of 2-methyl-branched chain fatty acids and of the bile acid intermediates di- and trihydroxycoprostanoic acids are oxidized by one single peroxisomal branched chain acyl-CoA oxidase in human liver and kidney. *J Biol Chem* **268**: 10335–10344.

Van Maldergem L, Espeel M, Wanders RJA, et al (1992) Neonatal seizures and severe hypotonia in a male infant suffering from a defect in peroxisomal β-oxidation. *Neuromusc Disord* **2**: 217–224.

Van Veldhoven PP, Vanhove G, Vanhoutte F, et al (1991) Identification and purification of a peroxisomal branched chain fatty acid acyl-CoA oxidase. *J Biol Chem* **266**: 24676–24683.

Wanders RJA, Van Roermund CWT, Van Wijland MJA, et al (1987a) Peroxisomal very long chain fatty acid β-oxidation in human skin fibroblasts: activity in Zellweger syndrome and other peroxisomal disorders. *Clin Chim Acta* **166**: 255–263.

Wanders RJA, van Roermund CWT, van Wijland MJA, et al (1987b) Peroxisomal fatty acid β-oxidation in relation to the accumulation of very long chain fatty acids in peroxisomal disorders. *J Clin Invest* **80**: 1778–1788.

Wanders RJA, van Roermund CWT, van Wijland MJA, et al (1988) Direct demonstration that the deficient oxidation of very long chain fatty acids in X-linked adrenoleukodystrophy is due to an impaired ability of peroxisomes to activate very long chain fatty acids. *Biochem Biophys Res Commun* **153**: 618–624.

Wanders RJA, Schelen A, Feller N, et al (1990a) First prenatal diagnosis of acyl-CoA oxidase deficiency. *J Inher Metab Dis* **13**: 371–374.

Wanders RJA, van Roermund CWT, Schelen A, et al (1990b) Bifunctional protein with deficient activity: identification of a new peroxisomal disorder. *J Inher Metab Dis* **13**: 375–379.

Wanders RJA, Denis S, Jakobs C, ten Brink HK (1992a) Identification of pristanoyl-CoA oxidase as a distinct clofibrate non-inducible enzyme in rat liver peroxisomes. *Biochim Biophys Acta* **1124**: 199–202.

Wanders RJA, Denis S, Van Roermund CWT, Jakobs C, Ten Brink HJ (1992b) Characteristics and subcellular localization of pristanoyl-CoA synthetase in rat liver. *Biochim Biophys Acta* **1125**: 274–279.

Wanders RJA, Van Roermund CWT, Brul S, Schutgens RBH, Tager JM (1992c) Bifunctional enzyme deficiency: identification of a new type of peroxisomal disorder in a patient with an impairment in peroxisomal β-oxidation of unknown aetiology by means of complementation analysis. *J Inher Metab Dis* **15**: 385–388.

Wanders RJA, van Roermund CWT, Lageweg W, et al (1992d) X-linked adrenoleukodystrophy: biochemical diagnosis and enzyme defect. *J Inher Metab Dis* **15**: 634–644.

Wanders RJA, Denis S, Dacremont G (1993a) Studies on the substrate specificity of the inducible and non-inducible acyl-CoA oxidases from kidney peroxisomes. *J Biochem (Tokyo)* **43**: 577–582.

Wanders RJA, Schutgens RBH, Barth PG, Tager JM, Van den Bosch H (1993b) Postnatal diagnosis of peroxisomal disorders: a biochemical approach. *Biochemie* **75**: 269–279.

Watkins PA, Chen WW, Harris CJ, et al (1989) Peroxisomal bifunctional enzyme deficiency. *J Clin Invest* **83**: 771–777.

Wolvetang EJ, Tager JM, Wanders RJA (1990) Latency of the peroxisomal enzyme acyl-CoA:dihydroxyacetonephosphate acyltransferase in digitonin-permeabilized fibroblasts: the effect of ATP and ATPase inhibitors. *Biochem Biophys Res Commun* **170**: 1135–1143.

J. Inher. Metab. Dis. 18 Suppl. 1 (1995) 125–134

Activity measurements of acyl-CoA oxidases in human liver

P. P. VAN VELDHOVEN
Katholieke Universiteit Leuven, Camus Gasthuisberg, Afdeling Farmakologie,
Herestraat, B-3000 Leuven, Belgium

Summary: Peroxisomal β-oxidation is involved in the degradation of different fatty acids or fatty acid derivatives including eicosanoids (prostaglandins, leukotrienes, thromboxanes), dicarboxylic fatty acids, very long-chain fatty acids, pristanic acid, bile acid intermediates (di- and trihydroxycoprostanoic acids), and xenobiotics. Separate β-oxidation systems are probably active inside peroxisomes, each acting on a distinct set of substrates, as suggested by the discovery of multiple acyl-CoA oxidases.

Using specific substrates or selective conditions, we can distinguish in rat liver the action of acyl-CoA oxidases (type I and II), a pristanoyl-CoA oxidase and a trihydroxycoprostanoyl-CoA oxidase and, in human liver, of acyl-CoA oxidase (type I and II) and a branched-chain acyl-CoA oxidase. When incubated with suitable CoA-esters, these different oxidases can be measured in a similar fashion by following fluorimetrically the dimerization of homovanillic acid, catalysed by peroxidase in the presence of hydrogen peroxide. The optimal assay conditions and possible pitfalls in this type of coupled assay are discussed. This knowledge can be used to reveal the existence of peroxisomal disorders in which only one acyl-CoA oxidase is deficient.

In mammals, peroxisomes are most abundant in liver and kidney. They play an important role in cellular lipid metabolism, e.g. the biosynthesis of ether lipids, cholesterol and dolichols and the degradation of fatty acids (and derivatives) via β-oxidation (see Mannaerts and Van Veldhoven 1992; van den Bosch et al 1992). Physiologically important substrates for the latter pathway include very long-chain fatty acids, dicarboxylic fatty acids, eicosanoids (prostaglandins, thromboxanes, leukotrienes), pristanic acid (an isoprenoid-derived 2-methyl branched-chain fatty acid), the bile acid intermediates di- and trihydroxycoprostanoic acid, and xenobiotics (Mannaerts and Van Veldhoven 1992). Except for the last, accumulation of the other carboxylates has been reported in peroxisomal disorders. Prior to their degradation, these carboxylates have to be activated to their CoA-esters. This process is catalysed by differed synthetases that are found at different cellular sites (mitochondria, peroxisomes, endoplasmic reticulum) (Mannaerts and Van Veldhoven 1992). The CoA-derivatives are shortened inside the peroxisomes in four consecutive steps. The first step is a desaturation reaction catalysed by an acyl-CoA oxidase generating 2-*trans*-enoyl-CoA and hydrogen peroxide.

In addition to the well-known palmitoyl-CoA oxidase (EC 1.3.3.-; ACox-I; see below), purified originally by the groups of Hashimoto (Osumi et al 1980) and Leighton (Inestrosa et al 1980) from rat liver, two additional oxidases have been isolated from this tissue. The three enzymes appear to have a characteristic substrate spectrum and tissue distribution (Schepers et al 1990; Van Veldhoven et al 1991, 1992). ACox-I acts on the CoA-esters of long-chain and very long-chain fatty acids, short-chain and long-chain dicarboxylic acids and prostaglandins (Van Veldhoven et al 1992). It has a native molecular mass of 139 kDa, is composed of 51 kDa and 23 kDa subunits and is several-fold induced by treatment of rats with hypolipidaemic drugs. A second oxidase, named pristanoyl-CoA oxidase (EC 1.3.3.-; PCox), acts on the CoA-esters of long-chain 2-methyl branched-chain fatty acids like pristanic acid, but also on those of straight-chain fatty acids (Van Veldhoven et al 1991, 1992). PCox is not induced by peroxisome proliferators, is present in liver and kidney, has a molecular mass of 513 kDa (by native gel electrophoresis) and consists of 70 kDa subunits (Van Veldhoven et al 1994a). A third acyl-CoA oxidase, trihydroxycoprostanoyl-CoA oxidase (EC 1.3.3.-; THCCox), is also not induced by proliferators and only found in liver. It acts on trihydroxycoprostanoyl-CoA, a bile acid intermediate, and is a dimer of 69 kDa subunits (Schepers et al 1990; Van Veldhoven et al 1992, 1994b). The identity of a possible fourth acyl-CoA oxidase, reported to act on valproyl-CoA (Vamecq et al 1993), remains to be proved.

In man, so far only two enzymes have been isolated (Casteels et al 1990; Vanhove et al 1993a). The first, with respect to substrate specificity and molecular structure, resembles the inducible rat ACox-I (human acyl-CoA oxidase; hACox; see below). The second oxidase acts on CoA-esters of substrates possessing a 2-methyl branch, such as pristanic acid and the bile acid intermediates, but it can also desaturate straight-chain acyl-CoA esters. It has been named branched-chain acyl-CoA oxidase (EC 1.3.3.-; BRCACox) and resembles the rat THCCox (Vanhove et al 1993a). Whether in man glutaryl-CoA desaturation is catalysed by a separate enzyme (Bennett et al 1991) remains questionable (Vanhove et al 1993a).

An additional confusing fact is that the ACox gene in both rat and man gives rise to two mRNAs via differential splicings (Miyazawa et al 1987; Varanasi et al 1994; B. Fournier and B.T. Poll-Thé, personal communication). So far only the enzyme corresponding to mRNA type I has been isolated from rat liver and is therefore referred to as ACox-I. A study on engineered rat ACox-I and ACox-II did not reveal major differences in substrate specificity towards a limited set of straight-chain acyl-CoAs (Hashimoto 1992).

In this paper, the measurement of acyl-CoA oxidases in human liver is described. BRCACox can be measured selectively using the physiological substrates pristanoyl-CoA or trihydroxycoprostanoyl-CoA. We prefer to use 2-methylpalmitoyl-CoA, a synthetic analogue of pristanoyl-CoA that is much easier to synthesize and that results in higher activities (Van Veldhoven et al 1991; Vanhove et al 1993a). No selective substrates are available for measuring the hACox, but palmitoyl-CoA can be used if the activity of the BRCACox is blocked by alkylation with *N*-ethylmaleimide (NEM).

The assay is based on the peroxidase (EC 1.11.1.7)-catalysed dimerization of homovanillic acid (HVA) into a fluorescent product in the presence of hydrogen peroxide (Guilbault et al 1967) (see Figure 1). A number of potential hydrogen donors are available (Zaitsu and Ohkura 1980) and tested in our laboratory, but we still prefer to use HVA (for

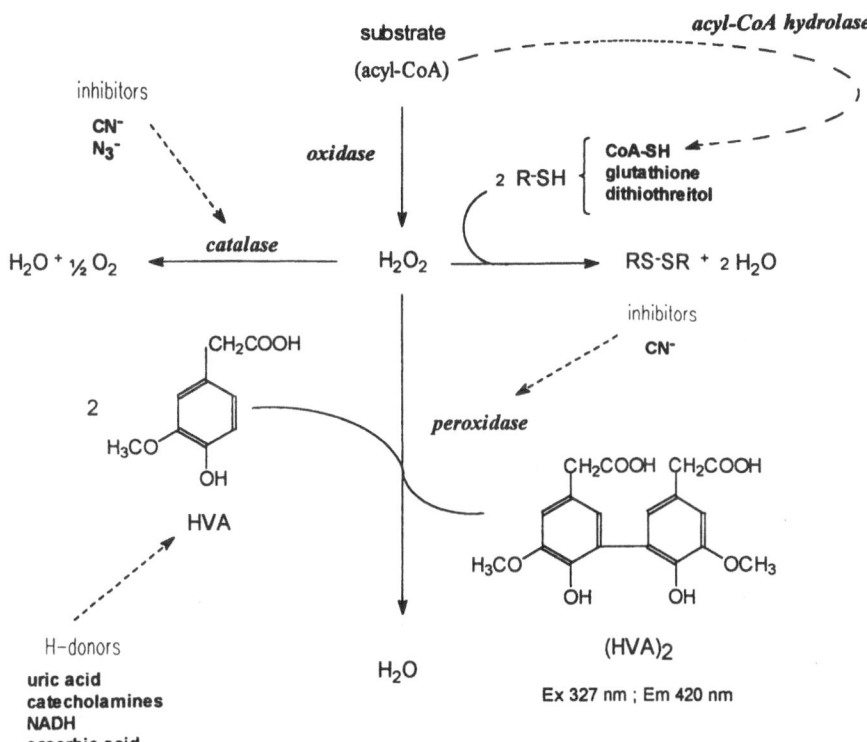

Figure 1 Possible pitfalls in fluorimetric hydrogen peroxide measurements

stability, sensitivity, price). Although the fluorescence yield of the HVA dimer is maximal at alkaline pH, it is possible to perform measurements in a kinetic mode at a lower pH, down to pH 7.5. Despite being somewhat more tedious, we advocate a (multiple) end-point assay. Firstly, the sample is not continuously UV-illuminated; this would lead to a drift in background fluorescence and pose a limit to the incubation time (Hirsch and Parks 1982). Secondly, measurements can be performed in small volumes, so only low amounts of (sometimes) costly substrates are consumed. Thirdly, a large number of samples can be processed simultaneously, which is especially useful when measuring low activities involving longer incubation times (1–2 h).

The fluorimetric assay can be applied to measure any (peroxisomal) oxidase (except monoamine oxidase (EC 1.4.3.4) since some catecholamines compete with HVA). Oxidases known to be present in mammalian peroxisomes act on a variety of substrates including polyamines, neutral D-amino acids, acidic D-amino acids, L-pipecolic acid, uric acid, xanthine, glycolate and alipatic 2-L-hydroxyacids (Mannaerts and Van Veldhoven 1992). With regard to the diagnosis of peroxisomal disorders, the measurement of L-pipecolate oxidase (EC 1.4.3.-) should be mentioned (Wanders et al 1989).

A number of pitfalls that have been encountered in the past 12 years we have been using this assay are shown in Figure 1 (see Van Veldhoven and Brees 1994). Firstly, the hydrogen peroxide produced can be destroyed chemically in homogenates by sulphydryl compounds (glutathione, dithiothreitol) or enzymatically by catalase (EC 1.11.1.6). Interference of the latter can be reduced by preincubating samples with azide. Although azide is also inhibitory to peroxidase, catalase is more sensitive. If the oxidase is stable, chemical interference can be removed by dialysis, or, if the activity is high, by dilution. If the enzyme is not affected, pretreatment of samples with *N*-ethylmaleimide removes the sulphydryl interference. The presence of hydrogen donors in sufficient amount to compete with HVA (NADH, ascorbic acid, serotonin, uric acid), is less likely in biological samples. With CoA-esters, one encounters another problem, i.e. hydrolysis (either chemically in the stock solutions or enzymatically during the assay) generating free CoA-SH that reacts with hydrogen peroxide.

The chemical interference, which is very pronounced and variable in human liver, is responsible for the lag phase in the production of peroxide when analysing crude homogenates (Casteels et al 1990), crude nuclear fractions and cytosolic fractions (unpublished data). If not recognized as such by the investigator, this lag phase might be interpreted as a deficiency (cf. a glutaryl-CoA oxidase deficiency syndrome (Vanhove et al 1993a; Bennett et al 1991)).

Finally, one should realize that a substrate-dependent hydrogen peroxide production in homogenates is not necessarily proof of a *peroxisomal* acyl-CoA oxidase. Indeed, mitochondrial acyl-CoA dehydrogenases (EC 1.3.99.3), especially those acting on short-chain (branched) acyl-CoA esters, are able to transfer electrons directly to oxygen under the described assay conditions, hence acting as an oxidase (Vanhove et al 1993b; Van Veldhoven et al 1994a).

EQUIPMENT, CHEMICALS AND SOLUTIONS

Equipment

Access to
- Balance (sensitivity ~1 mg)
- Motor-driven homogenization device (800–1500 rpm)
- Spectrofluorimeter (preferably with continuous spectrum adjustment) plus quartz cuvettes of 4 ml (or less)
- Low-speed centrifuge able to generate ~3000*g* and to accommodate 5 ml tubes, or an Eppendorf centrifuge with a rotor for 24 tubes
- Dispenser or automatic dispensing device (delivering between 1 and 3 ml)
- thermostated (shaking) water bath at 37°C
- vortexer
- 3 ml Potter–Elvehjem tissue grinder with Teflon pestle (clearance 0.004–0.006 inch; Kontes ref. K-886000-0020).
- Ice

On the bench
- Piece of Nylon or cotton wool gauze (~15×15 cm)+funnel (glass or plastic)
- Timers
- Conical shaped 5 ml plastic reaction tubes (~75×12 mm), or Eppendorf tubes of 1.5 ml plus suitable racks

- Glass or plastic reaction tubes of 14 ml (~100 × 16 mm) plus suitable racks
- Glass or plastic reaction tubes of 5 ml (~75 × 12 mm) plus suitable racks
- Adjustable 200 μl pipettes plus tips
- Adjustable 1000 μl pipettes plus tips
- Adjustable micropipette of 10 μl or 20 μl plus tips
- Coloured felt pens
- Ice buckets or containers

Chemicals and biological material

- Sample of human liver (~0.1–0.5 g) (stored at −80°C)
- Homovanillic acid (Aldrich ref. 14364-2; 1 g; 98%); peroxidase (horseradish, grade II, ~200 U/mg; Boehringer ref. 127361; 10000 U); palmitoyl-CoA (K-salt; Boehringer, Sigma, or Pharmacia); 2-methylpalmitoyl-CoA (not commercially available; contact author); trihydroxycoprostanoyl-CoA (not commercially available, contact author); uric acid (Merck 817); uricase (hog liver, 2 mg/ml; Boehringer ref. 127469; 2 mg); flavine–adenine nucleotide (Boehringer ref. 1102338; 200 mg); MOPS; *N*-ethylmaleimide; sodium bicarbonate; sodium carbonate; EDTA·Na$_2$; 70% perchloric acid; sodium azide; defatted bovine serum albumin.

Solutions

- 50 μmol/L uric acid (dissolve ~1 mg of uric acid in 100 ml of 50 mmol/L K-phosphate buffer, pH 7.5; measure absorbance at 293 nm ($\varepsilon = 12600$) and bring to 50 μmol/L; store at −20°C in aliquots)
- 1 mg/ml peroxidase (POD) (dissolve 10 mg in 10 ml of water; store at −20°C in 1 ml aliquots; stable, but avoid multiple freeze–thaw cycles since this will result in increased blank rates)
- 30 mmol/L homovanillic acid (HVA) (dissolve 55 mg in 10 ml of 0.1 mol/L HCl; store at −20°C in 1 ml aliquots; stable; warm before use to remove some haziness)
- 100 mmol/L NaN$_3$ (10 ml; store at −20°C
- 1 mmol/L FAD (1 ml; store at −20°C and shielded from light)
- 6% (w/v) defatted bovine serum albumin (BSA) (5 ml)
- 0.2 mol/L K-phosphate buffer, pH 8.3 (100 ml; store cold)
- 1 mmol/L palmitoyl-CoA (store frozen in aliquots; $A_{260} = 15.4$; 1 ml)
- 1 mmol/L 2-methylpalmitoyl-CoA (store frozen in aliquots; $A_{260} = 15.4$; 1 ml)
- 1 mmol/L trihydroxycoprostanoyl-CoA (store frozen in aliquots; $A_{260} = 15.4$; 1 ml)
- Homogenization medium (HM) (0.25 mol/L sucrose, 5 mmol/L MOPS, pH 7.5, 1 mmol/L EDTA, pH 7.5; 100 ml)
- HClO$_4$ 8% (w/v) (100 ml)
- 0.5 mol/L Na-carbonate buffer, pH 10.7, 10 mmol/L EDTA (bring 800 ml of 0.5 mol/L Na$_2$CO$_3$, 10 mmol/L EDTA to pH with approx 80 ml of 0.5 mol/L NaHCO$_3$, 10 mmol/L EDTA)
- 40 mmol/L *N*-ethylmaleimide (NEM) (dissolve 5 mg in 1 ml of water; prepare fresh)
- Urate oxidase (dilute Boehringer solution 3-fold in water just before use; place on ice)

ENZYME MEASUREMENTS

(1) For the specifications given in the following section, it is assumed that each oxidase activity, acting either on palmitoyl-, 2-methylpalmitoyl- or trihydroxycoprostanoyl-CoA, has to be determined in one sample.

(2) *Preparative work*
  • Label two tubes of 5 ml (A and B; *reaction-tubes*).
  • Label eight Eppendorf tubes (5 ml conical-shaped reaction tubes are also adequate) (A_i to A_j; B_i to B_j; the index corresponds to the stop-times mentioned below; *stop-tubes*); pipette $40 \mu l$ of 8% $HClO_4$ into each tube; place aside.
  • Label eight tubes (5 or 10 ml tubes are suitable) (A_i to A_j; B_i to B_j; the index corresponds to the stop-times; *reading-tubes*); add 1.4 ml of carbonate buffer to each tube. (Under the assay conditions described, the highest sensitivity is obtained at a 15-fold dilution. The final volume of 1.5 ml is sufficient to allow readings in a 4 ml cell in an Aminco SPF-500 fluorimeter.)

(3) Prepare 10% (w/v) homogenates of liver sample in cold homogenization medium (5× up and down strokes with Teflon pestle) and filter through nylon or cotton-wool gauze; store on ice.

(4) Prepare test and blank reaction mixtures ($400 \mu l$ for each sample to be analysed $+ 300 \mu l$ excess in order to prepare standard curves; see further).
  • *Test reaction mixture for palmitoyl-CoA oxidase activity:* 50 mmol/L K-phosphate buffer, pH 8.0, 0.1 mg/ml peroxidase, 0.15% (w/v) BSA, $125 \mu mol/L$ palmitoyl-CoA, 0.75 mmol/L HVA
  • *Test reaction mixture for 2-methylpalmitoyl-CoA oxidase activity:* 50 mmol/L K-phosphate buffer, pH 8.0, 0.1 mg/ml peroxidase, 0.3% (w/v) BSA, $125 \mu mol/L$ 2-methylpalmitoyl-CoA, 0.75 mmol/L HVA
  • *Test reaction mixture for trihydroxycoprostanoyl-CoA oxidase activity:* 50 mmol/L K-phosphate buffer, pH 8.0, 0.1 mg/ml peroxidase, 0.3% (w/v) BSA, $125 \mu mol/L$ trihydroxycoprostanoyl-CoA, 0.75 mmol/L HVA
  • *Blank reaction mixtures:* the composition of the corresponding blank reaction is identical except for the omission of the CoA-ester.

Please note the sequence of adding solutions: buffer first; BSA always before the CoA-ester to prevent binding of the substrate to the plastic walls; HVA just before starting the incubations to keep the fluorescence values low.

(5) Prepare preincubation mixtures ($40 \mu l$ for each sample to be analysed):
  • *Total palmitoyl-CoA oxidase activity:* 25 mmol/L NaN_3, $25 \mu mol/L$ FAD
  • *Human acyl-CoA oxidase-dependent palmitoyl-CoA oxidase activity:* 25 mmol/L NaN_3, $25 \mu mol/L$ FAD, 5 mmol/L NEM
  • *2-Methylpalmitoyl-CoA oxidase activity:* 25 mmol/L NaN_3, $50 \mu mol/L$ FAD
  • *Trihydroxycoprostanoyl-CoA oxidase activity:* 25 mmol/L NaN_3, $50 \mu mol/L$ FAD

(6) *Experimental work*
  • Dilute homogenates with homogenization medium to 1 g/100 ml (palmitoyl-CoA

oxidase and trihydroxycoprostanoyl-CoA oxidase) or 1 g/300 ml (2-methyl-palmitoyl-CoA oxidase).

- Place 80 μl of diluted homogenates at the bottom of the *reaction-tubes* A and B followed by 20 μl of the appropriate preincubation mixture; incubate at 0°C for 5 – 10 min (in order to inhibit catalase activity and to saturate the oxidase with its cofactor FAD); start reaction (timed) by adding 400 μl of appropriate warm blank reaction mixture to tube A and 400 μl of test reaction mixture to tube B; place tubes in shaking water bath at 37°C.
- Remove, at the stop-times indicated below, 100 μl from the blank and test incubations and deliver into the corresponding *stop-tubes* containing 40 μl of HClO$_4$; place on ice.
- *Stop-times for palmitoyl-CoA oxidase activity:* 8 – 16 – 24 – 32 min.
- *Stop-times for trihydroxycoprostanoyl-CoA oxidase activity:* 15 – 30 – 45 – 60 min.
- *Stop times for 2-methylpalmitoyl-CoA oxidase activity:* 10 – 20 – 30 – 40 min.
- Centrifuge *stop-tubes*, transfer 100 μl of supernatant to *reading-tubes* containing 1.4 ml of carbonate buffer; mix by vortexing and reading fluorescence after ~10 min (optimum setting: excitation 327 nm, slit 2 nm; emission 420 nm, slit 8 nm).
- Standardize fluorescence by means of uric acid and urate oxidase as follows. Add to two sets of four Eppendorf tubes 0, 4, 8, 16 μl of uric acid solution; adjust to 16 μl with water; add to one set 80 μl of blank reaction mixture, to the other set 80 μl of test reaction mixture (both mixtures were made in excess); start reaction with 4 μl of diluted urate oxidase; keep at 37°C; stop reaction after 20 min with 40 μl of 8% (w/v) HClO$_4$; proceed as described above for the test homogenates. The standard curves with and without substrate should possess the same slope and normally intercept the fluorescence axis at the same value. Some substrates, however, might possess an intrinsic fluorescence and will result in a different intercept. A significant difference in the slopes might be indicative of quenching or chemical interference due to the substrate.

RESULTS AND COMMENTS

The standard curves (H$_2$O$_2$ formed from uric acid/urate oxidase) normally show excellent correlation coefficients approaching 1. Also, the day-by-day variations of the increase in fluorescence per nanomole of uric acid are minimal (26.02±0.80; mean±SEM of 8 different experiments; Shimadzu RF-5001PC fluorimeter). Somewhat more variation is seen in the calculated oxidase activities. With palmitoyl-, 2-methylpalmitoyl- and trihydroxycoprostanoyl-CoA as substrate, activities amounted to 135±13 (n=6), 324±16 (n=6) and 51±16 (n=5) nmol/min per g tissue, respectively. More importantly, it has been clearly demonstrated that fluorescence readings have to be corrected for blank reactions, which can vary from tissue to tissue and which are more pronounced at low homogenate dilutions; one should also rely on multiple time points since lag phases in the appearance of the HVA dimer can be quite long, especially in human liver (Casteels et al 1990). The effects of substrate and albumin on the different acyl-CoA oxidases are complex and assay conditions should be carefully optimized (Casteels et al 1990; Van Veldhoven et al 1992;

Vanhove et al 1993a). The use of detergents is not recommended. Dependent on the enzyme to be measured and/or the composition of the assay mixtures, Triton X-100 can be stimulatory or inhibitory (unpublished data). Finally, the binding of the cofactor FAD to the different acyl-CoA oxidases can be rather weak (e.g. THCCox) to quite strong (e.g. PCox), hence the addition of various amounts of FAD to the preincubation mixtures (Van Veldhoven et al 1994a,b).

Compared to rat liver, human palmitoyl-CoA oxidase activity is ~5-fold less. It is important to stress that this activity reflects the action of two enzymes, namely human acyl-CoA oxidase and BRCACox. Depending on the assay conditions, the contribution of the latter can be considerable (20–50%). Since BRCACox is sensitive to sulphydryl reagents, the addition of 5 mmol/L NEM to the preincubation mixture (see measurement section) will result in a selective measurement of human acyl-CoA oxidase with palmitoyl-CoA as substrate.

Pristanoyl-CoA (or its synthetic analogue 2-methylpalmitoyl-CoA) and trihydroxy-coprostanoyl-CoA are only desaturated by BRCACox (Vanhove et al 1993a). Owing to the low activity with trihydroxycoprostanoyl-CoA, complicating the measurements in homogenates, 2-methylpalmitoyl-CoA is to be preferred as substrate. Pristanoyl-CoA, if available, is also useful but activities are ~2-fold lower compared to the synthetic analogue (Vanhove et al 1993a).

So far only a few cases with a likely isolated oxidase deficiency have been reported. Two patients described by Poll-The et al (1988) and showing accumulation of very long-chain fatty acids but normal levels of pristanic acid and bile acid intermediates might be deficient in acyl-CoA oxidase activity. A similar case was diagnosed prenatally (Wanders et al 1990). In some other patients presenting with abnormal bile acid intermediates but normal very long-chain fatty acid levels, BRCACox could be affected (Christensen et al 1990; Mandel et al 1992).

With regard to the assay itself, the following remarks can be made (Van Veldhoven and Brees 1994). Although the multiple steps complicate the assay, they were introduced for specific reasons. Direct alkalinization of the reaction mixtures is not possible since this results in a very rapid increase in the fluorescence that is not related to enzymatic activity. These increases are considerably smaller when proteins are first precipitated, followed by alkalinization of the acidic supernatant. The HVA dimer is very stable in the $HClO_4$ extracts, so that the actual analysis (i.e. alkalinization) can be performed later (even days later) without problem. Denaturation with TCA, however, should be avoided (it causes quenching of the HVA dimer fluorescence). Additional investigations revealed that the addition of EDTA or dithiothreitol to the carbonate buffer suppressed the time-dependent increase in fluorescence almost completely. A possible explanation could be that these compounds chelate divalent cations (probably Fe^{2+}), which can give rise to reactive oxygen-molecules under alkaline conditions, causing a chemical dimerization of HVA. In our hands, carbonate buffers are best with regard to sensitivity and low background fluorescence.

REFERENCES

Bennett MJ, Pallitt RJ, Goodman SI, Hale DE, Vamecq J (1991) Atypical riboflavin-responsive glutaric aciduria and deficient peroxisomal glutaryl-CoA oxidase activity: a new peroxisomal disorder? *J Inher Metab Dis* **14**: 165–173.

Casteels M, Schepers L, Van Veldhoven PP, Eyssen HJ, Mannaerts GP (1990) Separate peroxisomal oxidases for fatty acyl-CoAs and trihydroxycoprostanoyl-CoA in human liver. *J Lipid Res* **31**: 1865–1872.

Christensen E, Van Eldere J, Brandt NJ, Schutgens RBH, Wanders RJA, Eyssen HJ (1990) A new peroxisomal disorder: di- and trihydroxycholestanaemia due to a presumed trihydroxyco-prostanoyl-CoA oxidase deficiency. *J Inher Metab Dis* **13**: 363–366.

Guilbault G, Kramer DN, Hackley E (1967) New substrate for fluorimetric determination of oxidative enzymes. *Anal Chem* **39**: 271.

Hashimoto T (1992) Peroxisomal and mitochondrial enzymes. In Coates PM, Tanaka K, eds. *New Developments in Fatty Acid Oxidation.* New York: Wiley-Liss, 19–32.

Hirsch HE, Parks ME (1982) Fluorimetric oxidase assays: pitfalls caused by action of ultraviolet light on lipids. *Anal Biochem* **122**: 79–84.

Inestrosa NC, Bronfman M, Leighton F (1980) Purification of the peroxisomal fatty acyl-CoA oxidase from rat liver. *Biochem Biophys Res Commun* **95**: 7–12.

Mandel H, Berant M, Aizin A, et al (1991) Zellweger-like phenotype in two siblings: a defect in peroxisomal β-oxidation with elevated very long-chain fatty acids but normal bile acids. *J Inher Metab Dis* **15**: 381–384.

Mannaerts GP, Van Veldhoven PP (1992) Role of peroxisomes in mammalian metabolism. *Cell Biochem Funct* **10**: 141–151.

Miyazawa S, Hayashi H, Hajikata M, et al (1987) Complete nucleotide sequence of cDNA and predicted amino acid sequence of rat acyl-CoA oxidase. *J Biol Chem* **262**: 8131–8137.

Osumi T, Hashimoto T, Ui N (1980) Purification and properties of acyl-CoA oxidase from rat liver. *J Biochem* **87**: 1735–1746.

Poll-The BT, Roels F, Ogier H, et al (1988) A new peroxisomal disorder with enlarged peroxisomes and a specific deficiency of acyl-Coa oxidase (pseudo-neonatal adrenoleukodystrophy). *Am J Hum Genet* **42**: 422–434.

Schepers L, Van Veldhoven PP, Casteels M, Eysen HJ, Mannaerts GP (1990) Presence of three acyl-CoA oxidases in rat liver peroxisomes. An inducible fatty acyl-CoA oxidase, a non-inducible fatty acyl-CoA oxidase and a non-inducible trihydroxycoprostanoyl-CoA oxidase. *J Biol Chem* **265**: 5242–5246.

Vamecq J, Vallee L, Fontaine M, Lambert D, Poupaert J, Nuyts J-P (1993) CoA esters of valproic acid and related metabolites are oxidized in peroxisomes through a pathway distinct from peroxisomal fatty and bile acyl-CoA β-oxidation. *FEBS Lett* **322**: 95–100.

van den Bosch H, Schutgens RBH, Wanders RJA, Tager J (1992) Biochemistry of peroxisomes. *Annu Rev Biochem* **61**: 157–197.

Van Veldhoven PP, Brees C (1994) Acyl-Coa oxidase activity measurements. In Latruffe N, Bugaut MP, eds. *Peroxisomes.* Berlin: Springer Laboratory, 17–23.

Van Veldhoven PP, Vanhove G, Vanhoutte F, et al (1991) Identification and purification of a branched chain fatty acyl-CoA oxidase. *J Biol Chem* **266**: 24676–24683.

Van Veldhoven PP, Vanhove G, Asselberghs S, Eyssen HJ, Mannaerts GP (1992) Substrate specificities of rat liver peroxisomal acyl-CoA oxidases: palmitoyl-CoA oxidase (inducible acyl-CoA oxidase), pristanoyl-CoA oxidase (non-inducible acyl-CoA oxidase) and trihydroxyco-prostanoyl-CoA oxidase. *J Biol Chem* **267**: 20065–20074.

Vanhove G, Van Veldhoven PP, Fransen M, et al (1993a) The CoA-esters of 2-methyl-branched chain fatty acids and of the bile acid intermediates di- and trihydroxycoprostanic acid are oxidized by one single peroxisomal branched chain acyl-CoA oxidase in human liver and kidney. *J Biol Chem* **268**: 10355–10364.

Vanhove G, Van Veldhoven PP, Eyssen HJ, Mannaerts GP (1993b) Mitochondrial short chain acyl CoA dehydrogenase of human liver and kidney can function as an oxidase. *Biochem J* **292**: 23–30.

Van Veldhoven PP, Van Rompuy P, Fransen M, de Béthune B, Mannaerts GP (1994a) Large-scale purification and further characterization of rat pristanoyl-CoA oxidase. *Eur J Biochem* **222**: 795–801.

Van Veldhoven PP, Van Rompuy P, Vanhooren JCT, Mannaerts GP (1994b) Purification and further characterization of peroxisomal trihydroxycoprostanoyl-CoA oxidase from rat liver. *Biochem J* **304**: 195–200.

Varanasi U, Chu R, Chu S, Espinosa R, LeBeau MM, Reddy JK (1994) Isolation of the human peroxisomal acyl-CoA oxidase gene: organization, promoter analysis, and chromosomal localization *Proc Natl Acad Sci USA* **91**: 3107–3111.

Wanders RJA, Romeyn GJ, Schutgens RBH, Tager JM (1989) L-Pipecolate oxidase: a distinct peroxisomal enzyme in man. *Biochem Biophys Res Commun* **164**: 550–555.

Wanders RJA, Schelen A, Feller N, et al (1990) First prenatal diagnosis of acyl-CoA oxidase deficiency. *J Inher Metab Dis* **13**: 371–374.

Zaitsu K, Ohkura Y (1980) New fluorogenic substrates for horse radish peroxidase: rapid and sensitive assays for hydrogen peroxide and the peroxidase. *Anal Biochem* **109**: 109–113.

J. Inher. Metab. Dis. 18 Suppl. 1 (1995) 135–154

Immunocytochemical localization of peroxisomal proteins in human liver and kidney

M. ESPEEL and G. VAN LIMBERGEN

Department of Anatomy, Embryology and Histology, University of Gent, Godshuizenlaan 4, B-9000 Gent, Belgium

Summary: The sample preparation and immunocytochemical methods for investigating the presence and subcellular localization of peroxisomal proteins (catalase, the three β-oxidation enzymes, alanine:glyoxylate aminotransferase and a peroxisomal membrane protein) in human liver biopsies are described. We present a protocol for immunolabelling on ultrathin and semithin sections from the same tissue block, with protein A–colloidal gold as a reporter system. For this purpose, the tissue is embedded in Unicryl, a hydrophilic acrylic resin that is cured by ultraviolet illumination at 2°C. The limitations and possibilities of the methods are discussed together with methodological problems. Cryostat sections of prefixed material should be used for the visualization by light microscopy of cytoplasmic catalase. It is emphasized that immunolabelling for catalase in formalin-fixed archival liver samples and in liver autopsy tissue (in the latter also for the peroxisomal β-oxidation enzymes) permits visualization of peroxisomes; this can be helpful in diagnosing an index case retrospectively.

The impaired peroxisomal functions in patients with a peroxisomal disorder originate from the absence, mislocalization or deficient activity of one or more peroxisomal (enzyme) protein(s), or from the inability at the organelle level to assemble import-competent peroxisomes. The occurrence and subcellular localization of a protein can be determined by immunocytochemistry, which provides information about the tissue *in situ*. Here the method differs essentially from immunoblotting (see Wanders et al 1995), in which the presence of a protein is determined in tissue homogenates or in subcellular fractions. For the purpose of immunocytochemistry, the tissue architecture is kept as intact as possible via chemical fixation.

Several immunolocalization studies of peroxisomal matrix and membrane proteins have been performed in cultured skin fibroblasts from peroxisomal disorder patients. In the approach presented here, biopsied liver tissue is used. The argument is that cultured fibroblasts do not necessarily express the functional defect(s) observed in the patient (for recent examples see Mandel et al 1994; Schutgens et al 1994; Roels et al 1995a). In addition, peroxisomal enzymes may be expressed in the liver only, e.g. alanine:glyoxylate aminotransferase.

All immunocytochemical methods consist essentially of two main steps: (1) binding of the antibody to the antigen, based on the specific recognition of epitope(s) in the antigen against which the antibody is directed; (2) visualization of the bound antibody via a reporter system. The immunocytochemical methodology for the localization of catalase and the peroxisomal β-oxidation enzymes in human control liver and kidney at the light- and electron-microscope level was explored by Litwin et al (1987, 1988). They recommended protein A–colloidal gold as a detection method in combination with silver enhancement for light-microscopic visualization. The same detection system was used for the immunolocalization of catalase, the peroxisomal β-oxidation enzymes, alanine: glyoxylate aminotransferase and peroxisomal membrane proteins in the liver of peroxisomal disorder patients and in fetal human liver (Cooper et al 1988; Danpure et al 1989, 1993, 1994; Espeel et al 1990a,b, 1991a,b, 1993, 1995a,b; Hughes et al 1992, 1993) and kidney (Espeel et al 1991a).

Protein A occurs in the cell wall of almost all strains of *Staphylococcus aureus*. It has a high binding affinity for the Fc part of immunoglobulins belonging primarily to the IgG type. The binding affinity varies widely with the animal species from which the IgG is derived. All rabbit, pig, guinea pig and human IgGs are excellent protein A binders. Chicken IgG does not bind protein A and only a mild affinity is found for some mouse and rat IgG subclasses (Griffiths 1993).

Silver enhancement of colloidal gold particles is an application of the principle of 'autometallography' (Danscher and Noergaard, 1985; Danscher et al 1987). This principle is based on the property of several metals — including gold — to function as catalysts in the electron transfer for the reduction of silver ions to metallic silver. As a result, in the presence of an electron donor (hydroquinone), a shell of reduced (metallic) silver is deposited at the surface of the gold particles before silver precipitates in the medium. The metal particles visualize their own presence.

We describe here the procedures for immunolocalization at the light- and electron-microscope level of catalase (EC 1.11.1.6), the peroxisomal β-oxidation enzymes acyl-CoA oxidase (EC 1.3.99.3), bi(tri)functional enzyme (enoyl-CoA hydratase (EC 4.2.1.17 and 3-hydroxyacyl-CoA dehydrogenase (EC 1.1.1.35)) and 3-ketoacyl-CoA thiolase (EC 2.3.1.16), alanine:glyoxylate aminotransferase (EC 2.6.1.4) and a peroxisomal membrane protein (the latter at the ultrastructural level only) in liver samples embedded in Unicryl (BioCell, Cardiff, UK) — a recently introduced hydrophilic acrylic resin (Scala et al 1992) — from control subjects and from peroxisomal disorder patients. During the FEBS Advanced Course the protocol was demonstrated on the bench for the immunodetection of catalase in human control liver with polyclonal antibodies (IgG fraction) raised in rabbits against bovine liver catalase (Rockland Laboratories, code No. 200-4151). In Sections 1 and 2.4, several questions and remarks raised by the course participants are dealt with. The procedures for immunostaining for catalase and the peroxisomal β-oxidation enzymes at the light-microscope level (paraffin and cryostat sections) have been described previously (Litwin et al 1988; Espeel et al 1990a); they are not presented in this text, but the approach is dealt with in Section 2.4.

1. SPECIMEN PREPARATION

1.1. Fixation

The samples are fixed in 4% commercial formaldehyde in 0.12 mol/L sodium cacodylate (pH 7.3) containing 1% calcium chloride (w/v) at ambient temperature. The samples are sent to our laboratory by private express mail and, depending upon the duration of transport, the fixation time is usually around 24 h. For optimal tissue processing, it is preferred that the biopsy is taken and delivered in the first half of the week.

The choice of this fixative is determined by the fact that it is also used for the diamino-benzidine incubation to demonstrate catalase activity (see Roels et al 1995b) and that it is suited, according to Litwin et al (1987), for the immunocytochemical detection of the antigens mentioned in the introduction.

Upon arrival of the sample, small tissue pieces (less than 0.5 mm thick) are cut with a razor blade from the biopsy cylinder. These are further fixed in 0.5% glutaraldehyde in 0.1 mol/L sodium cacodylate (pH 7.3) containing 1% calcium chloride (w/v) for 1 h at 4°C. Thereafter they are rinsed in 0.1 mol/L sodium cacodylate (pH 7.3) containing 1% calcium chloride (w/v) at 4°C for at least 1 h. Depending on the time of arrival, they may be kept in this buffer at 4°C for an overnight period or over the weekend.

Remark: As a rule, part of the biopsy is divided into serial chopper sections while it is in the formaldehyde fixative. The sections are processed in alternating order for immuno-cytochemistry and for diaminobenzidine cytochemistry. The preparation of chopper sections implies that the tissue is enrobed in agar. The agar often sticks around the sections; it should be removed with a fine needle under a stereomicroscope before embedding in Unicryl.

1.2 Processing for Unicryl embedding

Specimens are rinsed for 1 h at 4°C in 0.1 mol/L sodium cacodylate (pH 7.3) without calcium chloride. The tissue samples are then immersed in ammonium chloride (50 mmol/L in phosphate-buffered saline (PBS, for preparation see Section 2.1) for 30 min at 4°C to block the free aldehyde groups. Dehydration follows at 4°C over ethanol at 50%, 70%, 90% (30 min each) and 100% (2×30 min). Specimens are then impregnated with pure Unicryl (2×1 h at 4°C); finally they are left in Unicryl overnight at 4°C. **Caution:** *Unicryl is supplied as a ready-to-use mixture. Handling unpolymerized Unicryl requires protective measures: wear acrylate-resistant gloves and work under an efficient fume hood.*

The next morning specimens are brought into BEEM capsules that are held in an aluminium rack. The rack is placed in an ultraviolet light polymerization chamber (Agar UVF 35; described by Glauert and Young 1989) at 2°C equipped with two UV lamps (Philips TL82/05; emitting at 360 nm). The distance between the capsules and the lamps is 10 cm. Polymerization takes 5 days.

1.3 Microtomy

Semithin sections (2 μm) are cut with a glass knife and mounted on silanated glass slides (see Section 3 for the preparation of silanated glass slides). Ultrathin sections (70–80 nm) are made with a diamond knife and collected on Formvar-coated nickel grids (300 mesh). The area from which the ultrathin sections are made is selected from semithin $2 \mu m$ sections and counterstained with toluidine blue (1% (w/v) in distilled water, containing 2.5% (w/v) sodium carbonate; prior to use the staining solution is filtered through a $0.2 \mu m$ filter).

During ultramicrotomy, care should be taken that the front of the specimen block does not come into contact with the water bath of the diamond knife, so the water level must be set to a lower level than usually used for epoxy resin sections. In our experience, the Unicryl resin is relatively easy to section (few wrinkles) and the sections are stable in the electron beam.

2. PROTOCOLS FOR IMMUNOSTAINING

The protocols presented are used for all antigens, but dilution of the primary antibody has to be adapted.

2.1 Semithin sections

The procedure is applied on $2 \mu m$ sections of human liver, mounted on silanated glass slides (for the preparation of silanated glass slides, see Section 3). Before starting, it is necessary to make a scratch with a diamond point around the sections; otherwise it becomes impossible to localize the sections once the reagents are on the slide. Take care that the sections do not become dry during the procedure. Unless stated otherwise, all steps are done at room temperature. The glassware in which the silver enhancement solutions are mixed must be perfectly clean.

Procedure

(1) Sections are soaked in phosphate-buffered saline (PBS) (see note (a)); 5 min.
(2) Treat with Triton X-100 (1% in PBS); 5 min.
(3) Rinse in PBS; 5 min.
(4) Blocking of free aldehyde groups with 0.1 mol/L glycine in PBS; 15 min.
(5) Reduction of non-specific binding with 1% bovine serum albumin ((BSA) Fraction V; Sigma) in PBS; 60 min.
(6) Incubation with the antibody diluted in PBS (see note (b)); overnight at 4°C (in a humid chamber).
(7) Rinse with PBS; 3×10 min.
(8) Incubation with protein A–colloidal gold (5 nm) (note (c)) diluted in PBS (note (b)) containing 0.1% BSA (Fraction V; Sigma); 60 min.
(9) Rinse with PBS; 3×10 min.
(10) Stabilization (chemical fixation) of antigen–antibody–protein A–colloidal gold complex with 2.5% glutaraldehyde in PBS; 15 min.
(11) Rinse with PBS; 2×5 min.

(12) Rinse with double-distilled water; 3×5 min.
(13) Silver enhancement with IntenSE M (Amersham Life Sciences, code no. RPN 491) at room temperature for 4–6 min. An equal number of droplets from solutions A and B are mixed in a clean vial, immediately before use. Bring some droplets of the mixture onto the slides with a clean Pasteur pipette and ensure that all the sections are immersed.
(14) Rinse with double-distilled water; 3×2 min.
(15) Remove non-reduced silver with commercial photographic fixer solution (e.g. Hypam or Agefix diluted 1/20 in double-distilled water); 1 min. *According to the manufacturer's instructions this step is not required; however, in our experience additional silver precipitation may occur later in the non-fixed preparations. The photographic fixer is also applied to the cryostat and paraffin sections.*
(16) (i) Rinse with double-distilled water.
Optional (note (d))
   (ii) Counterstain with toluidine blue (0.5% (w/v) solution in 0.01 mol/L HCl; pH 3.2) for 90 s at 45°C.
   (iii) Rinse in two baths of tap water; 2×1 min.
   (iv) Rinse with distilled water for 5 min.
(17) Dry the slides at 45°C (1 h) and mount (Fluoromount Mountant; Gurr Product No. 36098).

Notes: (a) Phosphate-buffered saline (PBS) is prepared as a 10× concentrated stock solution (80.0 g NaCl, 2.0 g KCl, 14.4 g $Na_2HPO_4.2H_2O$ and 2.0 g KH_2PO_4 in 1 litre distilled water). It can be stored at room temperature for several weeks. Prior to use, the solution is diluted 1/10 with distilled water. Check and adjust pH to 7.35. (b) The optimal dilutions have to be determined. (c) The protein A–colloidal gold is prepared by Dr J. W. Slot (Laboratory of Cell Biology, Utrecht, the Netherlands) according to the procedure described in Slot and Geuze (1985). Protein A–colloidal gold probes are commercially available. They remain stable over several months when stored at 4°C (do not freeze!). Always ensure that, when pipetting the required volume of the protein A–colloidal gold complex, the micropipette tip is clean. To prevent overall contamination, it is recommended to aliquot the stock. (d) Immunostained sections that have not been counterstained can be examined by dark-field illumination or phase-contrast microscopy to better visualize the tissue and/or the peroxisomes (Figures 3–6).

2.2 Ultrathin sections

The procedure, which is essentially the same as for the semithin sections, is applied on ultrathin (70–80 nm) sections collected on Formvar-coated nickel grids (300 mesh). Solutions of the different reagents are presented as droplets on a sheet of Parafilm. To prevent evaporation during the incubation with the primary antibody, a 'wet chamber' is used (the Parafilm lies on the bottom of a Petri dish containing a small vial filled with water). The grids float on top of the droplet; they are transferred using a loop of stainless steel wire. The back side of the grid should not become covered with any of the reagents and the side with the sections should not become dry during the procedure.

Procedure

(1) Soak with PBS; 2×5 min.
(2) Blocking of free aldehyde groups with 50 mmol/L NH₄Cl in PBS, 20 min, and with 100 mmol/L glycine in PBS, 20 min.
(3) Incubation with 2% BSA (Fraction V; Sigma) in PBS; 45 min.
(4) Primary antibody incubation; overnight at 4°C (in humid chamber).
(5) Rinse with PBS; 3×10 min.
(6) Incubation with protein A colloidal gold (15 nm) diluted in a 1:1 (v/v) mixture of PBS/5% (w/v) dry skim-milk powder in distilled water (adjust pH to 7.35); 45 min.
(7) Rinse with PBS; 3×10 min.
(8) Stabilization (chemical fixation) of antigen – antibody – protein A – colloidal gold complex with 2.5% glutardialdehyde in PBS; 15 min.
(9) Rinse with PBS; 5 min.
(10) Rinse with double-distilled water; 2×5 min.
(11) Conventional contrasting of the sections for electron microscopy with uranyl acetate (30 min) and lead citrate (5 min).

Negative controls

Incubation with primary antibody is replaced by incubation with normal rabbit serum or the IgG fraction of normal rabbit serum.

2.3 Results in control liver

Semithin sections (Figures 1 – 6; 13)

Bright-field illumination: In human control liver, peroxisomes are visualized as small, brown-to-black granules, dispersed throughout the cytoplasm of the parenchymal cells after immunostaining against the peroxisomal matrix enzymes catalase, acyl-CoA oxidase, trifunctional enzyme, 3-ketoacyl-CoA thiolase and alanine:glyoxylate aminotransferase. Use built-in controls as a criterion to evaluate the specificity of the reaction and background level: (i) there are no immunoreactive granules over the sinusoids and over strands of connective tissue; (ii) inside the parenchymal cells, immunoreactive granules are absent over lipid droplets and the nucleus.

By fine focusing it can be seen that the gold – silver deposit is *only on the surface of the section* and not inside the section.

Phase-contrast illumination: The peroxisomes are visualized as blue granules against a grey background. The granules are better delineated than in bright-field microscopy.

Dark-ground illumination: The peroxisomes are seen as bright spots against a dark background. The spots are already detectable at low magnifications that do not permit one to see the organelles in the other illumination modes.

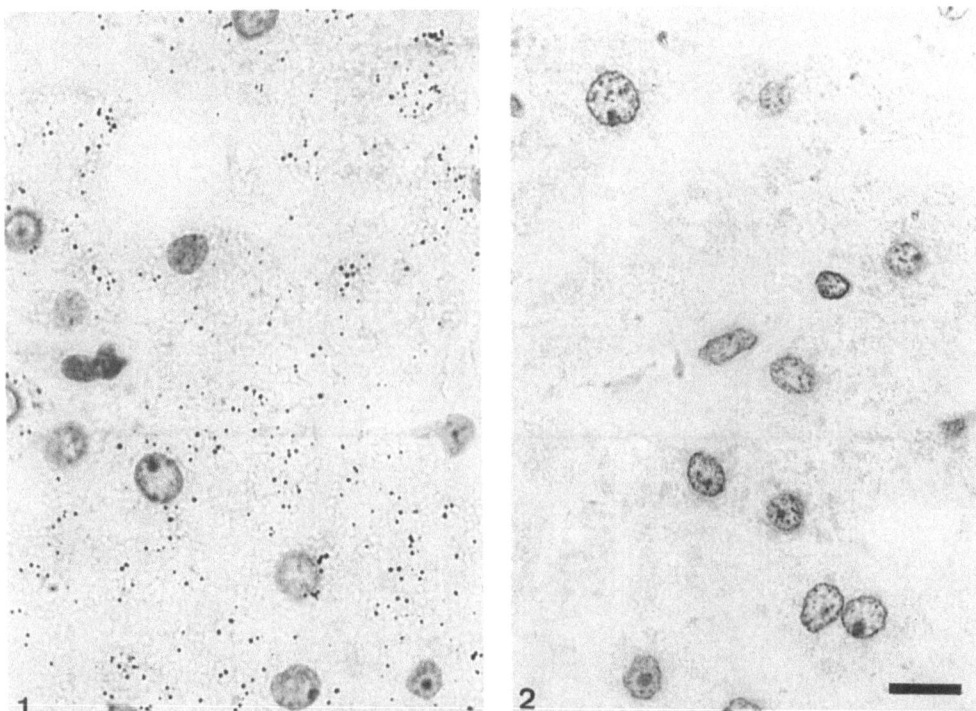

Figures 1 and 2 Semithin Unicryl section of human control liver, after immunostaining against catalase and counterstained with toluidine blue (**1**) and negative control incubation (**2**). The peroxisomes are visualized as small black granules throughout the parenchymal cells; the granules are absent in the negative control. Scale bar = 10 μm

Notes: (1) Especially in phase-contrast microscopy, granular structures can be seen not only at the surface of the section but throughout the whole section depth. These granules do not represent immunoreactive peroxisomes. They are also seen in the negative control incubations and in the sections counterstained with toluidine blue only. (2) Test the different illumination modes and compare their effect on peroxisome visualization. Dark-field illumination has an obvious enhancing effect on the visualization of the peroxisomes but concomitantly also of the background.

Ultrathin sections

In control liver, the colloidal gold particles are concentrated over the peroxisomal profiles after labelling for the above antigens (Figures 7 – 11). There is no significant label over any other subcellular structure. The negative control incubations do not show a labelling pattern; a few randomly dispersed gold particles are usually present throughout the section (Figures 12 and 20).

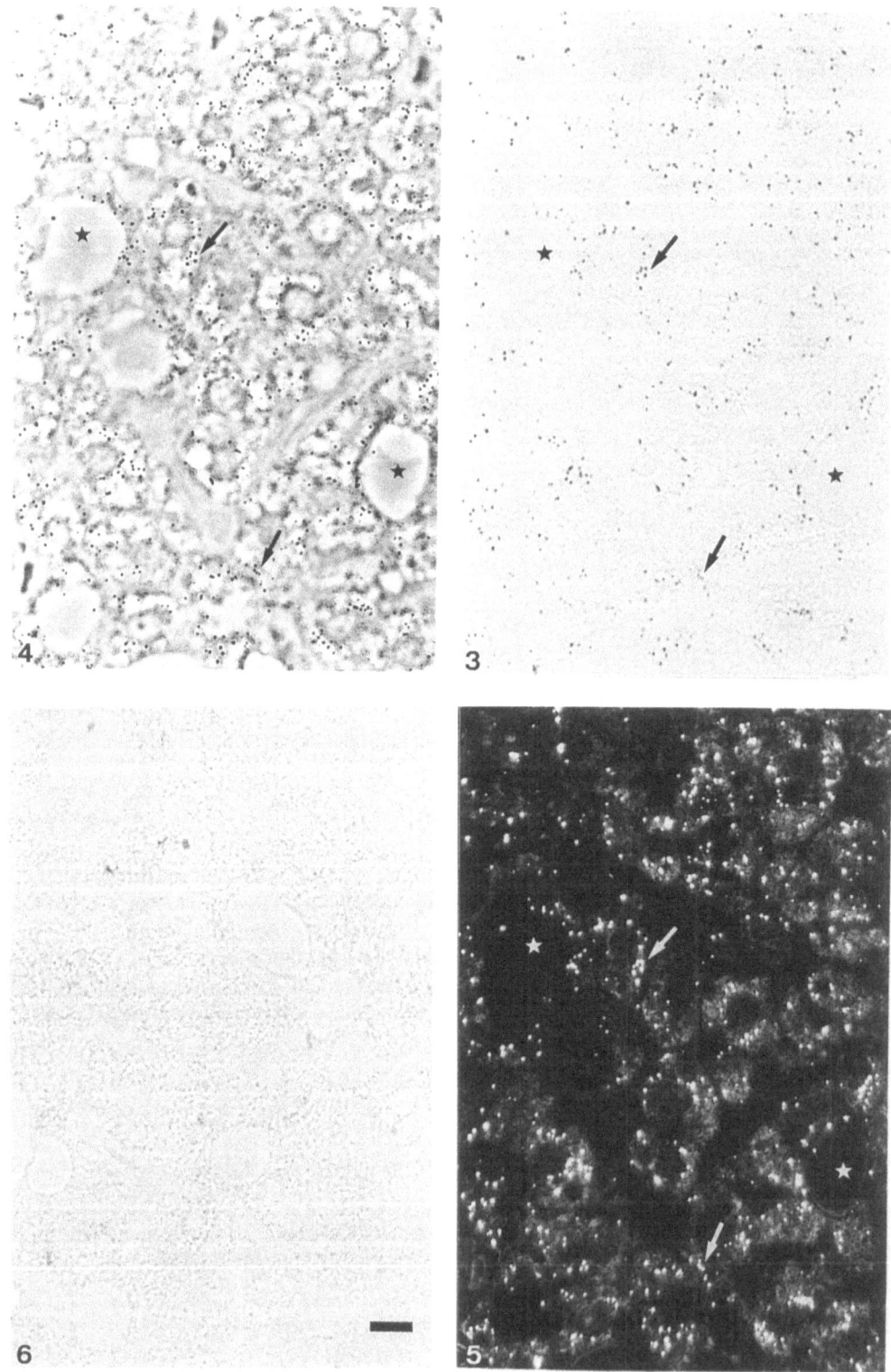

Remark: The ultrastructural detail in ultrathin acrylic resin sections differs markedly from that in conventional epoxy resin sections. The overall image has a lower contrast and the membranes of the organelles are poorly visualized, providing a less 'aesthetic' picture (term of Griffiths 1993); the mitochondrial cristae seem negatively stained after uranium and lead. In the liver, an additional characteristic is that the glycogen rosettes are not contrasted at all, leading to the image of an empty cytoplasm.

2.4 Comments and methodological problems

The use of semithin acrylic resin sections for immunodetection

Advantages: (1) Except for the silver enhancement on the semithin sections and the different diameter of the colloidal gold particles ($5\,\mu m$ versus $15\,\mu m$), the immunodetection method is similar for the semithin and ultrathin sections. As a result, the light-microscopic and ultrastructural images can be correlated at a detailed level. This has proved to be an advantage for the interpretation of some peculiar images: in some patients enlarged peroxisomes containing fat droplets and elaborate membrane invaginations are present; these inclusions, which are devoid of peroxisomal matrix, can be seen at the light-microscope level as a pale spot inside the organelle (compare Figures 14 and 15). Secondly, it is possible to first perform labelling on semithin sections and then to select an area for ultramicrotomy and proceed further on ultrathin sections.

(2) The labelling procedure does not require etching of the resin or trypsinization of the tissue, the latter being a critical step in unmasking the antigen in paraffin sections of formaldehyde-fixed liver (Litwin et al 1988). It seems that in the semithin acrylic resin sections the antigen is exposed — at the surface of the section — in a directly accessible and recognizable way for the antibody, without any pretreatment. This is an important advantage: on paraffin and cryostat sections of several control liver biopsies, peroxisomes were visualized only in the outer margin of the section, after immunostaining against catalase and the peroxisomal β-oxidation enzymes. This margin was sometimes only a few cells wide and the internal part of the section was unreactive. From diaminobenzidine cytochemistry in the same tissue block, it was clear that the peroxisomes were uniformly present in all the parenchymal cells. Litwin and co-workers have encountered similar problems in their formaldehyde-fixed samples, which they could resolve by adequate trypsinization (Jan Litwin, personal communication). In our experience, trypsinization did not show a beneficial effect in *all* the samples. Therefore, the phenomenon might lead to false negative results in the paraffin and cryostat sections in the case of a peroxisomal localization of the antigen. In the semithin and ultrathin sections the 'margin effect' was never observed. When the antigen is present in the *cytoplasm* (e.g. cytoplasmic catalase in

Figures 3–6 Semithin Unicryl section of human control liver (alcoholic liver steatosis) after immunostaining against alanine:glyoxylate aminotransferase without counterstaining, viewed in bright-field microscopy (**3**), phase-contrast microscopy (**4**), and dark-field microscopy (**5**). The asterisks indicate large fat droplets; the arrows indicate groups of peroxisomes for reference in the different illumination modes. There are no granules over the lipid droplets. The negative control incubation viewed in bright-field is presented in **Figure 6**; immunoreactive granules are absent. Scale bar = $10\,\mu m$

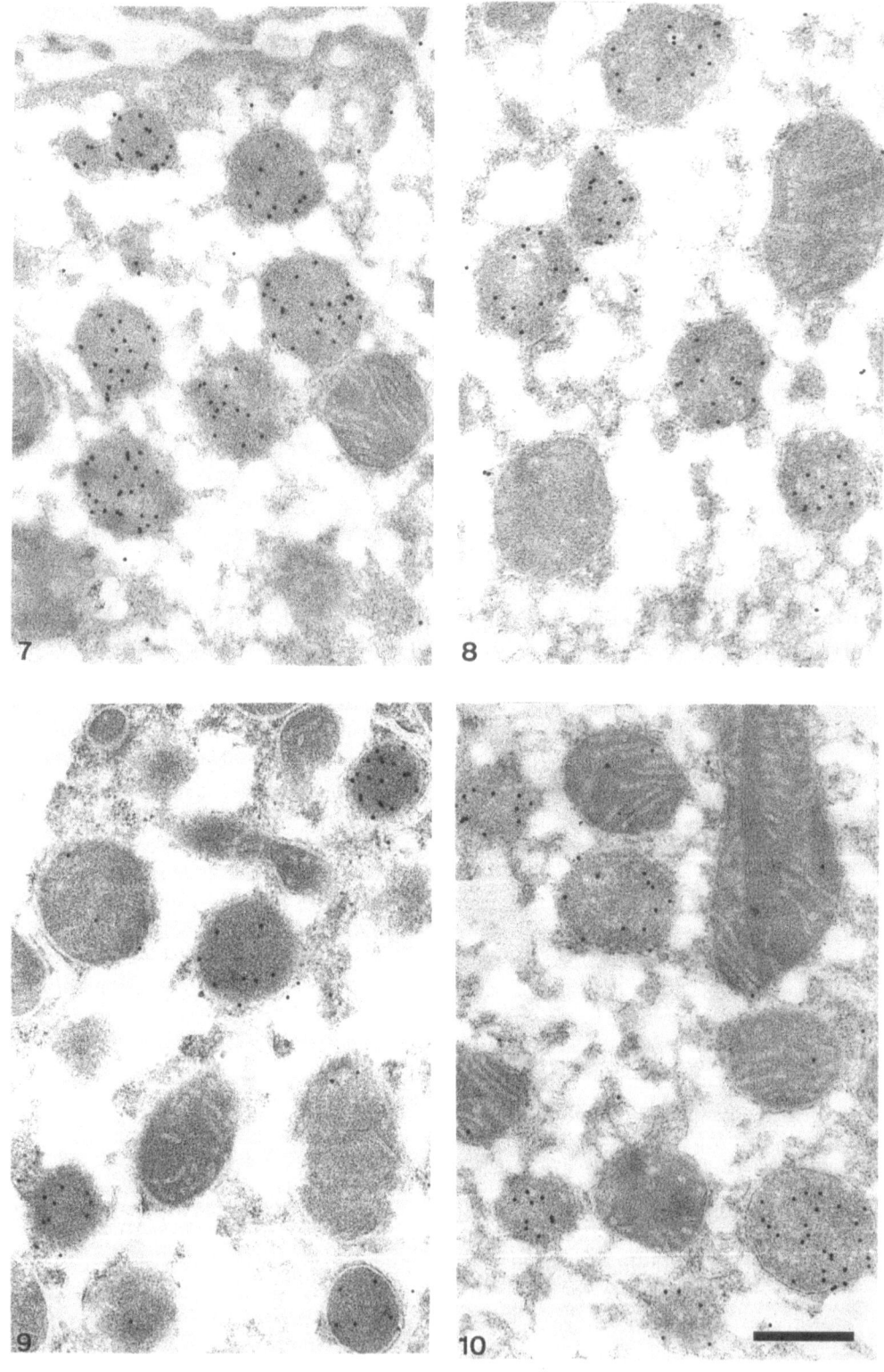

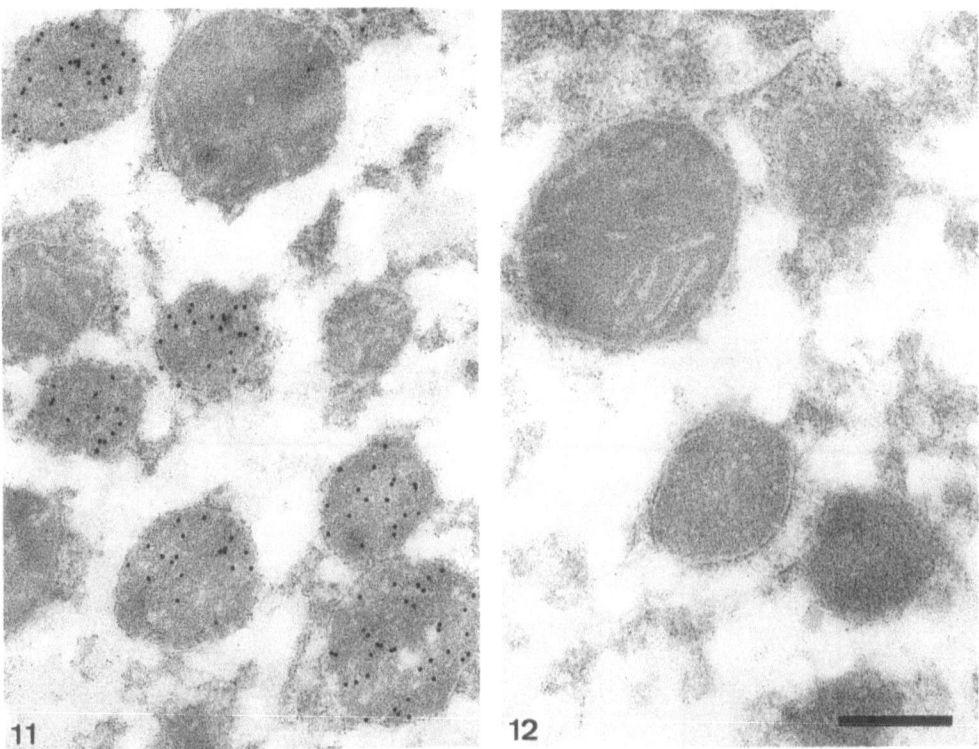

Figures 7–12 Immunodetection of catalase (**7**), acyl-CoA oxidase (**8**), trifunctional enzyme (**9**), 3-ketoacyl-CoA thiolase (**10**), and alanine:glyoxylate aminotransferase (**11**) in ultrathin Unicryl sections of human control liver, and a negative control incubation (**12**). In Figures 7–11 the peroxisomal profiles are labelled with gold particles; there is no distinct label in the negative control incubation (see also Fig. 17). Scale bar = 0.5 μm

the generalized peroxisomal disorder patients), the margin effect does not occur in the cryostat sections stained according to the previously published protocol (Espeel et al 1990a). Also in formalin-fixed archival samples and in autopsy samples, the margin effect was not observed.

(3) Since only peroxisomes at the section surface are detected, the number of visualized organelles does not depend on section thickness.

(4) In tissues embedded in acrylic resins, the antigenicity of many proteins is preserved over long periods; the resin behaves as a relatively inert medium. As a result, the embedded material can be examined retrospectively in view of new questions or with newly available antibodies. By embedding a control liver biopsy at regular intervals, closely matching positive control tissue is present throughout the sample collection.

(5) The mounted colloidal gold–silver immunostained sections (acrylic resin, paraffin and cryostat sections) remain stable over several years; there is no apparent loss or diffusion of the signal or of the tissue structure.

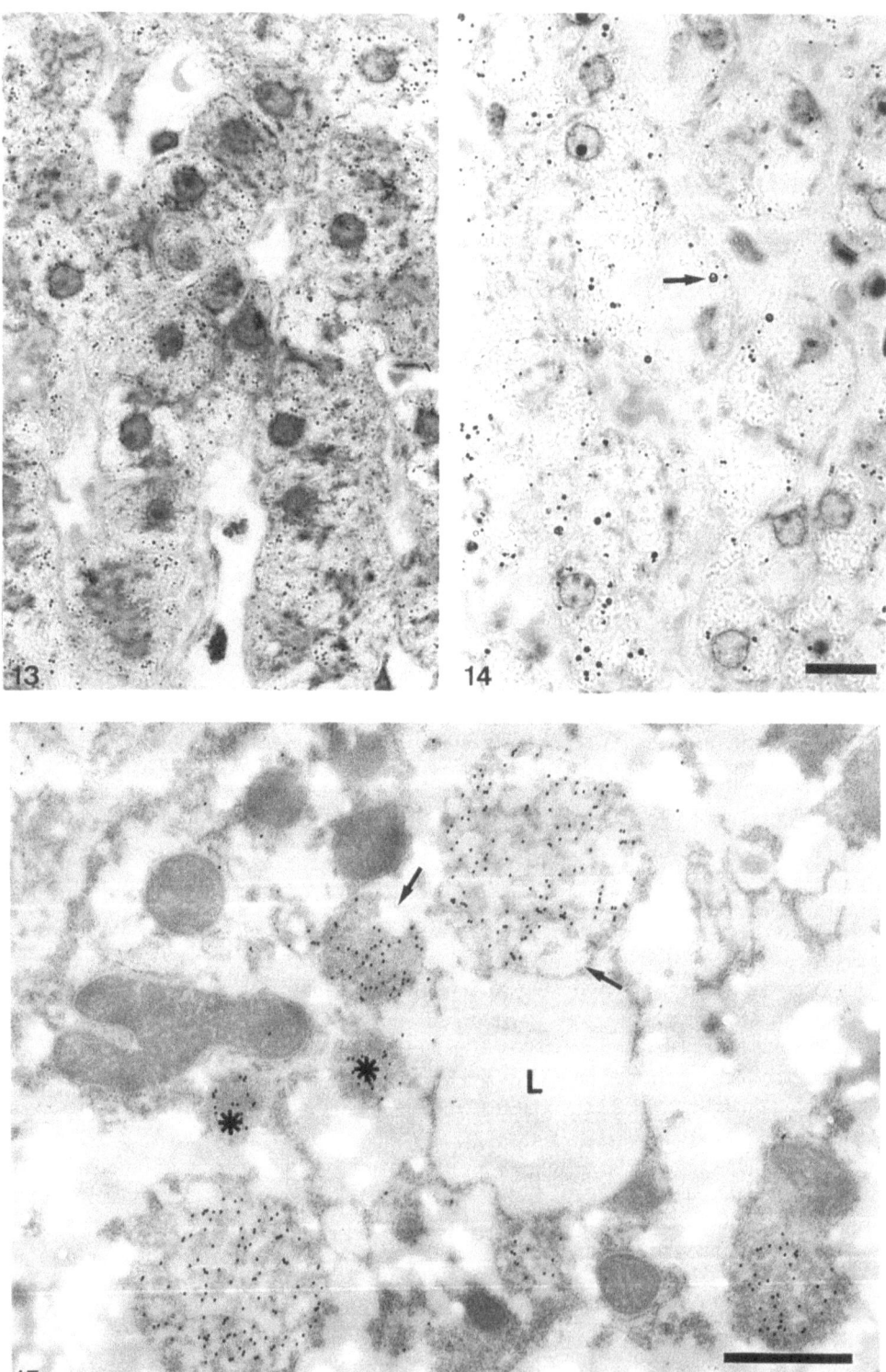

Disadvantages: (1) In the semithin 2 µm sections a weak to negative signal is obtained when an antigen is dispersed in the cytoplasm (e.g. catalase in patients with a generalized peroxisomal disorder). This is most probably related to the fact that the layer of colloidal gold particles (which are only on the section surface) is not dense enough for the silver to form a uniform deposit. In cryostat sections stained as described previously (Espeel et al 1990a), and in the ultrathin Unicryl sections, the cytoplasmic localization of catalase can be clearly demonstrated (Figures 16–20; Espeel et al 1993; Roels et al 1993); for its demonstration by staining for catalase activity see Roels et al (1995b).

(2) Because of the small size of the embedded tissue fragments, relatively small areas of tissue can be examined in one section. Several specimen blocks have to be sectioned.

(3) A silver/gold deposition on the resin, often just outside the tissue, may be seen occasionally. It does not occur in a consistent way: on the same slide it may be present on only some sections.

(4) The period between sampling the biopsy and the first immunostained semithin sections is about 2 weeks.

Limitations

The immunocytochemical detection relies on *the presence of the enzyme protein and not on its activity*. Even a small part of the molecule that still contains an epitope can give an immunopositive signal. Therefore, only the absence of a peroxisomal protein — as detected by immunocytochemistry — can be of diagnostic value. So far, we have found a normal peroxisomal localization of the three peroxisomal β-oxidation enzyme proteins in liver biopsies from three patients suffering from a peroxisomal β-oxidation defect who were clinically different from X-linked adrenoleukodystrophy (one case has been reported previously; in the kidney a peroxisomal localization of the three β-oxidation enzymes was also found (Espeel et al 1991a; Van Maldergem et al 1992)). Fibroblast complementation studies by Wanders et al (1992) revealed that in this patient the impaired peroxisomal β-oxidation was due to the inactivity of bi(tri)functional enzyme (R. J. A. Wanders, personal communication).

On the other hand, we could not detect immunoreactive enzyme protein in the peroxisomes of three cases: alanine:glyoxylate aminotransferase in two patients with primary hyperoxaluria type I, and acyl-CoA oxidase in a fetus from an interrupted pregnancy in the family described by Poll-The et al (1988). In a variant form of

Figures 13 and 14 Peroxisomes after immunolabelling for alanine:glyoxylate aminotransferase in semithin sections of a control liver (**13**) and of a chondrodysplasia punctata variant (**Fig. 14**; patient 4 in table 1 of Kerckaert et al 1995; biopsy at 11 years). In the patient's biopsy the peroxisomes are less numerous and many organelles are markedly enlarged. The arrow indicates an empty spot inside an enlarged peroxisome. Compare with Fig. 15 for the result at the ultrastructural level. Scale bar = 10 µm

Figure 15 Immunolabelling for alanine:glyoxylate aminotransferase in ultrathin sections of the specimen presented at the light-microscopic level in Fig. 14. The heterogenous size of the organelles, their reticular matrix and the empty places inside the matrix (arrows) are illustrated. Two normal-appearing organelles (asterisk) are also illustrated. L = lipid droplet. Scale bar = 1 µm

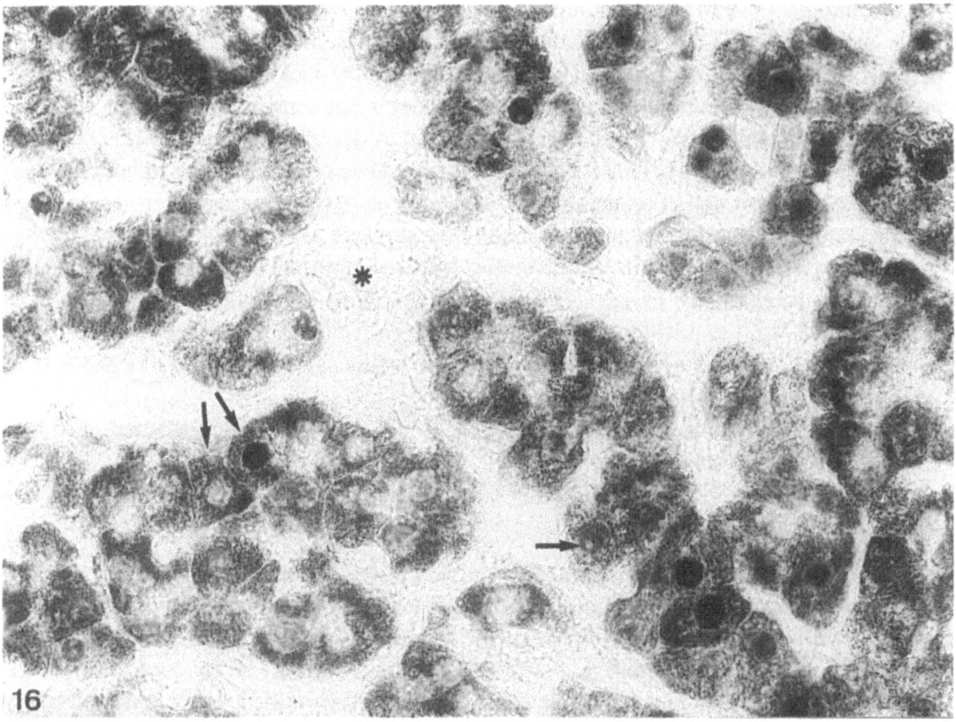

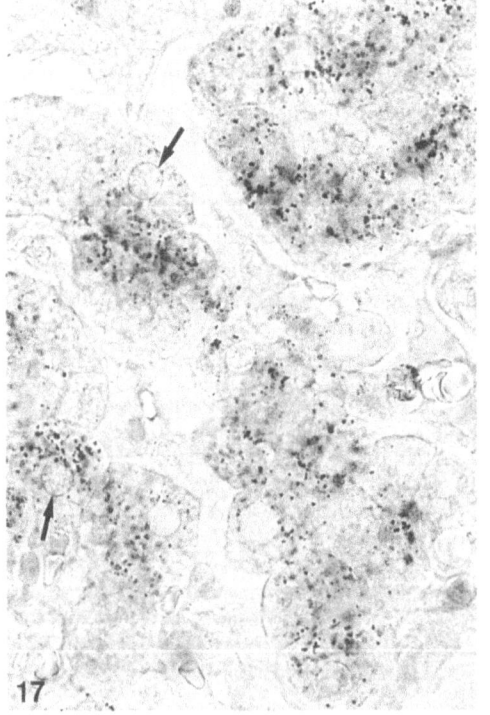

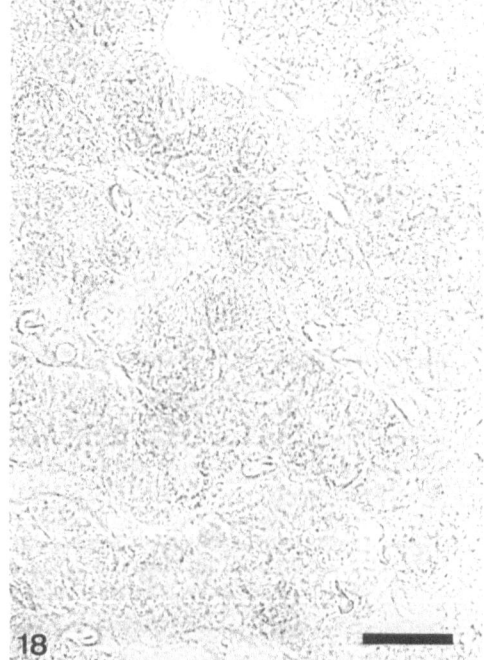

chondrodysplasia punctata described by Smeitink et al (1992) rare and extremely enlarged peroxisomes that were largely catalase-deficient were present in the liver biopsy. The organelles contained the three peroxisomal β-oxidation enzymes as well as alanine:glyoxylate aminotransferase; catalase was mainly localized in the cytoplasm and no normal catalase-containing peroxisomes were found (Espeel et al 1993). In two 'classical' rhizomelic chondrodysplasia punctata patients, Hughes et al (1992) also reported extremely enlarged and catalase-deficient peroxisomes in the liver.

In addition, no data are obtained by immunocytochemistry about the molecular processing of the reactive protein: for example, the unprocessed form of 3-ketoacyl-CoA thiolase (44 kDa) and the mature form (41 kDa) cannot be discriminated by our polyclonal antibodies.

Visualization of peroxisomes in autopsy and archival samples

It is useful to investigate archival and autopsy samples for the presence of peroxisomes, e.g. in order to document an index case retrospectively. In archival formaldehyde-fixed liver, immunocytochemistry for catalase antigen is recommended: peroxisomes could be identified in liver samples that had been kept in unbuffered formalin for 5 years. The presence of 'formalin-pigment', a characteristic brown-to-black deposit in tissues stored in formalin that becomes more pronounced after silver enhancement, may interfere with the immunopositive signal from the peroxisomes. Compare the image with unstained sections and with sections on which only the silver enhancement step was performed (Espeel et al 1991b; for images of formalin pigment see Roels et al, 1995b). Also, the immunoreactivity for alanine:glyoxylate aminotransferase may be preserved over several years (own unpublished observations).

In human autopsy liver, the three peroxisomal β-oxidation enzymes, in addition to catalase, are detectable up to at least 55 hours after death (Espeel et al 1990a). By immunolabelling against catalase and 3-ketoacyl-CoA thiolase, the hepatic peroxisomes could be visualized in a stillborn fetus at 26 weeks of gestation affected with X-linked recessive chondrodysplasia punctata (Van Maldergem et al 1991).

Human kidney

We have immunostained paraffin and cryostat sections of human kidney samples for catalase and the peroxisomal β-oxidation enzymes according to the protocol described previously (Espeel et al 1990a). On paraffin sections from a patient with a generalized

Figures 16–18 Immunolabelling for catalase (**16**) in 8 μm cryostat liver section of a generalized peroxisomal disorder patient. Catalase staining reveals a diffuse immunoreactivity in the cytoplasm of the parenchymal cells. The staining intensity varies between individual cells; there is no reaction in non-parenchymal tissue (asterisk). A positive reaction is also seen in a part of the nuclei; the reactivity differs between nuclei of adjacent cells (large arrows). The nucleolus is unreactive (small arrow). **Figure 17** shows catalase immunoreactivity in the peroxisomes (visualized as dark granules) of a control liver (cryostat section); there is no evidence for a reaction in the cytoplasm or in the nuclei (arrow). **Figure 18** shows the negative control incubation of the generalized peroxisomal disorder patient. Scale bar = 20 μm. The procedure for cryostat sections is described in Espeel et al (1990a); in addition, photographic fixer was applied (see protocol in Section 2.1, step 15)

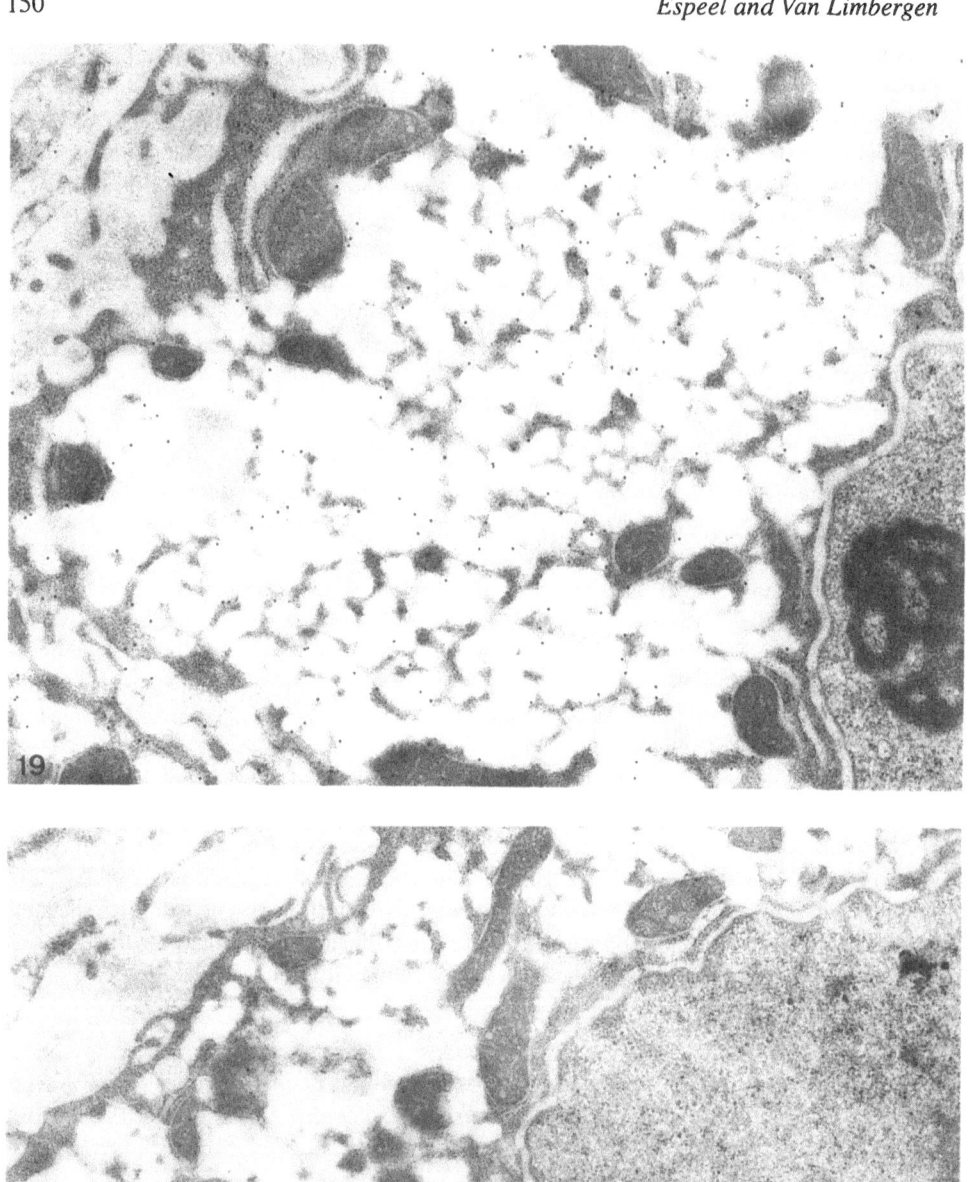

peroxisomal disorder, a distinct cytoplasmic localization of catalase together with a reaction in some nuclei was found in the epithelial cells of the proximal tubules (Figures 21 and 22). We found a normal peroxisomal localization of the three peroxisomal β-oxidation enzymes in the proximal tubules of the neonate with the peroxisomal β-oxidation defect (Espeel et al 1991a).

Also, in archival formalin-fixed kidney samples, the peroxisomes could be identified via immunostaining for catalase in cryostat and paraffin sections (unpublished observations).

Setup for biopsies from a patient with a suspected peroxisomal disorder

Each incubation experiment on semithin and ultrathin sections from a patient with a suspected peroxisomal disorder involves a simultaneous incubation of at least one control liver. For each analysis, we treat 10–12 slides/grids at most: primary antibodies for two antigens in two different dilutions and a negative control incubation in both patient and control samples. In case of an aberrant localization pattern or reaction, the procedures are repeated in combination with a parallel treatment of a human control sample. Other cytochemical stains are described by Roels et al (1995b).

3. PREPARATION OF SILANATED GLASS SLIDES

The immunostaining procedure requires a long incubation of the acrylate sections in aqueous solutions. Without the use of good adhesives, this results in the detachment of sections from the slides. Several products that provide maximal adhesion and minimal interference with the immunoreagents are commercially available.

A very efficient adhesion method that produces a low background is coating of the glass slides with organosilanes according to Henderson (1989). The binding mechanism is not known exactly; it is assumed that the organosilanes act as linker reagent (either ionically or covalently) between aldehyde, amino and ketone groups in the tissue and hydroxyl groups at the glass surface. There is no apparent reaction with silver solutions, cationic and anionic dyes, fat-soluble dyes, Schiff reagent or antibodies. Sections on silane-coated slides do not detach after digestion with amylase, trypsin, pronase, pepsin, protease or microwave staining. Coating with aminoalkylsilanes provides superior tissue adhesion — except for tissues fixed with mercury-containing solutions — and less background reaction than with slides coated with albumen, chrome-gelatin or poly-L-lysine (Henderson 1989).

Procedure

(1) Clean glass slides are dipped in a solution of 2% 3-aminopropyltriethoxysilane (APS; A-3648 Sigma, Munich) in acetone for 5 min at room temperature.

Figures 19 and 20 Immunolocalization of catalase in a patient with a generalized peroxisomal disorder; ultrathin Unicryl section (**19**). Labelled peroxisomes are absent. The gold particles are distributed over the cytoplasm and nucleoplasm but not the nucleolus; the perinuclear cisterna, mitochondrial profiles, sinusoidal lumen and space of Disse are relatively devoid of gold particles, similar to the negative control incubation (**20**). Scale bar = 1 μm

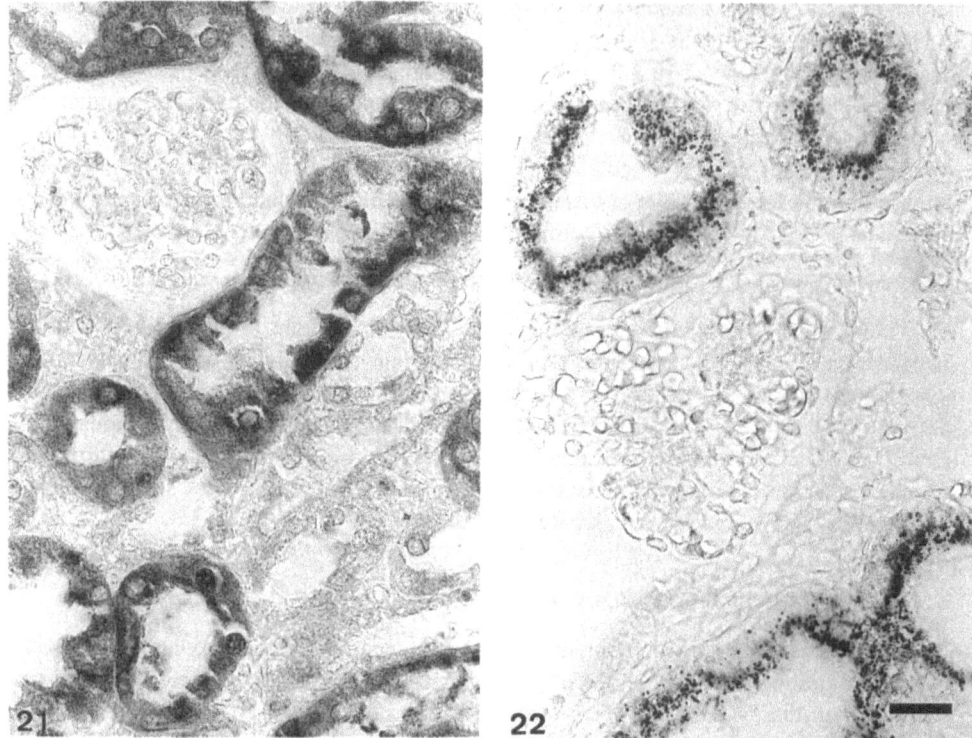

Figures 21 and 22 Immunostaining for catalase in the kidney cortex of a patient with a generalized peroxisomal disorder, paraffin section (**21**). A diffuse staining in the cytoplasm and some nuclei of the epithelial cells in the proximal tubules is observed; there is no punctate staining pattern reflecting a peroxisomal localization of catalase as in the control kidney (**22**; cryostat section). Scale bar = $20\,\mu$m

(2) The slides are immersed in deionized water for 2×5 min.
(3) Leave the slides to air dry or dry at 40°C for 2 days.
(4) When completely dry, the slides are stored dust-free at room temperature.

Four sections are placed in a droplet of distilled water on the glass slide. The slides are then left to dry at 40°C for at least 2 days prior to immunostaining.

Remark: The adhesive properties are not lost from slides stored for 6 months, or for sections picked up and stored for 6 months (Henderson 1989).

ACKNOWLEDGEMENTS

The assistance of Guido De Pestel, Robert De Smedt, Raf Mortier and Noël Verweire in preparing and performing the practical session during the training course is gratefully acknowledged. The control liver biopsies were provided by Professor Dr J. Versieck from the Gastroenterology Department of the Universitair Ziekenhuis, Gent.

We are indebted to Professor Dr J. Tager, Dr R. B. H. Schutgens, Dr R. J. A. Wanders (Amsterdam, the Netherlands), Professor Dr T. Hashimoto (Nagano, Japan), Dr M. Santos (Santiago, Chile) and Professor Dr A. Völkl (Heidelberg, Germany) for kindly providing the antibodies.

REFERENCES

Cooper PJ, Danpure CJ, Wise PJ, Guttridge KM (1988) Immunocytochemical localization of human hepatic alanine/glyoxylate aminotransferase in control subjects and patients with primary hyperoxaluria type I. *J Histochem Cytochem* 36: 1285–1294.

Danpure J, Cooper PJ, Wise PJ, Jennings PR (1989) An enzyme trafficking defect in two patients with primary hyperoxaluria type 1: Peroxisomal alanine/glyoxylate aminotransferase rerouted to mitochondria. *J Cell Biol* 108: 1345–1352.

Danpure CJ, Purdue PE, Fryer P et al (1993) Enzymological and mutational analysis of a complex primary hyperoxaluria type I phenotype involving alanine : glyoxylate aminotransferase peroxisome-to-mitochondrion mistargeting and intraperoxisomal aggregation. *Am J Hum Genet* 53: 417–432.

Danpure CJ, Fryer P, Griffiths S et al (1994) Cytosolic compartmentalization of hepatic alanine/glyoxylate aminotransferase in patients with aberrant peroxisomal biogenesis and its effect on oxalate metabolism. *J Inher Metab Dis* 17: 27–40.

Danscher G, Noergaard JO (1985) Ultrastructural autometallography: a method for silver amplification of catalytic metals. *J Histochem Cytochem* 33: 706–710.

Danscher G, Noergaard JO. Baatrup E (1987) Autometallography: tissue metals demonstrated by a silver enhancement kit. *Histochemistry* 86: 465–469.

Espeel M, Hashimoto T, De Craemer D, Roels F (1990a) Immunocytochemical detection of peroxisomal β-oxidation enzymes in cryostat and paraffin sections of human postmortem liver. *Histochem J* 22: 57–62.

Espeel M, Jauniaux E, Hashimoto T, Roels F (1990b) Immunocytochemical localization of peroxisomal β-oxidation enzymes in human fetal liver. *Prenat Diagn* 10: 349–357.

Espeel M, Roels F, De Craemer D et al (1991a) Peroxisomal localization of the immunoreactive β-oxidation enzymes in a neonate with a β-oxidation defect. Pathological observations in liver, adrenal cortex and kidney. *Virchows Archiv A Pathol Anat* 419: 301–308.

Espeel M, Rehder H, Hashimoto T, Roels F (1991b) Visualization of peroxisomes in archival human liver by catalase immunocytochemistry. *Micron Microsc Acta* 22: 259–260.

Espeel M, Heikoop JC, Smeitink JAM et al (1993) Cytoplasmic catalase and ghost-like peroxisomes in the liver from a child with atypical chondrodysplasia punctata. *Ultrastruct Pathol* 17: 623–636.

Espeel M, Roels F, Giros M et al (1995a) Immunolocalization of a 43 kDa peroxisomal membrane protein in the liver of patients with generalized peroxisomal disorders. *Eur J Cell Biol* 67: 319–327.

Espeel M, Mandel H, Poggi F et al (1995b) Peroxisome mosaicism in the livers of peroxisomal deficiency patients. *Hepatology* 22: 497–504.

Glauert M, Young RD (1989) The control of temperature during polymerization of Lowicryl K4M: there *is* a low-temperature embedding method. *J Microsc* 154: 101–113.

Griffiths G (1993) *Fine Structure Immunocytochemistry*. Berlin: Springer Verlag.

Henderson C (1989) Aminoalkylsilane: an inexpensive, simple preparation for slide adhesion. *J Histotechnol* 12: 123–124.

Hughes JL, Poulos A, Crane DI, Chow CW, Sheffield LJ, Silence D (1992) Ultrastructure and immunocytochemistry of hepatic peroxisomes in rhizomelic chondrodysplasia punctata. *Eur J Pediatr* 151: 829–836.

Hughes JL, Crane DI, Robertson E, Poulos A (1993) Morphometry of peroxisomes and immuno-localisation of peroxisomal proteins in the liver of patients with generalized peroxisomal disorders. *Virchows Arch A Pathol Anat* 423: 459–468.

Kerckaert I, De Craemer D, Van Limbergen G (1995) Practical guide for morphometry of human peroxisomes on electron micrographs. *J Inher Metab Dis* 18 (**Suppl. 1**): 172–180.

Litwin JA, Völkl A, Müller-Höcker J, Hashimoto T, Fahimi HD (1987) Immunocytochemical localization of peroxisomal enzymes in human liver biopsies. *Am J Pathol* **128**: 141–150.

Litwin J, Völkl A, Stachura J, Fahimi D (1988) Detection of peroxisomes in human liver and kidney fixed with formalin and embedded in paraffin: the use of catalase and lipid β-oxidation enzymes as immunocytochemical markers. *Histochem J* **20**: 165–173.

Mandel H, Espeel M, Roels F et al (1994) A new type of peroxisomal disorder with variable expression in liver and fibroblasts. *J Pediatr* **125**: 549–555.

Poll-The BT, Roels F, Ogier H et al (1988) A new peroxisomal disorder with enlarged peroxisomes and a specific deficiency of acyl-CoA oxidase (pseudo-neonatal adrenoleukodystrophy). *Am J Hum Genet* **42**: 422–434.

Roels F, Espeel M, Poggi F, Mandel H, Van Maldergem L, Saudubray JM (1993) Human liver pathology in peroxisomal diseases: a review including novel data. *Biochemie* **75**: 281–292.

Roels F, Espeel M, Mandel H et al (1995a) Cell and tissue heterogeneity in peroxisomal patients. In Wanders RJA, Schutgens RBH, eds. *Peroxisomal Disorders in Relation to Functions and Biogenesis of Peroxisomes.* Amsterdam: Elsevier, in press.

Roels F, De Prest B, De Pestel G (1995b) Liver and chorion cytochemistry. *J Inher Metab Dis* **18** (**Suppl. 1**): 155–171.

Scala C, Cenacchi G, Ferrari C, Pasquinelli G, Preda P, Manara GC (1992) A new acrylic resin formulation: a useful tool for histological, ultrastructural and immunocytochemical investigations. *J Histochem Cytochem* **40**: 1799–1804.

Schutgens RBH, Wanders RJA, Jakobs C et al (1994) A new variant of Zellweger syndrome with normal peroxisomal functions in cultured fibroblasts. *J Inher Metab Dis* **17**: 319–322.

Slot JW, Geuze HJ (1985) A new method of preparing gold probes for multiple-labeling cytochemistry. *Eur J Cell Biol* **38**: 87–93.

Smeitink JAM, Beemer FA, Espeel M et al (1992) Bone dysplasia associated with phytanic acid accumulation and deficient plasmalogen synthesis: a peroxisomal entity amenable to plasmapheresis. *J Inher Metab Dis* **15**: 377–380.

Van Maldergem L, Espeel M, Roels F et al (1991) X-Linked recessive chondrodysplasia punctata with XY translocation in a stillborn fetus. *Hum Genet* **87**: 661–664.

Van Maldergem L, Espeel M, Wanders R et al (1992) Neonatal seizures and hypotonia with elevation of very long chain fatty acids, normal bile acids, normal fatty acyl-CoA oxidase and intraperoxisomal localization of the three β-oxidation enzymes: a new peroxisomal disease? *Neuromusc Disord* **2**: 217–224.

Wanders RJA, Van Roermund CWT, Brul S, Schutgens RBH, Tager JM (1992) Bifunctional protein deficiency: identification of a new type of peroxisomal disorder in a patient with an impairment in peroxisomal β-oxidation of unknown etiology by means of complementation analysis. *J Inher Metab Dis* **15**: 389–391.

Wanders RJA, Dekker C, Ofman R, Schutgens RBH, Mooijer P (1995) Immunoblot analysis of peroxisomal proteins in liver and fibroblasts from patients. *J Inher Metab Dis* **18** (**Suppl. 1**): 101–112.

J. Inher. Metab. Dis. 18 Suppl. 1 (1995) 155–171
© SSIEM and Kluwer Academic Publishers.

Liver and chorion cytochemistry

F. Roels*, B. De Prest and G. De Pestel
Department of Human Anatomy, Embryology and Histology, University of Gent, Godshuizenlaan 4, 9000 Gent, Belgium

*Correspondence

Summary: Microscopic visualization of peroxisomes in chorionic villus cyto-trophoblast and in biopsy and autopsy samples of liver and kidney, the presence of enlarged liver macrophages containing lipid droplets insoluble in acetone and n-hexane as well as polarizing inclusions formed by stacks of trilamellar sheets are of diagnostic value in peroxisomal disorders. Methods are presented for evaluating these structures by light microscopy; trilamellar inclusions are only detected by electron microscopy. Macrophage features are preserved in archival paraffin blocks. In adrenal cortex, insoluble lipid, polarizing inclusions and trilamellar structures should be looked for. The stains are easily reproducible, and all reagents are commercially available.

VISUALIZATION OF PEROXISOMES

Aims and rationale

Peroxisomes are visualized by light and electron microscopy with the reaction for catalase activity, a peroxisomal marker enzyme. The reagent is diaminobenzidine (DAB), which has four reactive amino groups; when oxidized, the latter link DAB molecules and form an insoluble polymer that precipitates in position:

$$n\text{DAB} + n\text{H}_2\text{O}_2 \xrightarrow{\text{catalase}} \text{DAB polymer} + 2n\text{H}_2\text{O}$$

This reaction is catalysed by the *peroxidatic* activity of catalase, which is enhanced by prior fixation in aldehydes (Roels et al 1975). All peroxidases catalyse the same reaction, which explains why granules of blood neutrophils and eosinophils, which are often present in liver sections, also stain. Haemoglobin has peroxidatic activity (which is thermostable), and consequently erythrocytes always show reaction product (Figure 1). Sometimes lipofuscin granules (secondary lysosomes) demonstrate a thermostable reaction which may be due to traces of iron or copper. In addition, lipofuscin often binds osmium tetroxide, and by light microscopy such granules should be distinguished from peroxisomes. The latter are nearly always round, of similar size, and evenly distributed over the liver lobule. In contrast, lipofuscin granules are angular, larger, of various sizes and often display heterogeneous contents (Roels and Goldfischer 1979) and are easily recognized by electron microscopy; they are concentrated in one region, usually around the central vein (Figure 1).

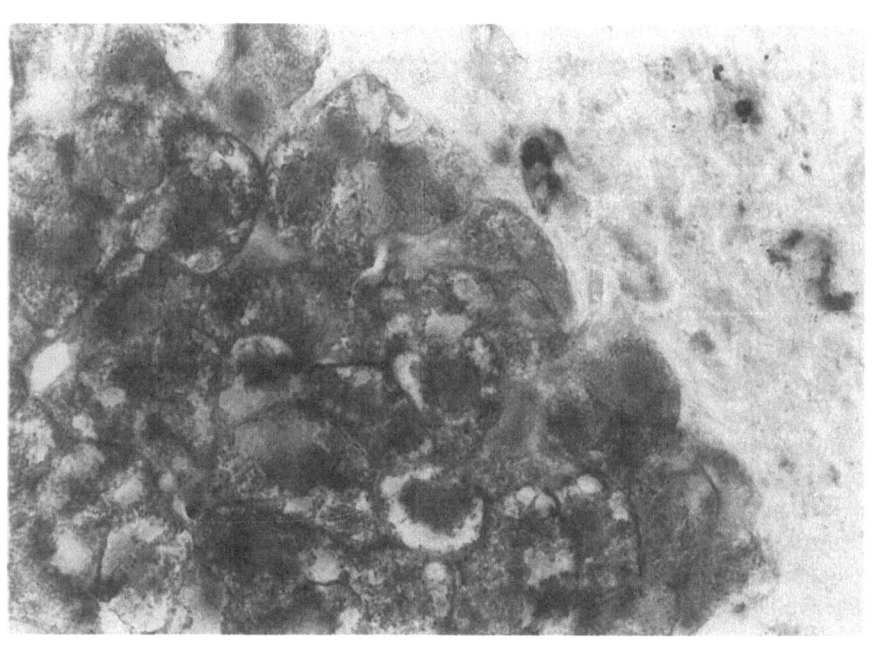

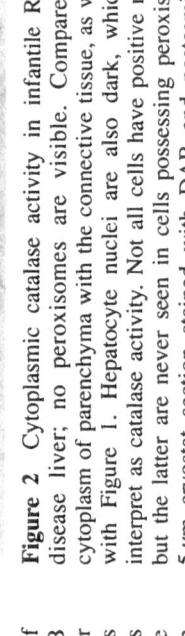

Figure 2 Cytoplasmic catalase activity in infantile Refsum disease liver; no peroxisomes are visible. Compare dark cytoplasm of parenchyma with the connective tissue, as well as with Figure 1. Hepatocyte nuclei are also dark, which we interpret as catalase activity. Not all cells have positive nuclei, but the latter are never seen in cells possessing peroxisomes. 5 μm cryostat section stained with DAB and postosmicated.

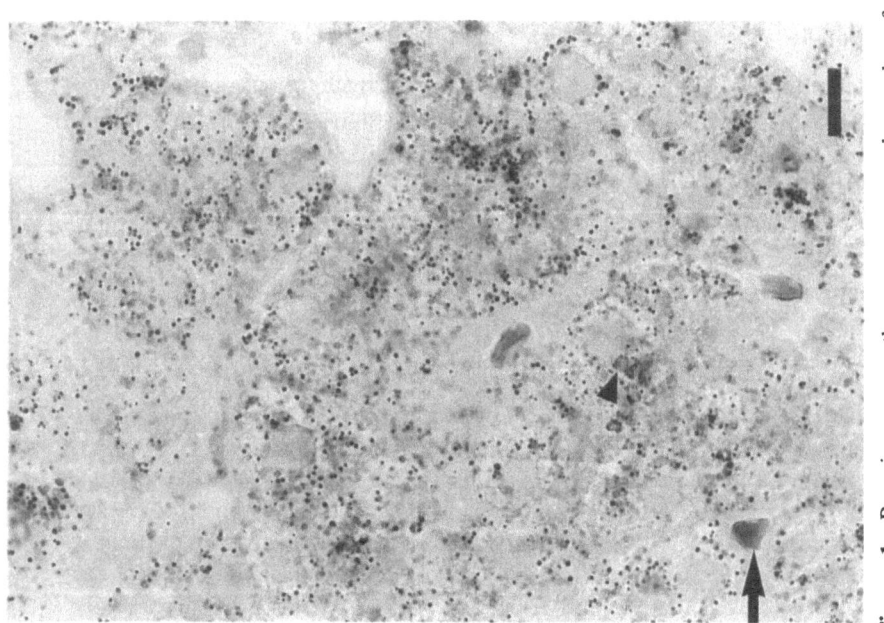

Figure 1 Peroxisomes are the numerous round granules of similar size seen in all hepatocytes of a control liver after DAB staining for catalase activity and post-osmication. Larger structures of irregular shape (arrowhead) are lipofuscin granules which are present in a few cells only; their number varies considerably between individuals but their distribution over the liver lobule is always uneven. Three stained erythrocytes are seen

In human kidney cortex, peroxisomes are large and numerous in the proximal convoluted tubule as well as in its pars recta. In the distal convoluted tubule and collecting segment, much smaller and fewer granules may be detected by light microscopy (Roels and Goldfischer 1979).

Fixation, in addition to improving the DAB reaction by catalase, immobilizes this enzyme; this is particularly important where catalase is free in the cytoplasm, as observed in liver parenchyma without peroxisomes (for example in the Zellweger syndrome (McKusick 214100) and infantile Refsum disease (McKusick 266510); Roels et al 1993b; Espeel et al 1993; Espeel and Van Limbergen, 1995). Freezing and prolonged storage in buffer favour diffusion of enzyme from fixed material.

Fast penetration of the fixative is the first important step, and requires very thin (< 1 mm) slices to be cut from a surgical biopsy; needle biopsies and chorionic villus samples can be immersed *in toto*. Glutaraldehyde is better than formaldehyde for immobilization of proteins but it penetrates slowly and decreases the sensitivity of catalase stain; it should not be used for this purpose in human material.

Catalase activity is lost by prolonged fixation and by embedding in paraffin or plastic; however, such liver and kidney specimens are still useful for immunolocalization of catalase (Espeel and Van Limbergen, 1995).

In autopsy material, catalase activity can be localized up to 48 h after death. Postmortem changes of peroxisomes have been studied by De Craemer et al (1990).

The DAB polymer is brown; it can bind osmium when treated with OsO_4, which increases the electron density and improves visibility on light microscopy also. DAB is also oxidized by cytochrome c + cytochrome oxidase, and may be used to visualize mitochondria (Novikoff and Goldfischer 1969; Roels 1974); this reaction is effectively inhibited by KCN (10^{-3} mol/L) added to the catalase medium.

The alterations of peroxisomal size, number and shape in peroxisomal disorders, as well as their absence in the Zellweger cerebrohepatorenal syndrome and in infantile Refsum disease, have been reviewed recently (Dimmick and Applegarth 1993; Roels et al 1993b, 1995). When catalase-containing organelles are absent, catalase is usually present in the cytoplasm; this can be visualized in many cases by DAB staining (Figure 2), but immunocytochemical localization in cryostat sections is more sensitive (Roels et al 1993b; Espeel and Van Limbergen, 1995). A third condition encountered is mosaicism of peroxisome distribution in the liver, i.e. normal cells adjacent to an area of parenchyma without any peroxisomes; the latter cells contain cytoplasmic catalase. Biochemically such patients present with a deficiency of multiple peroxisomal functions (Espeel et al 1995; Mandel et al 1994; Roels et al 1995). Obviously this condition can be recognized only when microscopy is performed.

Procedure

Fixative
Buffer:
    12.95 g sodium cacodylate (sodium dimethylarsenate, $(CH_3)_2AsO_2Na$, MW 214) in 450 ml double-distilled water
    5 g calcium chloride (final concentration = 1%)
Bring pH to 7.4 with 1 mol/L HCl and add double-distilled water to 500 ml (equivalent to 0.12 mol/L Na salt).

Formaldehyde, commercial solution 35%:   For 17.5 ml fixative, add 2 ml to 15.5 ml of buffer immediately before fixation (final concentration = 4%).

Fixation: Fixation at room temperature (22–24°C) for 18–24h is ideal for localization of catalase activity.

Liver and kidney: Cut unfrozen 50 μm chopper sections prior to rinsing; collect these in 1% $CaCl_2$ and incubate at once. For the preparation of cryostat sections, rinse a liver fragment briefly in 13% saccharose and freeze; collect 5–7 μm or 50 μm frozen sections in saccharose and incubate at once. Thick cryostat sections can be substituted for chopper sections if too little material is available.

Chorionic villi (first trimester):   Fixation time can be shorter, i.e. 12–24h, not longer. If the investigation is very urgent, fixation time can be reduced to 1h if 5% (v/v) absolute ethanol is added to the fixative.

After fixation, rinse briefly in saccharose. Villi are incubated *in toto*, but large fragments are reduced with a razor blade.

Incubation for localization of catalase activity
- *Buffer pH 10.5 (Theorell–Stenhagen):*

Phosphoric acid 0.33 mol/L	2 ml
Citric acid 0.33 mol/L	2 ml
Boric acid, H_3BO_3	70 mg
Sodium hydroxide 1 mol/L	7.3 ml
Distilled water	to 100 ml

  This mixture can be kept in a closed bottle for a few weeks.
- *Diaminobenzidine·4HCl:*   For 10 ml incubation medium, dissolve 20 mg in a few drops of double-distilled water; then add 10 ml of Theorell buffer. Check pH; if necessary bring to 10.5 with 1 mol/L and 0.1 mol/L NaOH.
- Add 1 mg KCN (cytochrome oxidase inhibitor) to 10 ml buffered DAB for the standard medium; to control medium (inhibition of catalase) add 200 mg KCN.
- Villi, or cryostat sections (5–7 μm), 50 μm unfrozen chopper sections or cryostat sections of formaldehyde-fixed liver or kidney are immersed in 10 ml DAB solution. In addition, each vial contains two cryostat sections (5 μm) of normal mouse liver, as positive control. Subsequently H_2O_2 is added: 0.1 ml of a 0.5% solution (final concentration = 0.005%) for villi and thin sections (which will be mounted); 0.2 ml of 0.5% solution (final concentration = 0.01%) for thick sections (which will be embedded).
- Incubate for 3h at room temperature (~22°C) while shaking.
- Rinse three times for 5 min each in 13% saccharose.
- Subsequent treatment is as follows:

Liver or kidney sections

Osmicate as follows.
- *Thin sections:* 1% OsO_4 in double-distilled water, 10 min or longer; rinse three times in double-distilled water; counterstain nuclei for 5s with 2% methyl green (purified with chloroform) in double-distilled water, dehydrate fast, and mount. Mouse liver does not require osmication; after double-distilled water, mount in water-soluble medium.

- *50 μm sections for light-microscopic demonstration of cytoplasmic catalase:* 1% aqueous OsO_4, 1–1.5 h; rinse, dehydrate and embed in plastic (Epon, LX). Examine 4 μm sections and compare parenchyma to bile duct and extracellular matrix; as well as to the parenchyma in the control stained in the presence of high cyanide concentration.
- *50 μm sections for electron microscopy:* Prepare before use:

4% OsO_4 in double-distilled water (stock)	1 ml
Na-cacodylate buffer, 0.134 mol/L, pH 7.4 (stock)	3 ml
$K_3Fe(CN)_6$	66 mg

Osmicate at 4°C overnight, up to 24 h; rinse three times in double-distilled water. This procedure gives excellent preservation of cellular membranes and also of glycogen. Dehydrate, embed in plastic.

Examine 2 μm sections by light microscopy: bright-field for evaluation of reaction intensity; phase-contrast for higher sensitivity to detect DAB-stained granules and for unstained structures; dark-ground illumination to visualize peroxisomes as bright granules even at low magnification (objective lens × 10, × 25). Glycogen gives a grey-blue colour to the cytoplasm, and for this reason this preparation is not suited for evaluating the presence of cytoplasmic catalase.

Counterstain ultrathin sections with lead only (no uranium).

Chorionic villi

Mount thin fragments *in toto*, after dehydration in ethanols and clearing; exert some pressure on the coverslip in order to flatten out the tubular villi.

Mount a larger fragment in the cryostat and cut 7 μm sections; mount on slides; apply aqueous osmium for 10 min, rinse, counterstain nuclei with methyl green (see above).

For electron microscopy and light microscopy of 4 μm plastic sections, osmicate in the presence of ferrocyanide, and embed.

Result: Trophoblast peroxisomes are very small, brown granules (black after osmication). Immersion lenses are necessary for phase-contrast microscopy and bright-field. With dark-ground illumination, peroxisomes are visualized as bright granules even at low magnification (objective lens × 25) (Roels et al 1987) (Figures 3–5). Bright-field microscopy is less sensitive, and not recommended unless peroxisomes have already been detected by phase-contrast; use a green filter. Peroxisomes are present exclusively in the deepest, cytotrophoblast layer, not in the syncytium. The latter often contains large, round fat droplets which are brown after osmication, and should not be mistaken for peroxisomes (Figure 6). As the cytotrophoblast layer is discontinuous, peroxisomes are not visible in all regions of a villus, even in healthy first-trimester chorion. The largest number is observed when the cytotrophoblast lies in the plane of the slide, and is viewed perpendicularly; peroxisomes often build groups around the nuclei (which are not visible unless stained with methyl green) (Roels et al 1987; and Figures 3–5).

Light microscopy of whole villi mounted *in toto* represent the fastest method, but do not show peroxisomes clearly in all samples of normal trophoblast; cryostat sections do. In case of doubt, 4 μm plastic sections or electron microscopy are reliable (Figure 6).

No peroxisomes are seen in first-trimester villi of a conceptus affected with the Zellweger

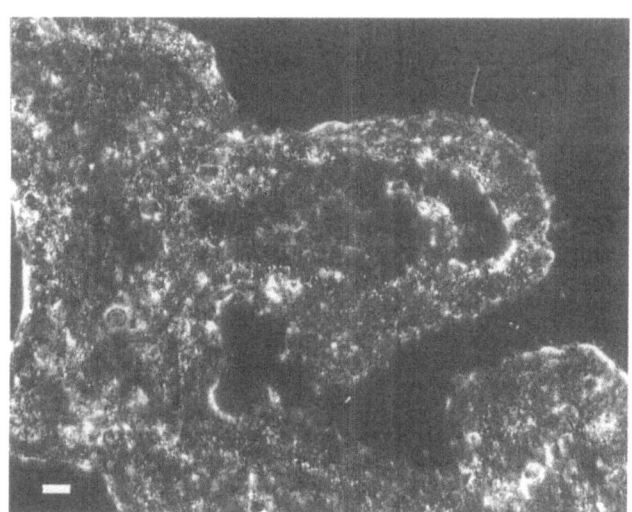

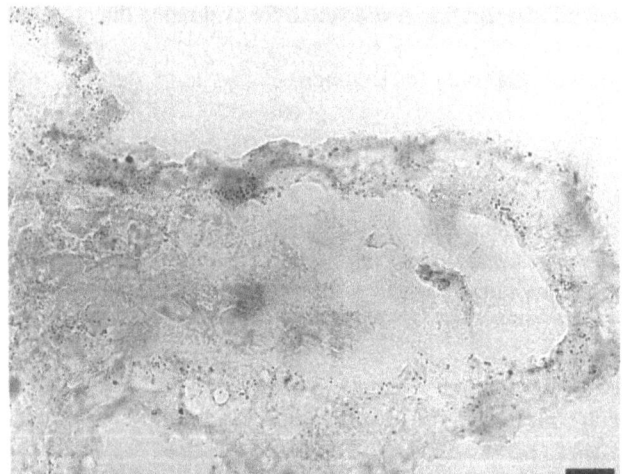

Figures 3–5 Chorionic villus sample from a normal pregnancy in first trimester, after visualization of peroxisomes by catalase activity.

Figure 3 (*top*) Dark-ground optics: peroxisomes stand out as numerous small bright granules in the deeper layer of the trophoblast, i.e. the cytotrophoblast. They build discrete groups reflecting the discontinuity of this layer; nuclei are unstained. The superficial layer limiting the intervillous space (syncytium) shows no peroxisomes. $7\,\mu$m cryostat section of villus stained *in toto*. ×25 objective lens. Bar = $10\,\mu$m

Figure 4 (*middle*) Same section. Bright-field optics and immersion lens of 63×, 1.4 NA: peroxisomes are very small and have low contrast (compare with liver in Figure 1). Same $7\,\mu$m cryostat section as in Figure 3. Bar = $10\,\mu$m

Figure 5 (*bottom*) Same section. Phase-contrast image: immersion lens ×100. Contrast is improved over that in bright field. Visualization is easier in the microscope by focusing up and down and by colour perception. Bar = $10\,\mu$m

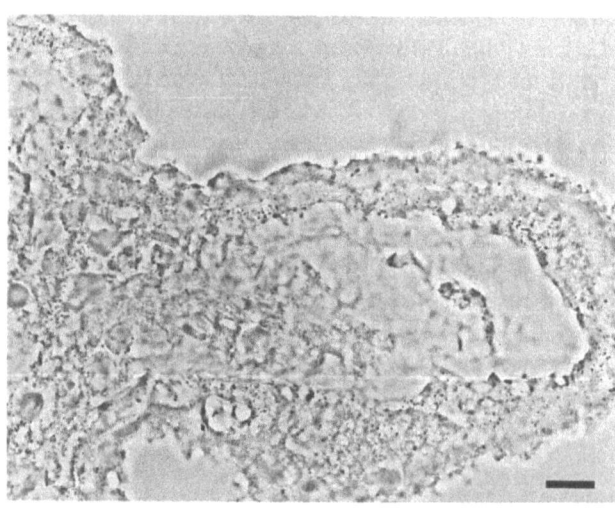

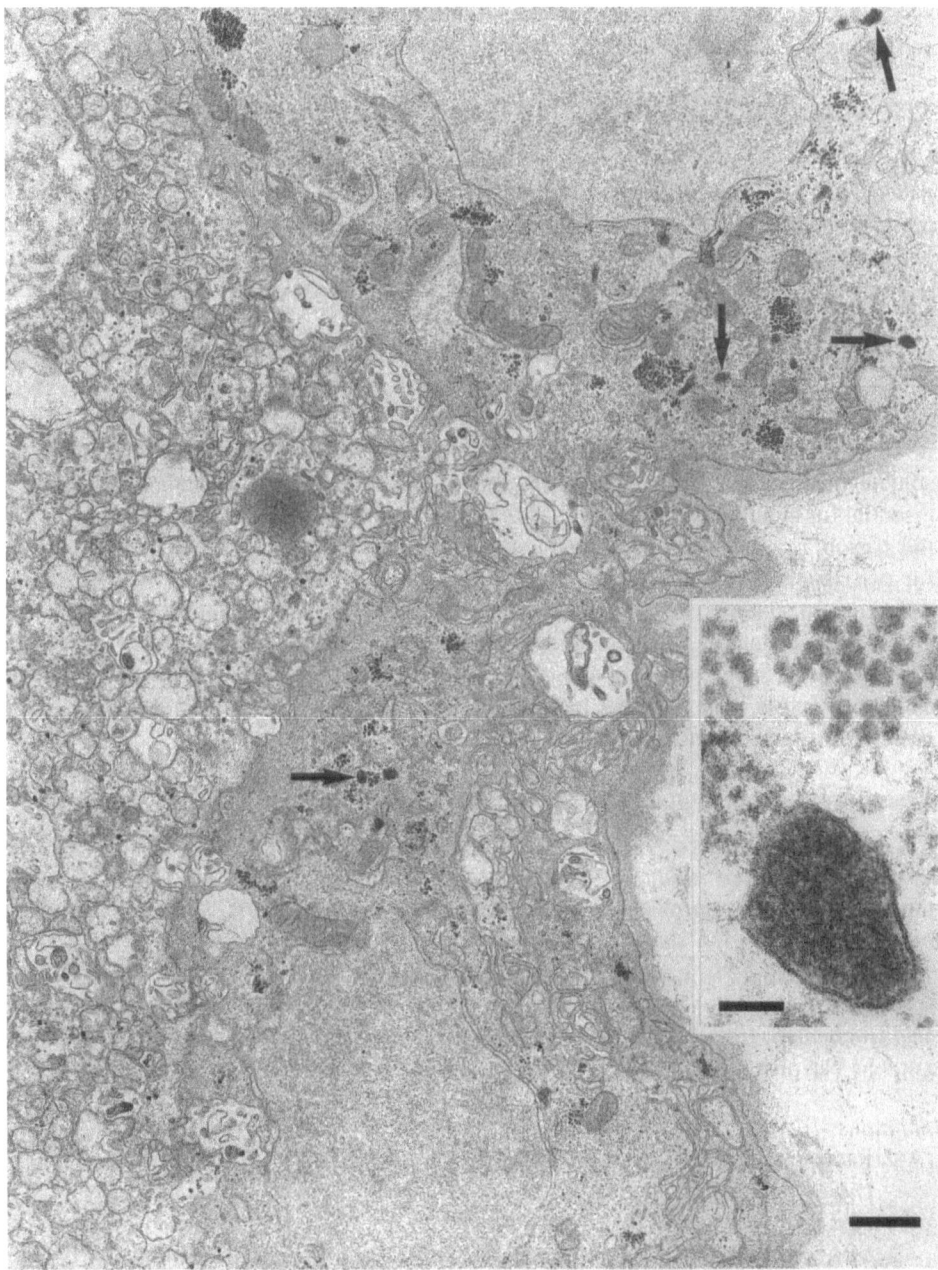

Figure 6 Ultrastructure of healthy first-trimester chorionic villus after staining for catalase activity. Two cytotrophoblast cells supported by the basal lamina are shown. They are characterized by the presence of flakes of monoparticulate glycogen (β-particles) (also in Zellweger trophoblast), and by the peroxisomes (arrows) (which are absent in a conceptus with ZS). Note the small size of these organelles in trophoblast. The adjacent syncytium shows no glycogen or peroxisomes but many endoplasmic reticulum membranes and a lipid globule that is much larger than the peroxisomes. Bar = 1 μm. *Inset:* the peroxisomal membrane should be visible in order to identify this organelle. Monoparticulate glycogen is recognized. Bar = 0.1 μm

syndrome (Roels et al 1993a). Contamination by maternal cells, which has happened in cell culture, is avoided because the trophoblast is easily identified under the microscope.

VISUALIZATION OF MACROPHAGES BY ACID PHOSPHATASE ACTIVITY

Enlarged liver macrophages (Kupffer cells) are observed in some peroxisomal deficiency patients, but are not in themselves specific (Roels et al 1986, 1991a, 1993b). Their presence should initiate search for trilamellar inclusions by electron microscopy. Acid phosphatase is localized in lysosomes, which also contain the inclusions; enzyme activity is higher in macrophages than in liver parenchymal lysosomes (Figure 7). Normal macrophages are thin cells alongside the sinusoids.

The sensitive method for light microscopy uses naphthol AS-TR-phosphate as substrate, which is hydrolysed by the enzyme; free naphthol AS-TR binds to the leukoform of *p*-rosaniline and the bright red complex precipitates on the spot. However, it is partly soluble in ethanol and embedding media. Cryostat sections, which can be mounted in water-soluble media and examined as such, should be used for this technique. Fixation is essential for immobilization of the enzyme, but the enzyme is also slowly inactivated. For this reason, fixation at 0°C is preferable, with times shorter than 24h. The same fixative as for peroxisomal staining is perfectly adequate, but the biopsy should be divided for different temperatures. Seven hours at room temperature is equally satisfactory.

Acid phosphatase activity is lost on paraffin embedding. However, enlarged macrophages usually stain strongly with periodic acid–Schiff, also in paraffin sections. This approach is used to select an area for osmication and electron microscopy in the search for trilamellar inclusions.

Procedure: Lojda method for acid phosphatase activity

Fixation: The same buffered formaldehyde–calcium fixative as used for peroxisome visualization, can be used. Because 24h fixation at room temperature results in strong inhibition of acid phosphatase activity, overnight rinsing in 0°C in three changes of cacodylate buffer+8% sucrose is necessary, as well as a long incubation (2h) at 37°C. After a rinse in 13% sucrose as a cryoprotective, freeze the fragment, make cryostat sections of $6-7\,\mu$m, which are thawed onto glass slides coated with APS (see Espeel and Van Limbergen, 1995). Sections should adhere to the slides during the full procedure; if not, one can pick them up with a pipette or glass hook.

Solutions
(A)   Pararosaniline chloride (= Basic Red 9, = Basic Fuchsin)   400 mg
        Dissolve in double-distilled water   8 ml
        Add HCl 36%   2 ml
        Mix well with magnetic stirrer until dissolved and filter. This solution can be kept for months in the refrigerator.

(B) $NaNO_2$, 4%: Can be stored at 4°C for 1 week.

Hexazotized pararosaniline: Mix thoroughly equal volumes of solutions A and B just before use. The mixture should become a transparent brown.

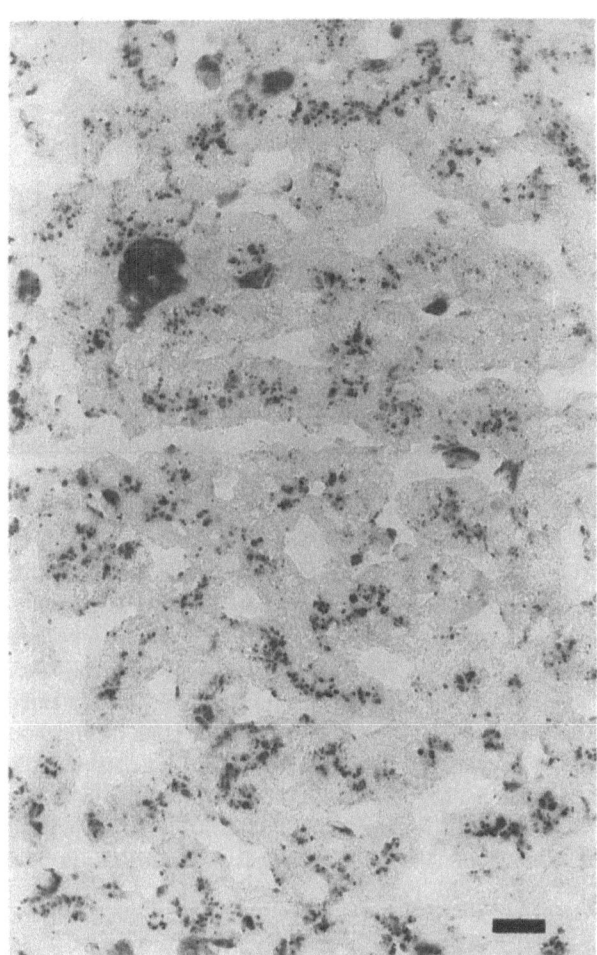

Figure 7 Acid phosphatase stain reveals enlarged macrophage in the liver of an infantile Refsum disease patient. Parenchymal cells contain small and normal-looking lysosomes forming rows along the bile canaliculi. Parenchymal lysosomes are less reactive and may not stain after fixation at room temperature; this liver was fixed at 0°C for 24 h. 5 μm cryostat section, Lojda method. Bar = 20 μm

Buffered hexazotized pararosaniline: (prepare just before use)
2 ml hexazotized pararosaniline
60.67 ml sodium acetate 0.82% (0.82 g/100 ml)

Incubation medium
Naphthol AS-TR phosphate	20 mg
Dissolve in *N,N*-dimethylformamide	1 ml
Buffered hexazotized pararosaniline	50 ml
Adjust to pH 5 with NaOH (1 mol/L)	
Filter	

Incubation
- 15 min, 30 min, 1 h and 2 h at room temperature, or at 37°C, depending on fixation time and temperature.
- Rinse in double-distilled water.
- Postfix in 10% formaldehyde with 1% $CaCl_2$ in double-distilled water overnight.
- Rinse in tap water.
- *Optional:* counterstain nuclei with haemalun (but do not dehydrate). Mount in Gurr Aquamount.
- *Optional:* Control incubation medium: as above plus NaF 0.01 mol/L (4.2 mg/10 ml).

LIPID INSOLUBLE IN ACETONE (LIVER AND ADRENAL CORTEX)

Lipid that stains with Oil Red O or Sudan Black B, and is insoluble in acetone, is pathognomonic for a peroxisomal β-oxidation defect (in X-ALD (McKusick 300100): Johnson et al 1976; Espeel et al 1991; Roels et al 1986, 1993b). Although the chemical nature of the material is still under investigation (Kerckaert et al 1988; Roels et al 1993b), it constitutes a significant diagnostic test. The lipid appears as droplets, usually very small, in non-parenchymal cells. Because the liver parenchyma often contains considerable numbers of fat globules, as normally does the adrenal cortex, the specific lipid may remain unnoticed unless the sections are treated with acetone prior to staining. Acetone dissolves the parenchymal fat and, after staining, the specific lipid stands out (Figure 8). The latter is seen even in paraffin sections stained with Sudan Black B. As usual, sections stained with fat stains should be mounted in water-soluble media, otherwise the stain disappears.

The solubility tests are performed on cryostat sections of formaldehyde–calcium fixed material; this fixative preserves lipids, and is also optimal for peroxisomal visualization.

Insoluble lipid and polarizing inclusions may be present in the same or adjacent non-parenchymal cells (Figure 9); the latter being large macrophages visualized with acid phosphatase or PAS.

Procedure: Oil Red O staining for lipids

Fixation: 24 h buffered formaldehyde with 1% $CaCl_2$; i.e. the same fixative used for peroxisome visualization, is best for all lipids. Glutaraldehyde fixed material can also be used.
Rinse liver fragments briefly in 13% sucrose; and freeze for cryostat.

Stock solution: Dissolve 0.5 g Oil Red O (ORO) in 100 ml isopropyl alcohol 98% (v/v). This solution can be kept for a long time in a dark bottle. *Before use:* Dilute 6 ml of the ORO stock solution with 4 ml double-distilled water. Filter the solution, which should be clear red.

Technique: Cut cryostat sections 5–7 μm. Mount the sections in double-distilled water on a microscope slide and dry. Immerse the slide with the dried sections into isopropyl alcohol 60% (v/v). Stain the slide in the filtered ORO solution for 10 min. Wash 10 s in isopropyl alcohol 60%. Wash in double-distilled water. Mount in Aquamount (Gurr). *Note:* this stain is not permanent. After a few days, the red oil forms crystal-like needles.

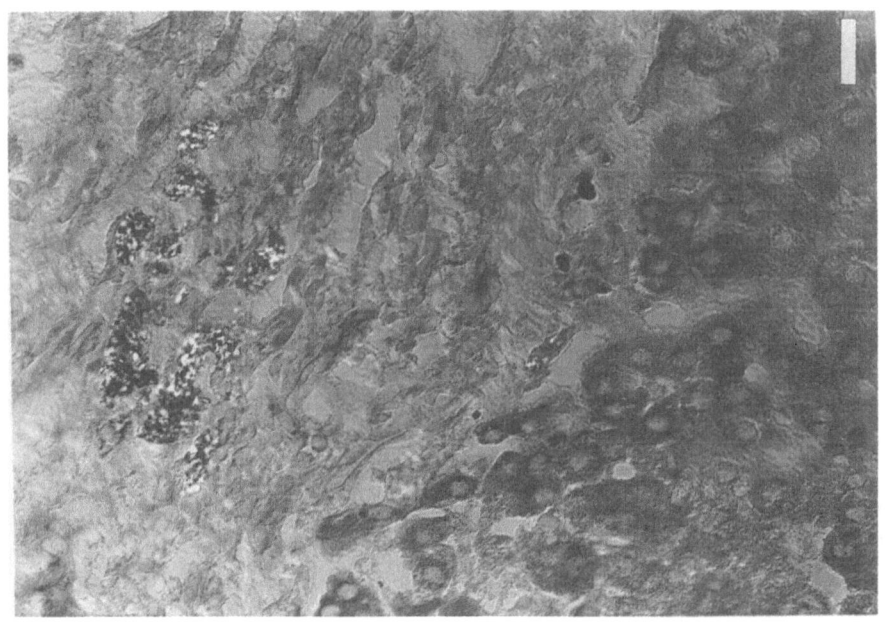

Figures 8–9 Fine droplets of lipid insoluble in acetone, together with small polarizing inclusions in macrophages in the liver of an IRD patient. Sudan Black B after acetone and hexane treatment. Bar = 20 μm.

Figure 8 (*left*) in parenchymal cells the lipid is dissolved by acetone, but erythrocytes remain black.

Figure 9 (*right*) Same region in polarized light: inclusions are seen as bright spots in between lipid droplets

Solubility test: Prepare cryostat sections from formaldehyde – calcium fixed block. Immerse sequentially in: absolute acetone for 15 min; absolute n-hexane for 30 min; again in absolute acetone 15 min; water. Apply Oil Red O or Sudan Black B stain.

Alternative: Immerse in two baths of absolute acetone, 15 min each; water. Apply Oil Red O or Sudan Black B stain.

Procedure: Sudan Black B staining for lipids

Fixation: see above.

Stock solution: Make a saturated solution of Sudan Black B (SBB) (approx. 4 g) in 50 ml ethanol 70% (v/v). Kept in a dark bottle, the solution is stable for a long time. Filter before use.

Technique: Cut cryostat sections 5 – 7 μm. Mount the sections in double-distilled water on a microscope slide and dry at room temperature. Immerse the slide with the dried sections into ethanol, 70% (v/v). Stain in the filtered and saturated SBB solution for 10 min. Wash quickly in ethanol, 70%. Wash in double-distilled water. Mount in Aquamount (Gurr). Stained lipids are blue-black; background is dark grey. Erythrocytes are always black, even after acetone extraction.

POLARIZING (BIREFRINGENT) MATERIAL

Fat globules in liver parenchyma stained or not with Oil Red O often show a birefringent rim in polarized light. This lipid, however, dissolves in acetone or during paraffin embedding and the birefringency disappears; it is not specific for a peroxisomal disorder. Polarizing inclusions resistant to all solvents (including n-hexane and xylene) are pathognomonic; they are localized predominantly in enlarged macrophages in liver and adrenal cortex but, when abundant, also in parenchymal lysosomes in small amounts. The macrophage inclusions may be huge and fill most of the cells (Roels et al 1986; Kerckaert et al 1988). They are often needle-like or fusiform (Figure 10). It is a most important feature of these inclusions that they do not take any stain, although their visibility may be obscured by a dark Sudan Black, PAS or acid phosphatase reaction surrounding them. Electron microscopy has shown that acid phosphatase activity is localized outside the parallel lamellae, which are the probable causes of the birefringence (Roels et al 1993b). As a consequence, the inclusions cannot be detected without polarized light.

They should be carefully distinguished from 'formaldehyde pigment' seen after prolonged (weeks) immersion in fixative; this pigment is yellow-brown and crystalline, and also birefringent; it is found in all tissues, intra- and extracellularly (Figures 11, 12) (formaldehyde pigment is also shown in Figures 3, 5 and 6 of Kerckaert et al (1988); it was unidentified at that time).

Polarizing inclusions are not found in all types of peroxisomal disorders (see details in Roels et al 1991a, 1993b); in addition, their accumulation and visibility under the light microscope are age-dependent. As a consequence, their absence does not exclude a peroxisomal deficiency, especially in children under 2 years of age.

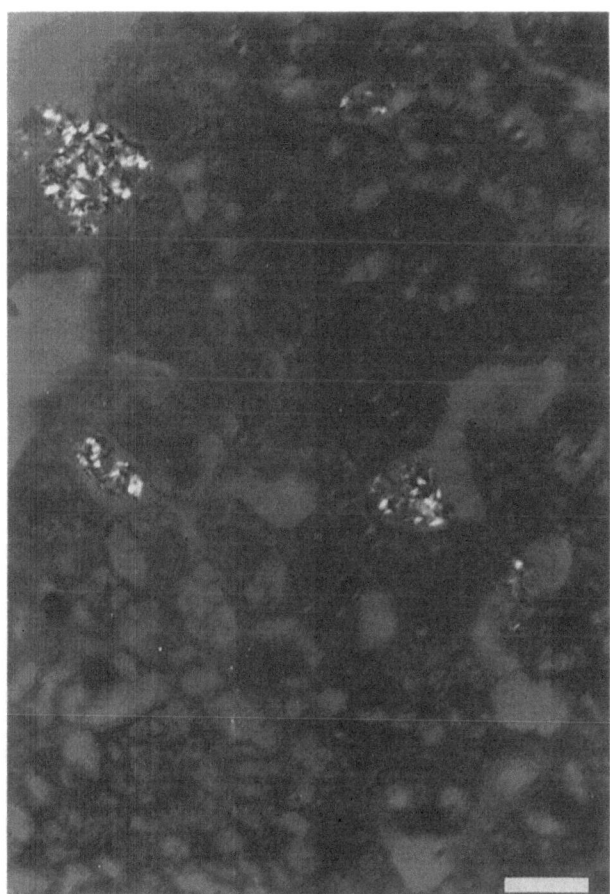

Figure 10 Numerous birefringent inclusions in globular macrophages between liver parenchymal cells of an infantile Refsum disease patient. They are angular, needle- or crystal-like. 4 μm plastic section observed in polarized light. The inclusions are invisible in bright field but have resisted dehydration in ethanol and embedding; they also persist after xylene and paraffin. Fragment stained for catalase activity; there are no peroxisomes but some cytoplasmic reaction product is present. Bar = 20 μm

The polarizing inclusions in Fabry disease (α-galactosidase A deficiency; McKusick 301500) are reminiscent of those in peroxisomal disorders; but their ultrastructure differs (Lake 1992). In erythropoietic protoporphyria (ferrochelatase deficiency; McKusick 177000) the polarizing inclusions in liver are constituted by brown pigment (Wolff et al 1975).

Bundles of collagen are normally birefringent.

TRILAMELLAR INCLUSIONS

Images are shown in Dingemans et al (1983), Espeel et al (1991), Gatfield et al (1968), Ghatak et al (1981), Haas et al (1982), Hughes et al (1990), Kerckaert et al (1988), Manz et al (1980), Mooi et al (1983), Pfeifer and Sandhage (1979), Poulos et al (1984), Roels et al (1986, 1991a, 1993b, 1995), Schaumburg et al (1974, 1975, 1977), Scotto et al (1982), Ulrich et al (1978), Vamecq et al (1986), Van Hoof and Roels (1989), Van Maldergem et al (1992).

Being pathognomonic for a peroxisomal disorder, trilamellar inclusions represent the

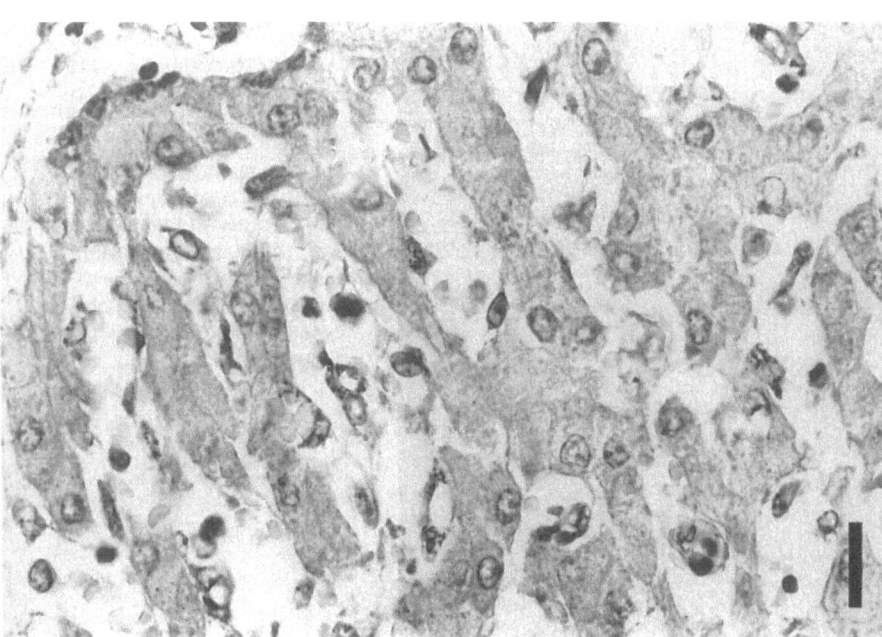

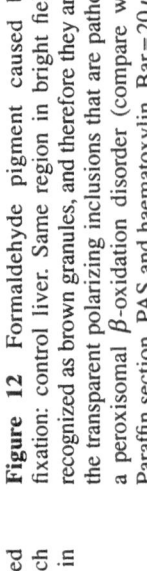

Figure 12 Formaldehyde pigment caused by prolonged fixation: control liver. Same region in bright field. Pigment is recognized as brown granules, and therefore they are distinct from the transparent polarizing inclusions that are pathognomonic for a peroxisomal β-oxidation disorder (compare with Figure 8). Paraffin section. PAS and haematoxylin. Bar=20 μm.

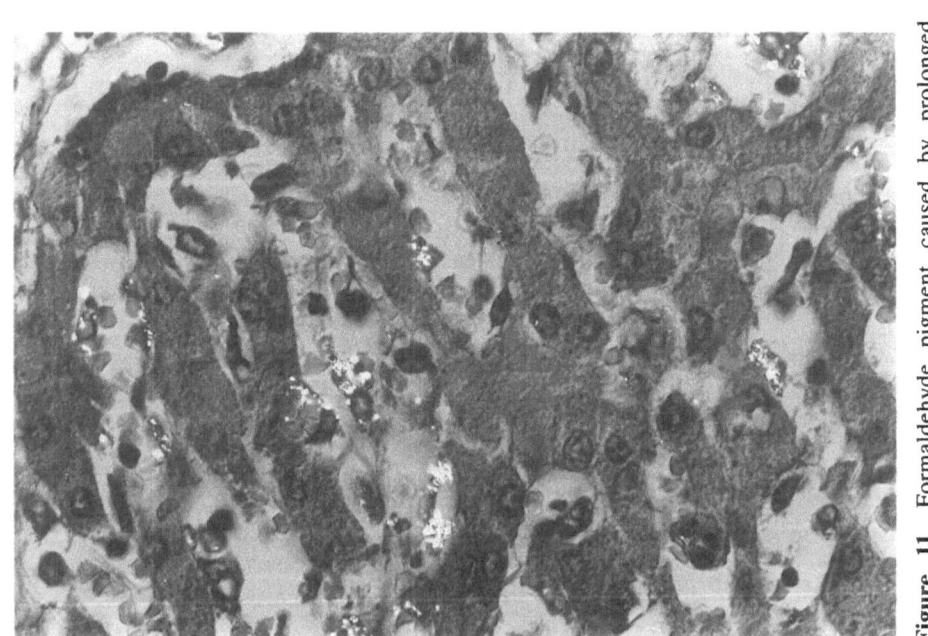

Figure 11 Formaldehyde pigment caused by prolonged fixation: control liver. Polarized light: bright inclusions which may be inside as well as outside cells. Compare bright field in next figure.

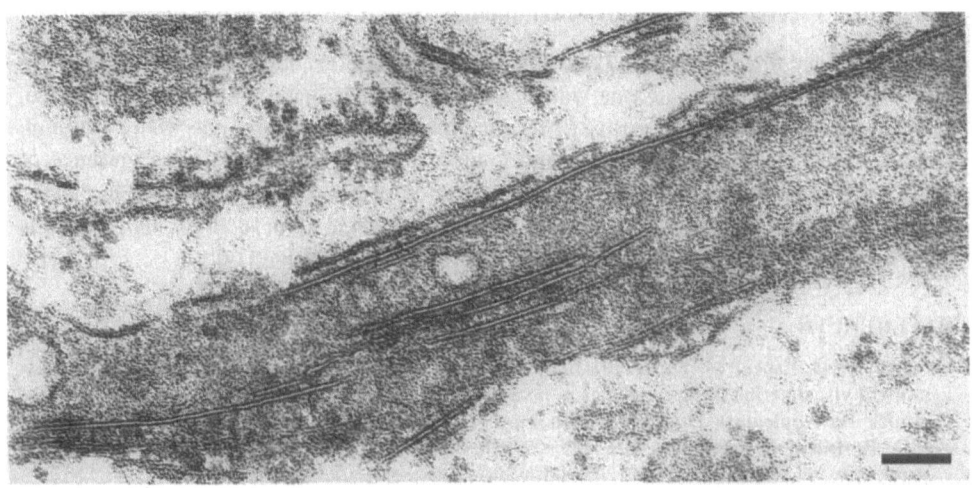

Figure 13 Ultrastructure of angulate lysosome containing only few trilamellar structures. The section was treated with acetone and n-hexane prior to osmication with ferricyanide. Trilamellar structures are unaltered while the lysosomal membrane appears damaged. Bar=0.1 μm

ultrastructural substratum of birefringence, and are thus predominant in macrophage lysosomes, and rare in parenchyma. Small stacks, or a single trilamellar structure, cannot be detected by light microscopy, but they can be seen by electron microscopy when the organelles, i.e. lysosomes in (enlarged) macrophages, are examined at higher magnification. Because macrophages are relatively few in comparison to liver parenchymal cells, they should be looked for by light microscopy of 2 μm sections, and this region should then be selected for ultrathin sectioning. Not all macrophages are necessarily positive, and the trilamellar structures are not distributed evenly over the liver (Roels et al 1993b); caution is required before one may conclude that they are absent. Their number increases with the age of the patient; we have not found any in the fetus.

The same fixation as used for peroxisome cytochemistry is excellent for the study of these inclusions. A peculiar and most interesting characteristic is their resistance to solvents and paraffin embedding (Figure 13); samples from archival paraffin blocks can be processed for electron microscopy and yield detailed high-resolution images of the trilamellar structures (Kerckaert et al 1988).

The same type of inclusion is described in macrophages in the brain and adrenal cortex of patients with peroxisomal disorders (Espeel et al 1991; Martin, this issue; Van Maldergem et al 1992; see the references mentioned above, and in Roels et al 1991a), in fact in some disorders (e.g. X-ALD, McKusick 300100) they are more abundant here than in the liver.

It is remarkable that Kupffer cells of patients with erythrohepatic protoporphyria (ferrochelatase deficiency) also contain angulate lysosomes with trilamellar structures that are ultrastructurally very similar to those in peroxisomal disorders; however, they are brown in ordinary light (Wolff et al 1975).

Procedure: Electron microscopy of trilamellar structures in paraffin blocks

Locate enlarged macrophages in PAS-stained paraffin sections; look for birefringence. Cut $20-30\,\mu$m sections from an area with macrophages, or if macrophages are not visible, select a portal triad with fibrosis. Bring the free floating sections into water (through xylene, or other appropriate solvent, and graded alcohols). Osmicate in the presence of ferrocyanide overnight at $0-4°C$ (see 'Visualization of peroxisomes'); embed in plastic. Look for macrophages (in phase-contrast) or birefringence in $2\,\mu$m sections, and select this area. Counterstain ultrathin sections with uranium and lead, or lead only.

REFERENCES

De Craemer D, Espeel M, Langendries M, Schutgens RBH, Hashimoto T, Roels F (1990) Post-mortem visualization of peroxisomes in rat and in human liver tissue. *Histochem J* **22**: 36–44.

Dimmick JE, Applegarth DA (1993) Pathology of peroxisomal disorders. In Landing BH, Haust MD, Bernstein J, Rosenberg HS, eds. *Genetic Metabolic Diseases*, Perspectives in Pediatric Pathology, vol. 17. Basle: Karger, 45–98.

Dingemans KP, Mooi WJ, van den Bergh Weerman MA (1983) Angulate lysosomes. *Ultrastruct Pathol* **5**: 113–122.

Espeel M, Van Limbergen G (1995) Immunocytochemical localization of peroxisomal proteins in human liver and kidney. *J Inher Metab Dis* **18 (Suppl. 1)**: 135–154.

Espeel M, Roels F, De Cramer D, et al (1991) Peroxisomal localization of the immunoreactive β-oxidation enzymes in a neonate with a β-oxidation defect. Pathological observations in liver, adrenal cortex and kidney. *Virchows Arch A Pathol Anat* **419**: 301–308.

Espeel M, Heikoop JC, Smeitink JAM, et al (1993) Cytoplasmic catalase and ghost-like peroxisomes in the liver from a child with atypical rhizomelic chondrodysplasia punctata. *Ultrastr Pathol* **17**: 623–636.

Espeel M, Mandel H, Poggi F, et al (1995) Peroxisome mosaicism in the livers of peroxisomal deficiency patients. *Hepatology*, **22**: 497–504.

Gatfield PD, Taller E, Hinton GG, Wallace AC, Abdelnour GM, Haust MD (1968) Hyperpipe-colatemia: A new metabolic disorder associated with neuropathy and hepatomegaly. *Can Med Assoc J* **99**: 1215–1233.

Ghatak NR, Nochlin D, Peris M, Myer EC (1981) Morphology and distribution of cytoplasmic inclusions in adrenoleukodystrophy. *J Neurol Sci* **50**: 391–398.

Haas JE, Johnson ES, Farrell DL (1982) Neonatal-onset adrenoleukodystrophy in a girl. *Ann Neurol* **12**: 449–457.

Hughes JL, Poulos A, Robertson E, et al (1990) Pathology of hepatic peroxisomes and mitochondria in patients with peroxisomal disorders. *Virchows Arch A, Pathol Anat* **416**: 255–264.

Johnson AB, Schaumberg HH, Powers JM (1976) Histochemical characteristics of the striated inclusions of adrenoleukodystrophy. *J Histochem Cytochem* **24**: 725–730.

Kerckaert I, Dingemans KP, Heymans HSA, Vamecq J, Roels F (1988) Polarizing inclusions in some organs of children with congenital peroxisomal diseases (Zellweger's, Refsum's, chondrodysplasia punctata (rhizomelic form), X-linked adrenoleukodystrophy). *J Inher Metab Dis* **11**: 372–386.

Lake BD (1992) Lysosomal and peroxisomal disorders. In Adams JH, Duchen LW, Edward Arnold, eds. *Greenfield's Neuropathology*, 5th edn. London: Hodder and Stoughton, 709–810.

Mandel H, Espeel M, Roels F, et al (1994) A new type of peroxisomal disorder with variable expression in liver and fibroblasts. *J Pediatr* **125**: 549–555.

Manz HJ, Schuelein M, McCullough DC, Kishimoto Y, Eiben RM (1980) New phenotypic variant of adrenoleukodystrophy. Pathologic, ultrastructural, and biochemical study in two brothers. *J Neurol Sci* **45**: 245–260.

Mooi WJ, Dingemans KP, van den Bergh Weerman MA, Jöbsis AC, Heymans HSA, Barth PG (1983) Ultrastructure of the liver in the cerebrohepatorenal syndrome of Zellweger. *Ultrastruct Pathol* **5**: 135–144.

Novikoff AB, Goldfischer S (1969) Visualization of peroxisomes (microbodies) and mitochondria with diaminobenzidine. *J Histochem Cytochem* **17**: 675–680.

Pfeifer U, Sandhage K (1979) Licht- und elektronenmikroskopische Leberbefunde beim Cerebro-Hepato-Renalen Syndrom nach Zellweger (Peroxisomen-Defizienz). *Virchows Archiv A, Pathol Anat Histol* **384**: 269–284.

Poulos A, Pollard AC, Mitchell JD, Wise G, Mortimer G (1984) Patterns of Refsum's disease. Phytanic acid oxidase deficiency. *Arch Dis Child* **59**: 222–229.

Roels F (1974) Cytochrome and cytochrome oxidase in diaminobenzidine staining of mitochondria. *J Histochem Cytochem* **22**: 442–446.

Roels F (1991) *Peroxisomes: A Personal Account*. Brussels: VUB Press, 1–151. ISBN 90-70289-94-6.

Roels F, Goldfischer S (1979) Cytochemistry of human catalase: the demonstration of hepatic and renal peroxisomes by a high temperature procedure. *J Histochem Cytochem* **27**: 1471–1477.

Roels F, Wisse E, De Prest B, van der Meulen J (1975) Cytochemical discrimination between catalases and peroxidases using diaminobenzidine. *Histochemistry* **41**: 281–312.

Roels F, Cornelis A, Poll-The BT, et al (1986) Hepatic peroxisomes are deficient in infantile Refsum disease: A cytochemical study of 4 cases. *Am J Med Genet* **25**: 257–271.

Roels F, Verdonck V, Pauwels M, et al (1987) Light microscopic visualization of peroxisomes and plasmalogens in first trimester chorionic villi. *Prenat Diagn* **7**: 525–530.

Roels F, Espeel M, De Craemer D (1991a) Liver pathology and immunocytochemistry in peroxisomal disorders: A review. *J Inher Metab Dis* **14**: 853–875.

Roels F, Espeel M, Pauwels M, De Craemer D, Egberts HJA, Van der Spek P (1991b) Different types of peroxisomes in human duodenal epithelium. *Gut* **32**: 858–865.

Roels F, Espeel M, Lissens W, et al (1993a) Fast prenatal diagnosis of Zellweger cerebro-hepato-renal syndrome (ZS) by microscopy of first trimester chorionic villi (CV). Abstracts of the 31st SSIEM Annual Symposium. Sheffield: SSIEM; W27.

Roels F, Espeel M, Poggi F, Mandel H, Van Maldergem L, Saudubray JM (1993b) Human liver pathology in peroxisomal diseases: a review including novel data. *Biochimie* **75**: 281–292.

Roels F, Espeel M, Mandel H, et al (1995) Cell and tissue heterogeneity in peroxisomal patients. In Wanders RJA, Schutgens RBH, Tabak HF, eds. *Peroxisomal Disorders in Relation to Functions and Biogenesis of Peroxisomes*. Amsterdam: Elsevier, in press.

Schaumberg HH, Powers JM, Suzuki K, Raine CS (1974) Adreno-leukodystrophy (sex-linked Schilder disease). Ultrastructural demonstration of specific cytoplasmic inclusions in the central nervous system. *Arch Neurol* **31**: 210–213.

Schaumburg HH, Powers JM, Raine CS, Suzuki K, Richardson Jr EP (1975) Adrenoleukodystrophy. A clinical and pathological study of 17 cases. *Arch Neurol* **32**: 577–591.

Schaumburg HH, Powers JM, Raine CS, et al (1977) Adrenomyeloneuropathy: a probable variant of adrenoleukodystrophy. II. General pathologic, neuropathologic, and biochemical aspects. *Neurology* **27**: 1114–1119.

Scotto JM, Hadchouel M, Odièvre M, et al (1982) Infantile phytanic acid storage disease, a possible variant of Refsum's disease: three cases including ultrastructural studies of the liver. *J Inher Metab Dis* **5**: 83–90.

Ulrich J, Hershkowitz N, Heitz Ph, Sigrist Th, Baerlocher P (1978) Adrenoleukodystrophy. Preliminary report of a connatal case. Light- and electron microscopical, immunohistochemical and biochemical findings. *Acta Neuropathol* **43**: 77–83.

Vamecq J, Draye J-P, Van Hoof F, et al (1986) Multiple peroxisomal enzymatic deficiency disorders. A comparative biochemical and morphologic study of Zellweger cerebrohepatorenal syndrome and neonatal adrenoleukodystrophy. *Am J Pathol* **125**: 524–535.

Van Hoof F, Roels F (1989) Liver ultrastructure and diagnosis of inborn metabolic disorders. *Micron Microscopica Acta* **20**: 59–62.

Van Maldergem L, Espeel M, Wanders RJA, et al (1992) Neonatal seizures and hypotonia with elevation of very long chain fatty acids, normal bile acids, normal fatty acyl-CoA oxidase and intraperoxisomal localization of the three β-oxidation enzymes: a novel peroxisomal disease? *Neuromusc Disord* **2**: 217–224.

Wolff K, Wolff-Schreiner E, Gschnait F (1975) Liver inclusions in erythropoietic protoporphyria. *Eur J Clin Invest* **5**: 21–26.

J. Inher. Metab. Dis. 18 Suppl. 1 (1995) 172–180
© SSIEM and Kluwer Academic Publishers.

Practical guide for morphometry of human peroxisomes on electron micrographs

I. KERCKAERT*, D. DE CRAEMER[1] and G. VAN LIMBERGEN
Departments of Anatomy, Embryology and Histology, University of Gent; [1]Department of Anatomy, Free University of Brussels, Belgium

*Correspondence: Dr I. Kerckaert, Department of Anatomy, Embryology and Histology, Section of Human Anatomy, University of Gent, Godshuizenlaan 4, B-9000 Gent, Belgium

Morphometry of peroxisomes is performed on electron micrographs of ultrathin sections after staining for catalase activity with diaminobenzidine; specific peroxisomal labelling is preferred to guarantee recognition. Peroxisomal number, size, axial ratio and volume parameters are determined and compared to control values. Results from 19 patients with loss of peroxisomal functions are listed. In many patients alterations in peroxisomal morphometric features are found. A brief guideline for interpreting morphometric data is included. Diagnostically relevant morphometric alterations are summarized.

SELECTION OF SECTIONS FOR MORPHOLOGICAL ANALYSIS

In $2\mu m$ sections of DAB-stained liver, zones are selected by light microscopy. This selection is based on the quality of the section (preservation of cells; optimal DAB-staining) and on the region to be measured, i.e. a portal or a central area. It is important to select comparable regions in the biopsy material from patients and controls, to exclude possible differences in peroxisomal number and size between regions in the liver tissue. Previous counts and measurements in rat liver (Loud 1968) revealed differences in numbers of peroxisomes but no changes in size between periportal, midcentral and central regions; counts in human liver (Roels et al 1983; Kerckaert et al 1989) revealed a variable difference between patients between periportal and central zones.

From these selected areas, 750 Å sections are cut and counterstained with lead only for 20 min.

ELECTRON MICROGRAPHS

Electron micrographs are taken at random in the four corners of the squares of the copper grids at a magnification of ×9000. About 20 pictures are taken from one biopsy, usually representing approximately 100 peroxisomes for each patient or condition. If this number is not reached, more pictures must be taken. Microscope magnification is calibrated with a grating replica of 2157 lines/mm for each series of micrographs. The micrographs are enlarged photographically 2.5 times; final magnification being ~×22 500.

This is suitable for normal-sized organelles. Recently, much smaller peroxisomal ghosts were found in human liver by Espeel; a final magnification of ×36 000 was necessary.

SEMI-AUTOMATED MEASUREMENTS

Morphometry is done from the micrographs using a semi-automatic image analysis system with manual delineation by means of a pen and a digitizer tablet. The boundaries of peroxisomes and parenchymal cells are traced.

From the parenchymal cells the area is measured; sinusoidal lumina and non-parenchymal cells are excluded. From each peroxisomal profile the perimeter, area and form-ellipse are measured. The form-ellipse (the axial ratio) is the ratio between the shortest and the longest diameter of an object. It is an expression of the roundness of the organelle. This shape parameter has a maximum value of 1, which represents a circle. Values below 1 indicate a deviation from the spherical shape. Mean, maximum and minimum form-ellipse and area values are noted. The digital tablet usually delivers the data in centimetres or millimetres. These values are divided by the final magnification and then expressed in micrometres. The number of peroxisomal profiles is also counted. Because our tablet is not connected to a PC, we enter the values manually into a database on the PC for further calculations. When a fully automated image analysis system is available, a number of parameters can be measured directly (Kerckaert et al 1989).

The point counting method can be used when no semi-automated analysis system is available. Manual counting of peroxisomes is then carried out on light photographs (magnification about ×1200) obtained by Nomarsky interference contrast light microscopy. The peroxisomes appear as bas-relief rounded hillocks. The area of the parenchymal cells is drawn with a pencil, excluding sinusoidal lumina and non-parenchymal cells, and measured in arbitrary units using a grid. Peroxisomal size or shape parameters cannot be measured on these micrographs. On electronmicrographs, using a grid, one can obtain the area.

FURTHER CALCULATIONS

Size parameters

Other size parameters are calculated from the area (expressed in μm^2) of the measured object. These parameters are:

- *d-Circle* (μm): the diameter of the circle with the same area as the measured profile.

$$d\text{-circle} = 2\sqrt{\frac{\text{area}}{\pi}}$$

- Mean above 95th percentile: the mean of the d-circle values of the 5% largest organelles. When only part of the peroxisomes are enlarged, the overall mean d-circle may not show this but the 5% mean will.

These size parameters can be measured directly on a digital tablet. For all parameters the minimum, maximum and mean values are given, with SD or SEM.

Correction for section thickness

In ultrathin sections, spherical objects are not always cut at the level of the equator. In order to eliminate underestimation of mean size due to the effect of sectioning, the corrected mean d-circle is calculated (Abe et al 1983).

$$\text{Corrected mean area} = \frac{3(2R + T)}{4R + 3T} \times \text{measured mean area}$$

where R = radius = measured d-circle/2; T = section thickness.

$$\text{Corrected mean d-circle} = 2\sqrt{\frac{\text{corrected mean area}}{\pi}}$$

Volume parameters

Formulae are used according to Weibel et al (1966) and Weibel (1969, 1979). Volume density (V_V), numerical density (N_V) and surface density (S_V) are calculated.

V_V is the total volume of a compartment (in our case the peroxisomal compartment) expressed as a fraction of cellular volume. The volume density is calculated as

$$V_V = \frac{A_{AP}}{A_T K} \qquad \text{(expressed in } \mu\text{m}^3/\mu\text{m}^3 \text{ or in } \%)$$

where A_{AP} = the total peroxisomal area; A_T = the total area of the parenchymal cells (reference area); K is a correction factor for the sectioning effect, calculated from

$$K = 1 + \frac{3T}{2D}$$

where T represents section thickness and D the mean profile d-circle.

N_V is the number of organelles per unit of cellular volume. The numerical density is defined as

$$N_V = \frac{\kappa N_a^{3/2}}{\beta V_V^{1/2}} \qquad \text{(expressed in } \mu\text{m}^{-3})$$

where V_V = volume density; N_a is the ratio between the number of peroxisomes and the area of the parenchymal cells (number of profiles per unit of reference area). The coefficient β relates to the shape of the organelles. In the literature, a correlation between the axial ratio of ellipsoids and the coefficient β is given (Weibel 1969). We use $\beta = 1.4$. The value of the size distribution coefficient κ approximates to 1.0 in these experiments.

S_V is the total peroxisomal surface area expressed as a fraction of cellular volume. The surface density is calculated as

$$S_V = \frac{4B_A}{\pi + 4T/D} \qquad \text{(expressed in } \mu\text{m}^{-1})$$

where B_A is the ratio between the sum of the measured profile perimeter and A_T is the total parenchymal cell area.

RESULTS AND DISCUSSION

The results of morphometric analysis in tissue of the patient are compared to control values. In Table 1 morphometric measurements of hepatic peroxisomes from patients suffering from a peroxisomal defect are shown, as well as control values. Before concluding that an altered value of a new patient is significant, the normal range of control values,

Table 1 Morphometry of peroxisomes

	d-Circle (μm)				V_v (%)	N_v (μm⁻³)	S_v (μm⁻¹)	Axial ratio (elongation)	
	Mean measured	Corrected	max.	95th[a]				(Mean)	(Min)
(1) Multiple peroxisomal defects									
12 y	0.714	0.907	1.441	1.351	0.444	0.016	0.033	0.824	0.681
(2) Acyl-CoA oxidase deficiency									
(a) 2.5 y	0.660	0.835	1.230	1.190	0.527	0.106	0.186	0.830	0.527
(b) at 3 w	0.681	0.851	1.389	1.104	5.269	0.208	0.400	0.841	0.436
at 2.3 y	0.698	0.877	1.233	1.170	2.956	0.107	0.234	0.865	0.441
(c) fetus 18 w	0.701	0.905	1.608	1.467	1.759	0.092	0.162	0.824	0.506
(3) NALD-like									
(a) 14 m	0.657	0.821	1.214	1.142	1.200	0.075	0.119	0.807	0.501
(b) 7 m	0.570	0.706	1.083	0.960	1.178	0.104	0.127	0.824	0.564
(c) 8 m	0.490	0.613	0.924	0.857	1.802	0.197	0.197	0.702	0.181
(4) Peroxisomal β-oxidation defect									
(a) 15 d	0.569	0.712	1.230	1.016	0.799	0.070	0.087	0.831	0.336
(b) 7 m	0.578	0.846	1.163	1.095	4.175	0.153		0.738	0.167
(c) 4 m	0.715	0.898	1.237	1.191	1.600	0.049			
(5 RCDP									
at 8 m	0.719	0.918	1.862	1.375	1.040	0.043	0.077	0.911	0.554
at 2 y 9 m autopsy	0.739	0.936	1.560	1.376	3.800	0.099			
(6) 'Atypical' RCDP									
at 9 y	1.119	1.440	2.767	2.310	0.422	0.003	0.018	0.818	0.430
at 11 y	0.934	1.172	1.897	1.648	2.350	0.030			
(7) Thiolase deficiency									
7 d	0.513	0.659	1.065	0.997	1.576	0.189	0.180	0.747	0.324
(8) IRD									
11 y	0.472	0.581	0.828	0.753	0.378	0.047	0.045	0.705	0.199

continued

Table 1 *continued*

| | d-Circle (µm) | | | V_v | N_v | S_v | Axial ratio (elongation) | |
	Mean measured	Corrected max.	95th[a]	(%)	(μm^{-3})	(μm^{-1})	(Mean)	(Min)
(9) Primary hyperoxaluria type I								
(a) 15 y	0.431	0.522	0.637	0.609	0.597	0.103	0.078	0.848	0.374
(b) 8 m	0.324	0.395	0.533	0.508	0.485	0.209	0.083	0.769	0.445
(c) 8 y	0.401	0.492	0.754	0.673	1.220	0.249		0.531	0.161
(10) Peroxisomal defects + mosaicism								
(a) 2 y	0.436	0.552	0.919	0.802	1.030	0.147		0.731	0.413
(b) 9 y	0.551	0.682	0.945	0.869	1.790	0.132			
(11) Controls								
7 adults:								
mean	0.525	0.643	0.940	0.768	1.047	0.100	0.110	0.861	0.410
SD	0.046			0.036	0.274	0.028	0.028		
$p = 0.05$ confidence limits	0.433–				0.499–	0.044–	0.054–		
	0.617				1.595	0.156	0.166		
Infant (6 w)	0.445	0.555	0.848	0.769	0.712	0.110	0.085	0.808	0.436
Infant (4 m)	0.518	0.640	1.027	0.848	1.183	0.128	0.131	0.884	0.569

[a]Mean of 5% highest values

Patients: **(1)** Elevated levels of pipecolic acid, phytanic acid and bile acid intermediates, but normal VLCFA and plasmalogens (Roels et al 1993; De Craemer 1994). **(2)** Girl (a) and boy (b) (Poll-The et al 1988; Roels 1991), and aborted fetus from the same parents (c), in amniocytes western blots showed absence of all three components of acyl-CoA oxidase protein; in amniotic fluid $C_{26:0}$ levels and C_{26}/C_{22} ratio increased; absence of acyl-CoA oxidase antigen confirmed in liver (De Craemer et al 1991b; Roels et al 1991). **(3)** Clinical picture and neuropathology of NALD, increased VLCFA, absence of enzymatic data (Aubourg et al 1986; Roels et al 1988, 1991; Roels 1991). **(4)** Accumulation of VLCFA; in liver presence of the peroxisomal proteins catalase (not tested in 4a), alanine : glyoxylate aminotransferase (not tested in 4a), bifunctional enzyme, acyl-CoA oxidase and thiolase (not tested in 4b) inside the organelles (immunocytochemistry performed by Espeel); oxidase activity normal in 4a and c (4a described by Espeel et al (1991) and Van Maldergem et al (1992)). **(5)** Impaired plasmalogen synthesis; elevated phytanic acid level in plasma; thiolase in precursor form in cultured skin fibroblasts (De Craemer et al 1991b; Roels et al 1994). **(6)** Biochemistry as in (5); absence of symmetric shortening of the proximal parts of the limbs (Espeel et al 1993); second biopsy (11 y) after 2 years of treatment (three plasmaphoreses performed at 4- and 5-day intervals followed by a low-phytanic acid diet); clinical improvement (Smeitink et al 1992) was accompanied by change in peroxisomal morphology. **(7)** Described by Goldfischer et al (1986) and Roels et al (1988, 1991); similar observations in Hughes et al (1990). **(8)** Classical IRD case (Scotto et al 1982; Roels et al 1986, 1991;Poll-The et al 1987; De Craemer 1994); not all IRD patients possess peroxisome-like structures; it is a heterogeneous group (see also Beard et al 1986; Hughes et al 1990). **(9)** Alanine : glyoxylate aminotransferase deficiency (9a,b from De Craemer et al 1989). **(10)** 'Mosaicism': parenchymal cells with and without peroxisomes in the liver; the organelles are measured and volume parameters calculated taking into account the area of the peroxisome-containing cells only; **(10a)** increased levels of VLCFA, pipecolic acid and bile acid intermediates (Espeel et al 1995a; Roels et al 1994); **(10b)** multiple peroxisomal defects (Mandel et al 1994; Espeel et al 1995a). **Controls:** De Craemer et al (1991a), De Craemer (1994)

Table 2 Alterations in size and number of peroxisomes[a]

	Size	
Number	Enlarged	Smaller
Normal	Acyl-CoA oxidase deficiency NALD-like	Primary hyperoxaluria type 1[b]
Increased	Acyl-CoA oxidase deficiency NALD-like[b] Thiolase deficiency[b] Peroxisomal β-oxidation deficiency[b]	Primary hyperoxaluria type 1[b]
Decreased	Multiple loss of peroxisomal functions RCDP Atypical RCDP Peroxisomal β-oxidation deficiency[b]	IRD[b,c] NALD

[a]Not including peroxisomal ghosts, i.e. very rare organelles devoid of five matrix enzymes
[b]Decreased axial ratio: elongation of peroxisomes
[c]In a minority of patients; most IRD livers have no peroxisomes

which is quite considerable, should be taken into account. Table 2 displays a survey of diagnostically relevant morphometric alterations. Results of this procedure can also be found in De Craemer et al (1991a,b, 1993a,b and in this issue), in Espeel et al (1991, 1993) and in Roels et al (1991, 1993, 1994).

In many patients with loss of peroxisomal functions, an increase in size of the hepatic peroxisomes is observed (Table 1). As a consequence the ratio between the surface (membrane) area and the volume of the organelles will be smaller, which will influence the import, activity and turnover of the peroxisomal enzymes and substrates (Roels 1991; Roels et al 1993). This results in a diminished exchange over the peroxisomal membrane relative to volume, i.e. to total enzyme content.

The ratio between surface area and volume should increase for augmentation of metabolic efficiency (Roels and Cornelis 1989; De Craemer 1994). This is the case when an increased number of smaller organelles is present (as in primary hyperoxaluria type 1). In conditions in which peroxisomal β-oxidation was induced (e.g. rats treated with thyroid hormone; Fringes and Reith 1982; Kerckaert et al 1989), a proliferation of peroxisomes together with a decrease in size was measured.

A decrease of the form-ellipse (axial ratio) value indicates elongated peroxisomes or abnormal forms. Elongation means an increase in surface area for a constant volume. This may contribute to increase the rate of metabolism (De Craemer 1994). Elongation was striking in peroxisomal disorder patients suffering from infantile Refsum disease (IRD), peroxisomal thiolase deficiency, neonatal adrenoleukodystrophy (NALD)-like syndrome, peroxisomal β-oxidation defects and primary hyperoxaluria type I. Not all peroxisomes within a total population are elongated or irregularly shaped. The minimum axial ratio illustrates better the severity of shape disturbance.

Hughes and Poulos (1992) published a three-type classification of hepatic peroxisomal alterations in patients with peroxisomal disorders, according to their morphology. De

Craemer (1994) distinguished five hepatic types. We make a distinction between seven types as follows.

(1) Extremely large organelles with a flocculent matrix as seen in RCDP and in multiple peroxisomal defects patients.
(2) Peroxisomes, with a number of them being enlarged (as reflected in the mean above the 95th percentile) or elongated, or with an increase in number: in acyl-CoA oxidase deficiency, peroxisomal thiolase deficiency, NALD-like syndromes, peroxisomal β-oxidation defects.
(3) Small (mean d-circle 0.472 μm) peroxisome-like structures in IRD patient 8 in Table 1.
(4) Small peroxisomes (mean diameter below 0.5 μm) with a number of them being elongated, or with an increased peroxisomal number in primary hyperoxaluria type I.
(5) Very small, dense peroxisomes (0.15–0.24 μm diameter) with a clear halo beneath the limiting membrane, as in some IRD (Beard et al 1986; Hughes et al 1990) and NALD (Goldfischer et al 1985; Vamecq et al 1986) patients.
(6) 'Empty' vesicles smaller than 0.5 μm diameter (Espeel et al 1995b) in patients with loss of multiple peroxisomal functions.
(7) Morphologically normal peroxisomes as in X-ALD.

Our results are based on DAB-marked peroxisomes. Recently Hughes et al (1993) published morphometric results of unlabelled peroxisomes: their volume parameter values were smaller than our control values but their diameter was higher. It is probable that the smallest peroxisomal contours were not taken into account because they were not recognized as peroxisomes without DAB staining, and on the other hand that vesicles of other kinds were mistaken for peroxisomes. Whatever the explanation, this experience shows that one cannot substitute data from unlabelled sections for those from DAB-stained peroxisomes, and vice versa. We always rely on specific peroxisomal markers (catalase activity with DAB; immunolocalization of membrane or matrix proteins). By ultra-structure alone, peroxisomal profiles may be missed even by experienced pathologists. We advise morphometry to be performed *after specific staining of the organelles* and to compare with control values obtained by the same procedures of staining and measurement.

STATISTICS

For comparison of groups Kruskall–Wallis and Mann–Whitney tests are used.

In order to evaluate the deviation of an individual patient, confidence limits of $p = 0.05$ or 0.01 are set around the control mean by adding/subtracting two or three times the standard deviation. This assumes that the parameter is normally distributed.

REFERENCES

Abe K, Matsushima S, Mita M (1983) Estimation of the mean size of spheres from the profiles in ultrathin sections: a stereological method. *J Electron Microsc* **32**: 57–60.
Aubourg P, Scotto J, Rocchiccioli F, Feldmann-Pautrat D, Robain O (1986) Neonatal adrenoleukodystrophy. *J Neurol Neurosurg Psychiatry* **49**: 77–86.

Beard ME, Moser AB, Sapirstein V, Holtzman E (1986) Peroxisomes in infantile phytanic acid storage disease: a cytochemical study of skin fibroblasts. *J Inher Metab Dis* **9**: 321–334.

De Craemer D (1994) Peroxisomes in acquired and congenital diseases: human and experimental studies. PhD thesis. Brussels, 133–155.

De Craemer D, Rickaert F, Wanders RJA, Roels F (1989) Hepatic peroxisomes are smaller in primary hyperoxaluria type 1. *Micron Microsc Acta* **20**: 125–126.

De Craemer D, Kerckaert I, Roels F (1991a) Hepatocellular peroxisomes in human alcoholic and drug-induced hepatitis: a quantitative study. *Hepatology* **14**(5): 811–817.

De Craemer D, Zweens MJ, Lyonnet S, et al (1991b) Very large peroxisomes in distinct peroxisomal disorders (rhizomelic chondrodysplasia punctata and acyl-CoA oxidase deficiency): novel data. *Virchows Archiv(A)* **419**: 523–525.

De Craemer D, Pauwels M, Roels F (1993a) Peroxisomes in cirrhosis of the human liver: a cyto-chemical, ultrastructural and quantitative study. *Hepatology* **17**: 404–410.

De Craemer D, Pauwels M, Hautekeete M, Roels F (1993b) Alterations of hepatocellular peroxisomes in patients with cancer. Catalase cytochemistry and morphometry. *Cancer* **71**(12): 3851–3858.

Espeel M, Roels F, Van Maldergem L, et al (1991) Peroxisomal localization of the immunoreactive β-oxidation enzymes in a neonate with a β-oxidation defect: pathological observations in liver, adrenal cortex and kidney. *Virchows Archiv(A)* **419**: 301–308.

Espeel M, Heikoop JC, Smeitink JAM, et al (1993) Cytoplasmic catalase and ghost-like peroxisomes in the liver from a child with atypical chondrodysplasia punctata. *Ultrastruct Pathol* **17**: 623–626.

Espeel M, Mandel H, Poggi F, et al (1995a) Peroxisome mosaicism in the liver of peroxisomal deficiency patients. *Hepatology*, **22**: 497–504.

Espeel M, Roels F, Giros M, et al (1995b) Immunolocalization of a 43 kDa peroxisomal membrane protein in the liver of patients with generalized peroxisomal disorders. *Eur J Cell Biol* **67**: 319–327.

Fringes B, Reith A (1982) Time course of peroxisome biogenesis during adaptation to mild hyperthyroidism in rat liver. A morphometric-stereologic study by electron microscopy. *Lab Invest* **47**: 19–26.

Goldfischer S, Collins J, Rapin I, et al (1985) Peroxisomal defects in neonatal-onset and X-linked adrenoleukodystrophies. *Science* **227**: 67–70.

Goldfischer S, Collins J, Rapin I, et al (1986) Pseudo-Zellweger syndrome: deficiencies in several peroxisomal oxidative activities. *J Pediatr* **108**: 25–32.

Hughes JL, Poulos A (1992) Ultrastructural examination of tissues from patients with peroxisomal disorders. In: Megias-Megias L, Rodriguez-Garcia MI, Rios A, Arias JM, eds. *Electron Microscopy 92: Biological Sciences.* Granada: University of Granada, 521–522.

Hughes JL, Poulos A, Robertson E, et al (1990) Pathology of hepatic peroxisomes and mitochondria in patients with peroxisomal disorders. *Virchows Archiv(A)* **416**: 255–264.

Hughes JL, Bourne AJ, Poulos A (1993) Establishment of a normal range of morphometric values for peroxisomes in paediatric liver. *Virchow's Archiv(A)* **423**: 453–457.

Kerckaert I, Claeys A, Just W, Cornelis A, Roels F (1989) Automated image analysis of rat liver peroxisomes after treatment with thyroid hormones: changes in number, size and catalase reaction. *Micron Microsc Acta* **20**(1): 9–18.

Loud AV (1968) A quantitative stereological description of the ultrastructure of normal rat liver parenchymal cells. *J Cell Biol* **37**: 27–46.

Mandel H, Espeel M, Roels F, et al (1994) A new type of peroxisomal disorder with variable expression in liver and fibroblasts. *J Pediatr* **125**: 549–555.

Poll-The BT, Saudubray JM, Ogier HAM, et al (1987) Infantile Refsum disease: an inherited peroxisomal disorder. Comparison with Zellweger syndrome and neonatal adrenoleukodystrophy. *Eur J Pediatr* **146**: 477–483.

Poll-The BT, Roels F, Ogier H, et al (1988) A new peroxisomal disorder with enlarged peroxisomes and a specific deficiency of acyl-CoA oxidase (pseudo-neonatal adrenoleukodystrophy). *Am J Hum Genet* **42**: 422–434.

Roels F (1991) Peroxisomes, a personal account. Brussels: VUB Press, 1–151. ISBN: 90-70289-94-6.

Roels F, Cornelis A (1989) Heterogeneity of catalase staining in human hepatocellular peroxisomes. *J Histochem Cytochem* **37**: 331–337.

Roels F, Pauwels M, Cornelis A, et al (1983) Peroxisomes (microbodies) in human liver. Cytochemical and quantitative studies of 85 biopsies. *J Histochem Cytochem* **31**: 235–237.

Roels F, Cornelis A, Poll-The BT, et al (1986) Hepatic peroxisomes are deficient in infantile Refsum disease: a cytochemical study of 4 cases. *Am J Med Genet* **25**: 257–271.

Roels F, Pauwels M, Poll-The BT, et al (1988) Hepatic peroxisomes in adrenoleukodystrophy and related syndromes. Cytochemical and morphometric data. *Virchows Archiv(A)* **414**: 275–285.

Roels F, Espeel M, De Craemer D (1991) Liver pathology and immunocytochemistry in congenital peroxisomal diseases: a review. *J Inher Metab Dis* **14**: 853–874.

Roels F, Espeel M, Poggi F, Mandel H, Van Maldergem L, Saudubray JM (1993) Human liver pathology in peroxisomal diseases: a review including novel data. *Biochimie* **75**: 281–292.

Roels F, Espeel M, Mandel H, et al (1994) Cell and tissue heterogeneity in peroxisomal patients. In Wanders RJA, Schutgens RBH, Tabak HF, eds. *Peroxisomal Disorders in Relation to Functions and Biogenesis of Peroxisomes*. Amsterdam; Elsevier, in press.

Scotto JM, Hadchouel M, Odièvre M, et al (1982) Infantile phytanic acid storage disease, a possible variant of Refsum's disease: three cases including ultrastructural studies of the liver. *J Inher Metab Dis* **5**: 83–90.

Smeitink JAM, Beemer FA, Espeel M, et al (1992) Bone dysplasia associated with phytanic acid accumulation and deficient plasmalogen synthesis: a peroxisomal entity amenable to plasmapheresis. *J Inher Metab Dis* **15**: 377–380.

Vamecq J, Draye JP, Van Hoof F, et al (1986) Multiple peroxisomal enzymatic deficiency disorders. A comparative biochemical and morphologic study of Zellweger cerebrohepatorenal syndrome and neonatal adrenoleukodystrophy. *Am J Pathol* **125**: 524–535.

Van Maldergem L, Espeel M, Wanders RJA, et al (1992) Neonatal seizures and severe hypotonia in a male infant suffering a defect in peroxisomal β-oxidation. *Neuromusc Disord* **2**: 217–224.

Weibel ER (1969) Stereological principles for morphometry in electron microscopic cytology. *Int Rev Cytol* **26**: 235–302.

Weibel ER (1979) *Stereological Methods*, vol. 1, *Practical Methods for Biological Morphometry*. London: Academic Press.

Weibel ER, Kistler GS, Scherle WF (1966) Practical stereological methods for morphometric cytology. *J Cell Biol* **30**: 23–38.

J. Inher. Metab. Dis. 18 Suppl. 1 (1995) 181–213
© SSIEM and Kluwer Academic Publishers.

Secondary alterations of human hepatocellular peroxisomes

D. DE CRAEMER

Menselijke Anatomie & Embryologie, Vrije Universiteit Brussel, Laarbeeklaan 103, B-1090 Brussels, Belgium

Summary: The morphological and morphometric characteristics of peroxisomes in normal human liver and the peroxisomal alterations in the liver of patients with acquired or congenital non-peroxisomal diseases are reviewed. Secondary peroxisomal changes are observed in steatosis, hepatitis and cirrhosis induced by various agents (viruses, alcohol, drugs, etc.), in cholestasis, in hepatomas, in extrahepatic cancer with or without liver metastasis, in extrahepatic inflammatory processes, in metabolic disorders affecting metabolism of carbohydrates, lipids and lipoproteins, glycoproteins, amino acids, bilirubin or copper, and in altered thyroid hormone levels. They are recognized as a proliferation of peroxisomes (increased in number and to a lesser extent in surface density and volume density) often accompanied by a minor reduction in size (at most to 68% of the mean diameter in control livers) but very rarely by an increase in mean peroxisomal diameter, and as proliferation-related changes in shape (tails, gastruloid cisternae, funnel-like constrictions, elongation, protrusions) in at least a few of the peroxisomes. These secondary alterations of the peroxisomes are clearly distinguishable from the primary changes in peroxisomes observed in the liver of patients with congenital peroxisomal disorders.

The morphological and morphometric characteristics of hepatic peroxisomes have been studied in many congenital peroxisomal disorders (Roels et al 1991, 1993). In contrast, ultrastructural reports on other congenital disorders and on acquired disorders rarely contain peroxisomal data; when included, they are mostly based on visual impressions. This lack of peroxisomal data is due in part to the inconspicuous morphology of human peroxisomes in routinely stained liver biopsy samples and gives the false impression that peroxisomal changes are rare in human pathology. It is our experience that the optimal way to look for peroxisomal alterations is a morphometric approach combined with the use of a peroxisome-specific stain that facilitates identification of the organelles. This procedure often leads to the discovery of significant alterations in size, number and shape of the peroxisomes in the liver of patients with acquired or congenital diseases; alterations that were undetectable in routinely stained sections and without morphometric analysis. In the older literature description of peroxisomes was omitted or they were considered normal. However, secondary peroxisomal alterations may be significant: (1) they may influence the activities of peroxisomal enzymes and the metabolic pathways in which these enzymes intervene; (2) they have a repercussion when the diagnosis of a primary peroxisomal disorder is suspected, since the observed peroxisomal alterations can be elicited by other disease conditions.

Current knowledge of the peroxisomal morphology in normal human liver and in the liver of patients with an acquired or congenital non-peroxisomal disease is summarized here. The implications of the observed peroxisomal alterations on the metabolic efficiency of the peroxisomal enzymes are discussed in a separate section. For information concerning peroxisomes in patients with peroxisomal disorders, the reader is referred to other chapters in this Handbook.

PEROXISOMAL MORPHOLOGY IN NORMAL HUMAN LIVER

A study of the peroxisomal pathology in human liver requires the knowledge of the normal morphological and morphometric characteristics of these organelles. In man, the notion of 'normal liver' usually refers to a biopsy sample that was performed for clinical and biochemical symptoms suggesting mild liver disease but in which no pathological alterations were found by light microscopy. One might argue that such patients are not truly normal as they may suffer from an unrecognized lesion, from a disease that might indirectly affect the liver, and have often taken medication prior to the biopsy sampling. In practice the microscopist has to deal with a range of normal ultrastructures in human liver biopsies (Phillips et al 1987). In this section, we discuss different aspects of the peroxisomal morphology in normal human liver.

Under the electron microscope, peroxisomes in control human livers have a homogeneous, finely fibrillar or granular, non-compartmentalized matrix surrounded by a single tripartite membrane (de la Iglesia 1969; Sternlieb 1979). Rodent peroxisomes normally display nucleoids or marginal plates. In contrast, peroxisomes with matrix inclusions (cores, nucleoids or marginal plates) have infrequently been described in a minority of normal human livers (Sternlieb 1970, 1979; Ma and Biempica 1971; Sternlieb and Quintana 1977, 1982; Phillips et al 1987). We observed amorphous inclusions in a few peroxisomes in one adult out of seven adult and two paediatric control liver biopsy samples (De Craemer, unpublished observation). According to de la Iglesia (1969), normal peroxisomes do not contain nucleoids as a general rule. In addition, de Duve and Baudhuin (1966) looked at 'numerous' human liver sections and did not detect one core-containing peroxisome. This is confirmed by Roels et al (1988), who reported only a single patient with peroxisomal marginal plates and one other patient with amorphous peroxisomal nucleoids in a total series of more than 150 biopsy samples studied by electron microscopy up to 1994 (Roels, personal communication). Espeel et al (1991) found an amorphous acentric nucleoid in the hepatic and renal peroxisomes of one patient with a peroxisomal disorder. Nucleoids were also found in liver peroxisomes of the peroxisomal disorder patient described by Vanhole et al (1994). We examined the ultrastructure of more than 80 liver biopsy samples of patients with an acquired disease affecting the liver; only in three cases were nucleoids observed in the matrix of peroxisomes. Taking into account the low incidence of peroxisomal matrix inclusions in human liver, we assume that peroxisomes normally lack these inclusions.

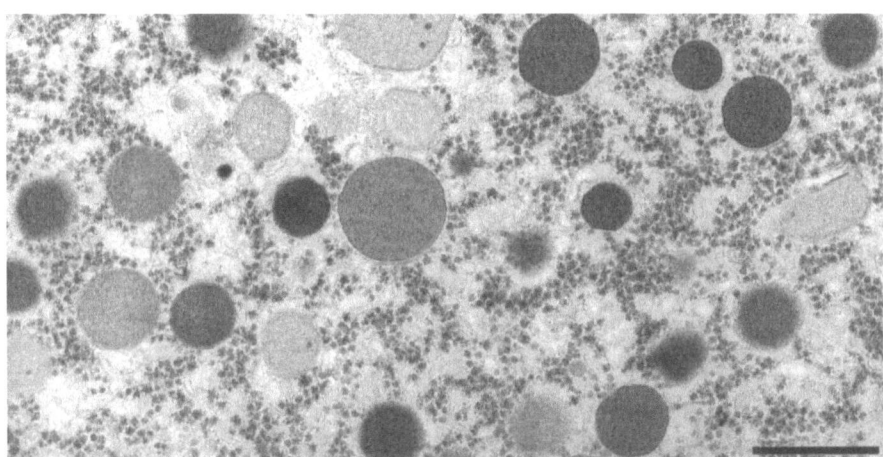

Figure 1 Electron micrograph after catalase staining shows peroxisomes in control human liver. Peroxisomes contain an electron-dense matrix filled with catalase reaction product. Tangential sections through peroxisomes are easily identified. The organelles form a typical cluster. The shape of the peroxisomes is regular; some are round to oval, others are slightly angular. The electron density of the matrix of each peroxisome is homogeneous; between the organelles, a heterogeneity in electron density is observed, reflecting differences in catalase concentration (Roels and Cornelis 1989). Glycogen rosettes are well contrasted. Bar = 1 μm

The identification of tangentially sectioned peroxisomes is facilitated after cytochemical staining for the activity of a peroxisomal enzyme. As a result, more peroxisomal profiles are observed when compared to routinely stained sections. Morphometric analysis of peroxisomes performed in routinely stained sections risks underestimating the real number of organelles present or, conversely, including profiles that are not peroxisomes. In addition, the mean size of peroxisomes also may be incorrect. Using catalase cytochemistry the peroxisomal matrix is filled with an electron-dense catalase reaction product (Roels and Goldfischer 1979; Beier and Fahimi 1986) (Figure 1) and the organelles become visible for light-microscopic evaluation (Figure 2). In recent years, immunocytochemical localization of proteins in the peroxisomal matrix and membrane has applied successfully in human tissues, mostly from patients with a (presumed) peroxisomal disorder (Cooper et al 1988; Litwin et al 1988; Danpure et al 1989, 1990; Espeel et al 1990a,b; Kamei et al 1993).

Peroxisomes in unaffected human liver have a random distribution or are arranged in clusters that are randomly distributed in the cytoplasm (Figure 2). In 1 of 10 control livers a perinuclear configuration of the peroxisomes (Figure 3) in several hepatocytes was reported (Roels et al 1983). As the perinuclear configuration of peroxisomes is related to proliferation (Roels et al 1983; De Craemer et al 1991a, 1993a,b), we assume that this phenomenon may not be found in real control livers. Therefore, the observation by Roels et al (1983) has to be considered as an example of the variation in 'normal' livers.

The shape of peroxisomes in normal liver is round, ellipsoid or slightly angular (Figure 1). The mean diameter of the peroxisomes varies between 0.487 and 0.620 μm in adult human liver (De Craemer et al 1993b) and between 0.51 and 0.63 μm in routinely stained

Figure 2 Light microscopy of hepatic peroxisomes visualized by their catalase activity. Peroxisomes are recognized as small dots in the cytoplasm of the hepatocytes. A relatively homogeneous distribution of the peroxisomes is observed in this control liver. Two inclusions in a macrophage also stain. Bar = 10 μm

paediatric liver from patients ranging in age from 3 months to 18 years (Hughes et al 1993). In the liver of a 6-week-old infant, mean peroxisomal d-circle was 0.445 μm (De Craemer et al 1991c). The mean area of the peroxisomal profiles in 15 control livers was 0.20 μm^2; this corresponds to a d-circle (diameter of the circle with the same area) of 0.505 μm (Hughes et al 1992).

Several different parameters are used to quantify the number of peroxisomes. In the liver of 5 non-alcoholics Chen et al (1987) found 8.3 peroxisomes per 100 μm^2 cytoplasm (range 5.9 – 10.1). In the livers of 4 young adults (medical students), the number of peroxisomes per cell was 1050 ± 14 (Rohr et al 1976). The numerical density (the number of organelles per unit volume) ranges between 0.052 and 0.132/μm^3 in adult liver (De Cramer et al 1993b) and between 0.057 and 0.188/μm^3 in paediatric liver (Hughes et al 1993).

The volume density of the peroxisomal compartment, i.e. the fraction of cellular volume occupied by peroxisomes, in hepatocytes of control livers ranged from 0.734% to 1.441% in liver of 7 adults (De Craemer et al 1993b) and from 1.7% to 3.2% in human paediatric

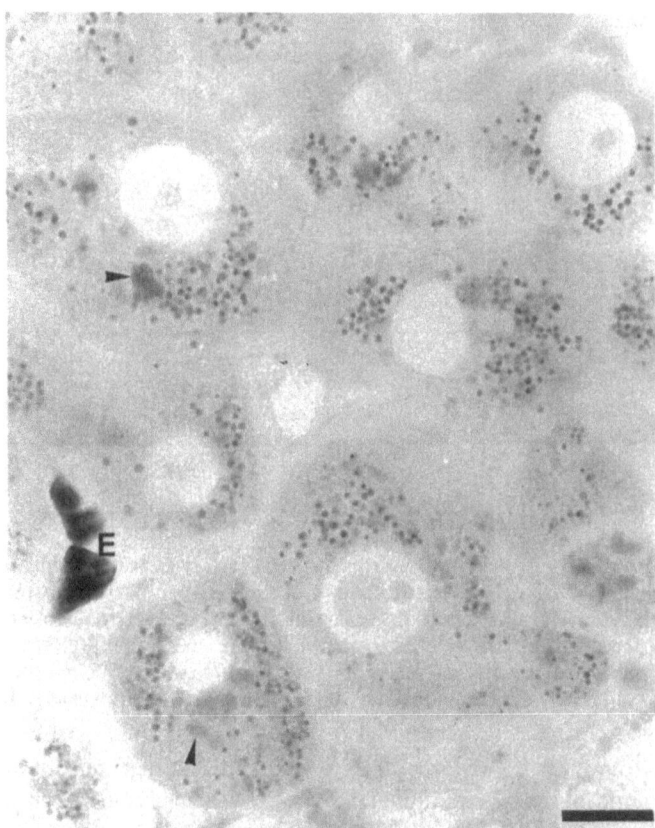

Figure 3 This light micrograph shows a perinuclear configuration of the peroxisomes in the liver of a patient with cirrhosis. Catalase cytochemistry. A perinuclear distribution is found in a variable number of hepatocytes in patients suffering from diverse diseases; in the remaining parenchymal cells peroxisomes reveal a normal homogeneous distribution. The number (numerical density) of peroxisomes in this liver was increased to 171% as shown by morphometry. Lipofuscin is also present in the hepatocytes (arrowheads). Large black structures are erythrocytes (E). Bar = 10 μm

liver (Hughes et al 1993). The mean volume density in routinely stained liver samples of 5 non-alcoholics was 1.64±0.99% (Chen et al 1987). This wide range can be ascribed to individual differences of the patients (age, drug treatments, diets, other conditions) as well as to different approaches in determining the reference volume (e.g. hepatocellular cytoplasm without lipid droplets (Oudea et al 1973; De Craemer et al 1991a,b, 1993a,b), complete hepatocyte (Oudea et al 1973; Rohr et al 1976) or liver tissue (Rohr et al 1976; Gariot et al 1986)). Finally, the morphometric technique applied and, most importantly, the use or absence of a cytochemical marker will also determine the results. A few reports also deal with other volumetric parameters: the mean peroxisomal volume was $0.13 ± 0.04 \, \mu m^3$ in the control liver of 3 children (Landrieu et al 1982) and the total peroxisomal volume per hepatocyte in 4 healthy young adult livers was $132 ± 10 \, \mu m^3$ (Rohr et al 1976). These two findings are in agreement with the number of peroxisomes per hepatocyte (1050)

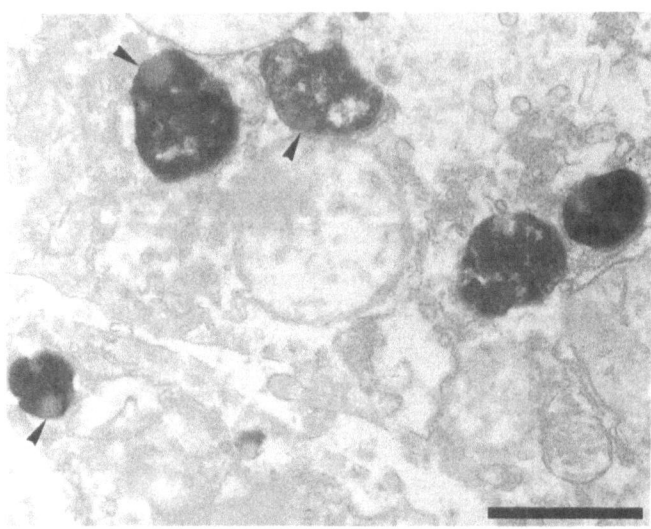

Figure 4   Electron micrograph of a parenchymal cell in the liver of a patient with acute heart failure and a normal liver histology. Catalase cytochemistry. Post-mortem alterations resulted at 4°C by postponing the fixation of liver for 48 h. In some peroxisomes, the electron-dense reaction product shows a heterogeneous dispersion in the matrix, leaving transparent (electron-lucent) spots. Heterogeneity in electron density of the matrix between peroxisomes is enhanced. Amorphous matrical inclusions devoid of catalase reaction product are observed (arrowheads). Bar = 1 μm

calculated by Rohr et al (1976). The mean peroxisomal surface density ranges between 0.07 and 0.11 μm^{-1} in normal adult human liver (De Craemer et al 1993b) and is slightly higher in paediatric liver (Hughes et al 1993).

Spontaneous post-mortem alterations of the peroxisomes lead to a decrease of peroxisomal catalase staining that occurs in a heterogeneous distribution throughout the liver (De Craemer et al 1990b). Ultrastructurally, the peroxisomes reveal a matrix filled with clumped catalase reaction product leaving transparent spots (Figure 4). The mean diameter and number of peroxisomes were slightly (~10%) decreased in samples fixed with a delay of 48 h (De Craemer et al 1990b). Evaluation of the condition caused by the pathology of the patient is possible up to 48 h *post mortem* because catalase staining remains effective on condition that the body or the tissue samples are conserved at a temperature of 4°C. Immunocytochemical localization of peroxisomal proteins is still possible 3 days *post mortem* (Espeel et al 1990a).

PEROXISOMAL MORPHOLOGY IN HUMAN PATHOLOGY

The most frequent alterations of peroxisomes in the liver of patients with a 'non-peroxisomal' disease are changes in number and size, and the presence of inclusions; they are summarized in Table 1. Other peroxisomal alterations, of matrix density, distribution and shape, in the liver of patients with specific diseases are also discussed in this section.

We make a distinction between data on number and size of peroxisomes obtained by morphometric analysis of random areas of sections specifically stained for peroxisomes,

Table 1 Peroxisomal alterations in size and number, and the presence of matrix inclusions in the liver of patients with different non-peroxisomal diseases as reported in the literature

Pathology	Reference (±)[a]	Peroxisomal size evaluated[b]		Peroxisomal number evaluated[b]		Inclusions	
		Visually[c]	Morphometrically	Visually[c]	Morphometrically	Core	Marginal plate
Viral hepatitis							
Chronic active	Sternlieb and Quintana 1982 (−)	Increase (1/1)					
	Sternlieb 1979 (−)						
	De Craemer et al 1991b (+)		Increase[d] (2/3)	Increase (3/3)		+	
Acute	de la Iglesia 1969 (−)	Increase	Normal (2/2)		Increase (1/2)		
	Schaffner 1966 (−)	Increase		Increase			
	De Craemer et al 1991b (+)		Decrease (1/1)		Increase (1/1)		
Chronic with immuno-suppression	Reinke et al 1988 (−)		Decrease[e] [22]	Decrease [22]			
Chronic persistent	De Craemer et al 1991b (+)	Increase[f] [8]	Decrease [8]		Increase (7/8)		
Cholestasis							
Chlorpromazine-induced	Cooper et al 1988 (+)			Increase (1/1)	Increase (1/1)		
Extrahepatic	Jezequel et al 1983 (−)			Increase		+	
Jaundice of pregnancy	Phillips et al 1987 (−)			Increase		+	
Unspecified	Roels et al 1983 (+)			Increase (3/11)			
Metabolic diseases							
Glycogen storage disease III	Phillips et al 1987 (−)	Decrease		Decrease			
Pyruvate dehydrogenase deficiency	De Craemer unpublished (+)		(Increase)[g] (1/1)		Increase (1/1)		
Abetalipoproteinaemia	Collins et al 1989 (−)		Increase[d] (1/1)			+	+
Hyperlipoproteinaemia type IIA	Gariot et al 1986 (−)		Increase [4]				
Hyperlipidaemia+CPIB	Hanefeld et al 1983 (−)		Decrease[h] [16]		Increase [16]		

continued

Table 1 (*continued*)

| Pathology | Reference (±)[a] | Peroxisomal size evaluated[b] | | Peroxisomal number evaluated[b] | | Inclusions | |
		Visually[c]	Morphometrically	Visually[c]	Morphometrically	Core	Marginal plate
Hyperlipidaemia + gemfibrozil	de la Iglesia et al 1982 (–)						+
Carnitine deficiency	Phillips et al 1987 (–)			Increase			
Congenital total lipo-dystrophy	Phillips et al 1987 (–)	Decrease (1/1)			Decrease (1/1)	+	
	Harbour et al 1981 (–)			Increase (1/1)		+	
	Klar et al 1987 (–)			Increase (1/1)			
	De Craemer et al 1992 (+)		Normal (2/2)		Increase (1/2)		
Medium- and long-chain fatty acid oxidation defect	Lake et al 1987 (–)	Increase (2/2)		Increase (3/3)			
Cerebrotendinous xanthomatosis	Salen et al 1978 (–)			Increase (2/2)			
	Goldfischer and Sobel 1981 (–)					+	
α-1-Antitrypsin deficiency	Phillips et al 1987 (–)	Increase		Increase			
Hypofibrinogenaemia	De Craemer unpublished (+)				Increase (1/1)	+	
Tyrosinaemia type 1	Phillips et al 1987 (–)	Increase					
Partial ornithine trans-carbamylase deficiency with acute hepatic failure	Landrieu et al 1982 (–)		Increase+, decrease[i] (1/1)		Decrease (1/1)		
Gilbert's syndrome + alkaptonuria	Brown and Smuckler 1970 (–)					+	
Idiopathic recurrent cholestasis	de la Iglesia 1969 (–)					+	
Rotor syndrome	Evans et al 1981 (–)			Increase (1/1)		+	+
Familial cholestatic cirrhosis	Sternlieb and Quintana 1977 (–)		Increase[d] (1/1)				

continued

Table 1 *(continued)*

Pathology	Reference (±)[a]	Peroxisomal size evaluated[b] Visually[c]	Peroxisomal size evaluated[b] Morphometrically	Peroxisomal number evaluated[b] Visually[c]	Peroxisomal number evaluated[b] Morphometrically	Inclusions Core	Inclusions Marginal plate
Wilson disease	Phillips et al 1987 (−)	Increase		Increase			
	Sternlieb 1970 (−)						+
	Sternlieb and Quintana 1977 (−)		Increase[d] (3/5), decrease[d] (1/5)			+	+
Menkes disease	Sternlieb and Quintana 1982 (−)	Increase (1/1)					
	Sternlieb and Quintana 1977 (−)						
Reye syndrome	Phillips et al 1987 (−)		Decrease[d] (2/2)	Increase (1/2)			
	Thaler et al 1970 (−)			Increase			
	Partin et al 1971 (−)			Increase (2/2)			
	Iancu et al 1972 (−)			Increase [10]			
	Partin et al 1984 (−)			Increase			
	Svoboda and Reddy 1980 (+)	Normal (6/6)		Increase (2/2)			
Hyperthyroidism	Kerckaert et al 1939 (+)			Increase (1/1)		+	
Alcoholic liver diseases							
Early acute alcoholism Alcoholic foamy degeneration	Rubin and Lieber 1967 (−)	Increase (2/2)	Increase (2/2)	Increase (2/2)			
	Uchida et al 1983 (−)			Decrease [21]			
Alcoholic fatty liver	De Craemer et al 1995 (+)	Decrease (2/19)	Decrease (8/12)	Increase (3/19)	Increase (10/12)		
Alcoholic hepatitis	De Craemer et al 1991a (+)	Increase[f] (3/4)	Decrease (2/4)		Increase (4/4)	+	
Ethylic cirrhosis	Roels and Cornelis 1989 (+)			Increase (1/1)			
	De Craemer et al 1993b (+)		Decrease (7/11)		Increase (11/11)		
Mixed alcoholic liver diseases	Phillips et al 1987 (−)	Increase		Increase		+	
	Horvath et al 1975 (−)			Increase (10/30)			

continued

Table 1 (continued)

Pathology	Reference (±)[a]	Peroxisomal size evaluated[b]		Peroxisomal number evaluated[b]		Inclusions	
		Visually[c]	Morphometrically	Visually[c]	Morphometrically	Core	Marginal plate
Malignant diseases							
Hepatoma (hepatocarcinoma)	Phillips et al 1987 (−)			Decrease			
	Ruebner et al 1967 (−)					+	
	Keeley et al 1972 (−)			Increase (1/1) Absent (2/5)			
	Ma and Blackburn 1973 (−)			Decrease			
Liver cell adenoma	Phillips et al 1987 (−)						
Normal liver parenchyma of patients with							
extrahepatic carcinoma	De Craemer et al 1993a (+)		Decrease (12/20)	Increase (21/39)[j]	Increase (17/20)	+	
colorectal cancer	De Netto et al 1991 (−)						+
hepatoma	De Craemer et al 1993a (+)						+
'malignancy'	Roels et al 1983 (+)			Increase (2/22)	Increase (1/2)		
Other liver diseases							
Non-alcoholic fatty liver	De Craemer et al 1995 (+)	Decrease (2/8)	Decrease (7/7)		Increase (7/7)		
Drug-induced hepatitis[k]	De Craemer et al 1991a (+)		Decrease[l] (4/6)		Increase (4/6)		
Drug and toxic liver injury	Phillips et al 1987[m] (−)	Increase +, decrease					
Non-alcoholic cirrhosis	De Craemer et al 1993b (+)		Decrease (1/3)		Increase (2/3)		
Cirrhosis of mixed aetiologies	De Craemer et al 1993b (+)	Decrease (11/32)		Increase (8/32)			

a With (+) or without (−) a peroxisome-specific marker
b The most reliable data on size and number are obtained by morphometric analysis and by the use of a peroxisomal marker. Figures in parentheses indicate the number of patients in whose liver the peroxisomal alteration was observed versus the total number of patients investigated; figures in square brackets indicate the total number of patients studied; if both are absent, no data on the number of patients investigated are available
c By light and/or electron microscopy. d Maximum diameter instead of the true diameter (d-circle) was investigated
e Smaller peroxisomes were only found in areas with proliferated endoplasmic reticulum containing HBsAg-particles
f A small number of peroxisomes appeared swollen
g Mean peroxisomal size in this 3-month-old girl was higher than in cytochemically stained liver of two controls (age 6 weeks and 4 months) described by De Craemer et al (1991c) but was smaller than in two routinely stained liver samples from infants (age 3 and 4 months) described by Hughes et al (1993)
h Estimated from the ratio numerical density/volume density. i Two subpopulations consisting of smaller and enlarged organelles
j In 13 liver samples, increase in peroxisomal number was restricted to small areas of liver parenchyma
k Hepatitis induced by amitryptiline, aprindine, clomipramine, methimazole, phenytoin or an unspecified drug
l A decrease in size when compared to 7 human adult livers (De Craemer et al 1993a). m See Table 5 in Chanter 4 of Phillips et al (1987)

and conclusions reached by visual evaluation of routinely stained samples. The latter are based on less reliable, subjective criteria, and are therefore inferior to the objective quantitative data obtained by morphometric analysis.

In the following we first describe the peroxisomal changes for each group of diseases or conditions. Subsequently each peroxisomal parameter (such as size, shape and number) is discussed in a separate section.

Viral hepatitis

The mean diameter of the overall population of peroxisomes in patients with chronic hepatitis B and on immunosuppressive treatment was unchanged (0.64μm) when compared to controls, while the diameter of the peroxisomes in proliferated areas of endoplasmic reticulum with HBsAg particles dropped to 0.49μm (Reinke et al 1988). Furthermore, the mean peroxisomal diameter was decreased in a patient recovering from acute viral hepatitis, and showed a tendency to decrease in 8 patients with chronic persistent hepatitis; in 2 patients with chronic aggressive hepatitis, peroxisomal size was unchanged (De Craemer et al 1991b).

Peroxisomal proliferation was evident in 9 of 11 patients with viral hepatitis (De Craemer et al 1991b). In contrast, fewer peroxisomes were reported in the liver of patients with chronic hepatitis B combined with immunosuppressive therapy (Reinke et al 1988); lytic decomposition of the organelles via partial or total destruction of the membrane and autophagocytosis were the observed mechanisms by which peroxisomes were broken down (Reinke et al 1988).

Peroxisomes revealed a condensed matrix in routinely stained livers of patients with chronic active hepatitis (Sternlieb and Quintana 1977). In contrast, peroxisomes with a rarified, less electron-dense matrix were observed in acute viral hepatitis (de la Iglesia 1969). Light-microscopic evaluation of liver sections stained for catalase activity revealed a decrease in peroxisomal staining in several patients (Roels et al 1983; De Craemer et al 1991b). A perinuclear configuration of the hepatocellular peroxisomes was sometimes evident (Figure 3); a normal homogeneous distribution of the organelles was observed in the other cells (De Craemer et al 1991b).

On ultrastructural analysis of these samples, a decrease in electron density of the catalase reaction product in the matrix was noticed in several but not all peroxisomes (De Craemer et al 1991b). The decrease in catalase activity showed two patterns: a homogeneous loss of electron density and a heterogeneous distribution of the catalase reaction product in the matrix (Figure 5). Some peroxisomes showed transparent spots (Figure 5).

In all 11 liver samples of patients with viral hepatitis, abnormally shaped peroxisomes were observed (De Craemer et al 1991b): some peroxisomes were curled or strongly angular (Figure 6), others contained a tail (Figure 7), a membrane loop (a so-called gastruloid cisterna) (Figure 8), or one or more cytoplasmic invaginations of the membrane (Figure 8). In 3 patients with chronic active hepatitis, the predominant profile of the peroxisomes was irregular (Sternlieb and Quintana 1977).

In an animal model of viral hepatitis, peroxisomal staining intensity indicated a progressive decrease in catalase activity; the number of peroxisomes was decreased by 62% after 72h of infection (De Craemer et al 1990a). This rapid loss of peroxisomes, probably via a

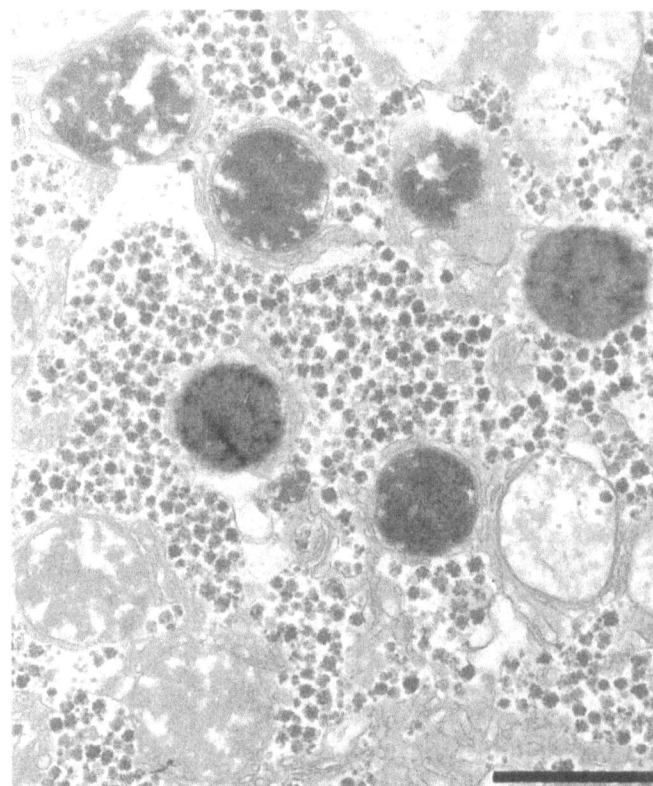

Figure 5 Electron micrograph of a liver sample stained for catalase activity in a patient with chronic persistent hepatitis. Catalase reaction product is heterogeneously distributed in the matrix of most peroxisomes, leaving transparent spots. Catalase reaction product also strongly differs between peroxisomes. Endoplasmic reticulum membranes close to the peroxisome sometimes create the impression of a double membrane. Bar = 1 μm

lytic process, was not observed in human viral hepatitis and may be ascribed to the more severe condition in the animals leading to death (De Craemer et al 1990a).

Based on visual evaluation only, an increase in peroxisomal size and number was reported in several liver biopsy samples of patients with viral hepatitis (Schaffner 1966; de la Iglesia 1969; Sternlieb 1979). The enlargement of the peroxisomes seems to be supported by a study of the maximum diameter of a restricted number of peroxisomes (n = 30) in 3 patients (Sternlieb and Quintana 1977). However, this parameter does not give information about the size or area of the measured profiles as it is increased in each elongated or irregularly shaped peroxisome.

Cholestasis

By visual evaluation after catalase cytochemistry, catalase activity in the hepatic peroxisomes was decreased in 10 out of 16 patients with cholestasis (Roels et al 1983). In 5

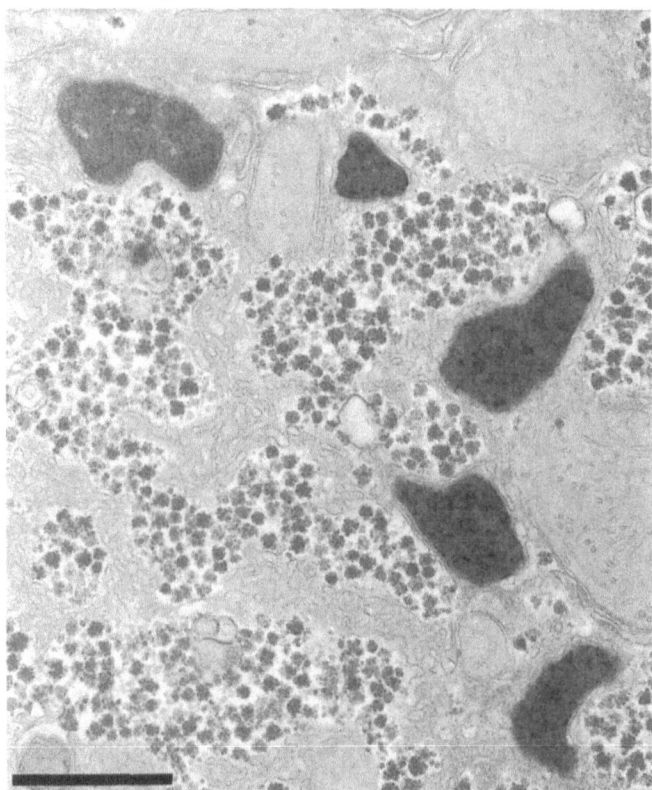

Figure 6 Irregularly shaped peroxisomes in the liver of a patient with chronic persistent hepatitis: some organelles are curled or strongly angular. The amount of catalase reaction product is comparable to that in control livers. Endoplasmic reticulum cisternae are close to the peroxisome, a feature seen in most livers. Bar = 1 μm

patients with extrahepatic cholestasis, peroxisomes with a dense amorphous core were occasionally observed; no alterations in peroxisomal volume and surface density were found when compared to 4 controls (Jezequel et al 1983). Light-microscopic analysis revealed a peculiar distribution of the peroxisomes around the nucleus in several cells of liver biopsy samples in 14 patients (Roels et al 1983).

An increase in peroxisomal number was visually evaluated in several patients with jaundice of pregnancy (Phillips et al 1987), but quantitative data supporting this observation are lacking.

Drug toxicity

Morphometric analysis revealed a decrease of the mean diameter of the peroxisomes during phenytoin treatment and a 2.4-, 2.8- and 4.1-fold increases in the number of peroxisomes in patients with hepatitis induced by, respectively, aprindine, methimazole or phenytoin (De Craemer et al 1991a). In a patient with chlorpromazine-induced cholestasis, the peroxisomal frequency (number of peroxisomes per area of cytoplasm) was more than

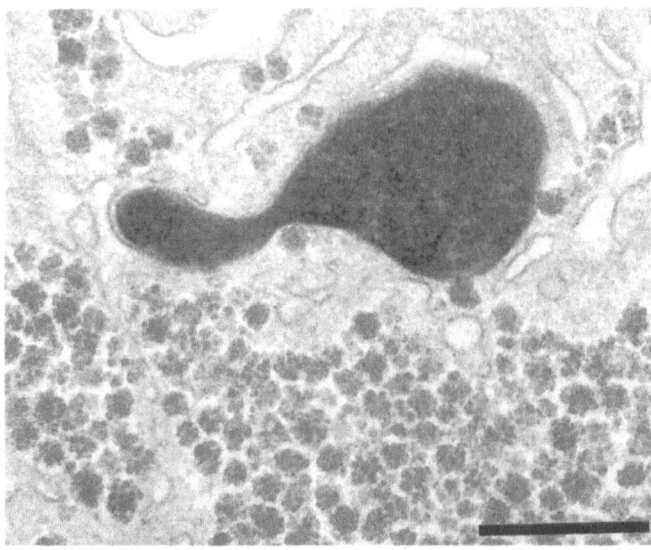

Figure 7   Tailed peroxisome in the liver of a patient with chronic persistent hepatitis. Catalase cytochemistry. This is rarely seen in human liver. Bar = 0.5 μm

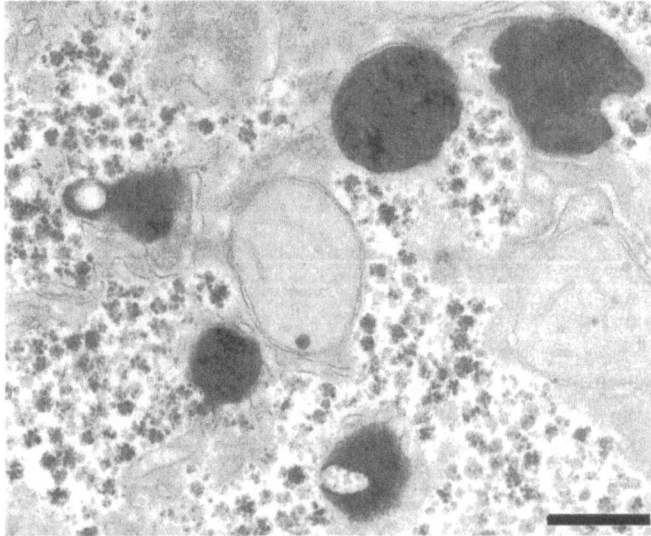

Figure 8   Cluster of five peroxisomes in a hepatocyte of a patient with chronic persistent hepatitis. One organelle has a normal shape; two others are irregularly shaped owing to an invagination of the membrane; one peroxisome contains a short protrusion; the fifth peroxisome reveals a catalase-positive gastruloid cisterna. Bar = 0.5 μm

doubled when compared to the mean frequency in 4 control livers (Cooper et al 1989). The proliferation of peroxisomes observed in the liver of mice treated with a high dose of chlorpromazine was ascribed to the ability of this drug to depress mitochondrial fatty acid oxidation (Vamecq et al 1987).

Peroxisomes with a condensed matrix were reported in the liver of a patient with isoniazid hepatitis (Sternlieb and Quintana 1977). A decrease in catalase staining of the peroxisomes was observed in several patients with toxic hepatitis (De Craemer et al 1991a). Amiodarone treatment induced crystalline inclusions in the peroxisomal matrix (Phillips et al 1987). Irregularly shaped peroxisomes were infrequently observed in toxic hepatitis (De Craemer et al 1991a).

In a patient suffering from methimazole hepatitis, peroxisomes showed a preferential localization around the cell nucleus (De Craemer et al 1991a). Peroxisomes were sometimes arranged around glycogen bodies in the liver of a patient on treatment with glucocorticoids (Phillips et al 1987).

By visual evaluation, an increase in peroxisomal number has been reported in patients treated with acetylsalicylic acid, anabolic steroids, chlorbiphenols, clofibrate, isoniazid, 6-mercaptopurine, oral contraceptives, trimethoprim plus sulfamethoxazole, and spirolactone. Fewer peroxisomes were found after treatment with orotic acid and glucocorticoids. See review by Phillips et al (1987). Unfortunately, no quantitative data are available to support these visually evaluated peroxisomal changes. In an electron micrograph of the liver of a patient on treatment with glucocorticoids, hyperplasia of the peroxisomes was evident: more than 340 peroxisomes are present in an area covering $390\,\mu m^2$ (Fig. D15 in Phillips et al 1987).

Peroxisome proliferators

Much attention is paid to a large number of pharmaceutical (including hypolipidaemic), agrochemical and industrial substances that cause peroxisome proliferation in the liver of male rats and mice. The peroxisome proliferators, recently classified into nine chemically structure-related groups (Bentley et al 1993), induce a dramatic (5–20-fold) increase in the number of peroxisomes and in activity of certain peroxisomal enzymes, especially those involved in β-oxidation of fatty acids, while catalase activity is only moderately increased (Reddy et al 1982; Conway et al 1989; Bentley et al 1993; Reddy and Mannaerts 1994). Peroxisomal size is *enlarged* in the liver of responsive animals (Roels 1991).

In contrast, hamsters show an intermediate response, while guinea pigs, monkeys and humans appear to be relatively insensitive or non-responsive at dose levels that produce a marked response in rats and mice (see Tables 4 and 5 in Bentley et al (1993)). This species-specific difference in peroxisome proliferation is also observed in cultured hepatocytes (Bentley et al 1993). In cultured human hepatocytes no response to peroxisome proliferators is observed (Bichet et al 1990; Blaauboer et al 1990; Bentley et al 1993). Hyperlipoproteinaemic patients under long-term treatment with hypolipidaemic drugs only occasionally show a minor (at most 50%) increase in peroxisomal number accompanied by a *reduction* in size (Staubli and Hess 1975; de la Iglesia et al 1982; Blumcke et al 1983; Hanefeld et al 1983; Gariot et al 1984; Bentley et al 1993).
[See also Note added in proof, p.208]

High-fat diets

Diets supplemented with large amounts of fatty acids ($\geq 10\%$ w/w) in the form of linol salad oil, partially hydrogenated marine oil, fish oil, oleic acid or semi-synthetic mixture, induce the peroxisomal β-oxidation capacity and catalase activity in mice and rats (Ishii et al 1980; Neat et al 1980, 1981; Christiansen et al 1981; Thomassen et al 1982, 1985; Vamecq et al 1987, 1993; Yamazaki et al 1987; Flatmark et al 1988; Rustan et al 1992; De Craemer et al 1993d, 1994). The number of hepatic peroxisomes is also increased (Christiansen et al 1981; Vamecq et al 1987, 1993; De Craemer et al 1994) but, in contrast to the effect of peroxisome proliferators, peroxisomal size remains unchanged or is decreased (Christiansen et al 1981; Vamecq et al 1987, 1993; Roels 1991; De Craemer et al 1994). To our knowledge, data on human liver peroxisomes are not available.

Metabolic abnormalities

In glycogen storage disease type III (McKusick 232400), a disorder of carbohydrate metabolism, peroxisomes were found to be decreased in size and number (Phillips et al 1987). No quantitative analysis was performed.

In the liver of a 3-month-old girl with pyruvate dehydrogenase deficiency (McKusick 312170) we found that the size (mean d-circle $0.526\,\mu$m), number, volume density and surface density of the peroxisomes were slightly increased as compared to the two age controls described by De Craemer et al (1991c). However, Hughes et al (1993) found a higher mean peroxisomal d-circle ($0.56\,\mu$m) in routinely stained control livers of two infants aged 3 and 4 months. When compared to the peroxisomes in 7 normal adult livers, only the numerical density of peroxisomes was slightly higher than the highest control value. The shape of the organelles was less regular, as reflected in the mean and minimum axial ratio (shortest over longest diameter) of the peroxisomes (De Craemer, unpublished observation).

Peroxisomes are more often described in disorders of lipid and lipoprotein metabolism. In one patient with abetalipoproteinaemia (McKusick 200100) the shape of most peroxisomes was pleiomorphic or irregular; the mean largest diameter of the peroxisomes was increased, indicating that the organelles were more elongated than normal, and dense nucleoids and marginal plates were present in certain peroxisomes (Collins et al 1989). Patients with hyperlipoproteinaemia type II (McKusick 144400) revealed a doubling of the peroxisomal volume density when compared to controls. The number of peroxisomes was not significantly altered, but the frequency distribution of the diameter of hepatic peroxisomes showed a shift to higher values (Gariot et al 1986). In contrast to the massive proliferation of the peroxisomes in rodent liver, hypolipidaemic drugs had only a minor or no effect on the peroxisomal number in liver of patients with different types of hyperlipoproteinaemia (Staubli and Hess 1975; de Iglesia et al 1982; Bentley et al 1993; Blumcke et al 1983; Hanefeld et al 1983; Gariot et al 1984). Peroxisomes with marginal plates were observed in hyperlipoproteinaemic patients under long-term gemfibrozil treatment (de la Iglesia et al 1982).

Four reports deal with the ultrastructural alterations of the liver in 5 patients with congenital total lipodystrophy (McKusick 269700) (Harbour et al 1981; Klar et al 1987; Phillips et al 1987; De Craemer et al 1992). Electron-dense aggregates in the peroxisomal matrix were observed in 2 patients (Harbour et al 1981; Phillips et al 1987). Catalase staining of the hepatic peroxisomes was normal in the 2 patients reported by De Craemer et al (1992), indicating that catalase activity was not changed in congenital total lipodystrophy. By morphometric analysis of the peroxisomes in these two patients, proliferation of the organelles was observed in the patient with massive steatosis, while in the patient with mild lipid accumulation peroxisomal number was normal. In both patients the mean peroxisomal size fell within the control range (De Craemer et al 1992). Contradictory results were obtained by visual impression of the peroxisomes in 3 other patients: the organelles seemed to be increased in number in 2 patients (Harbour et al 1981; Klar et al 1987) and appeared reduced slightly in size and number in a third patient (Phillips et al 1987). However, these observations were not confirmed by quantitative analyses.

In three patients with long- or medium-chain fatty acid oxidation defect, peroxisomes were 'numerous'; in 2 patients the peroxisomes contained a densely flocculent granular matrix (Lake et al 1987). The peroxisomal morphology in patients with very long-chain fatty acid oxidation defects is not discussed here as these patients suffer from a disorder of peroxisomal β-oxidation.

By visual evaluation, 'unusually increased numbers' of 'somewhat enlarged' peroxisomes were observed in patients with cerebrotendinous xanthomatosis (McKusick 213700) (Salen et al 1978). Hepatocellular peroxisomes sometimes contained a crystalline structure (Goldfischer and Sobel 1981), revealed an increased matrix density, and were often noted around crystals in the cytoplasm (Salen et al 1978).

In α-1-antitrypsin deficiency (McKusick 107400), a disorder of glycoprotein metabolism, an increase in number of hepatocellular peroxisomes was often reported, but morphometric data are lacking (Phillips et al 1987).

Recently, we had the opportunity to study a liver biopsy from a 6-year-old girl with hypofibrinogenaemia (McKusick 134800). This disorder of glycoprotein metabolism is characterized by fibrinogen storage in the endoplasmic reticulum of the hepatocytes (Pfeifer et al 1981). Morphometric analysis revealed a doubling of the volume density, numerical density (number) and surface density of the hepatocellular peroxisomes (De Craemer, unpublished observation); the size of the organelles was normal. In part of the liver parenchyma, the organelles revealed a very complex shape owing to the presence of several protrusions and dents (Figure 9). As a consequence, the median (0.806) and minimum (0.388) axial ratios (short over long diameter) were lower than in 9 controls (De Craemer, unpublished observation).

Peroxisomal alterations have been reported in two disorders of amino acid metabolism. In tyrosinaemia type 1 (McKusick 276700), peroxisomes may occasionally contain nucleoids or small lipid droplets (Phillips et al 1987). In ornithine transcarbamylase deficiency (McKusick 311250), variation in size and number of peroxisomes, a flocculent aspect of the peroxisomal matrix, and an increased peroxisomal breakdown have repeatedly been reported (Phillips et al 1987). A morphometric analysis of the peroxisomes in the less affected zones of the liver parenchyma was performed in only one patient with acute hepatic failure and partial ornithine transcarbamylase deficiency: the peroxisomes were 40% fewer than in 3 control livers (Landrieu et al 1982). Based on size, two subpopulations of the peroxisomes were recognized: the mean radius of the smaller

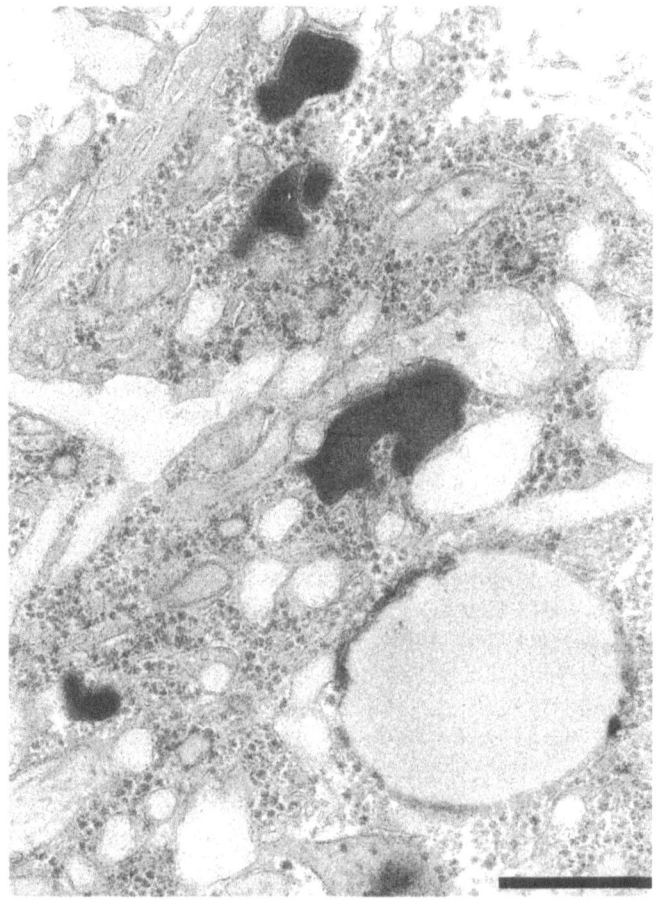

Figure 9   Electron micrograph from the liver of a patient with hypofibrinogenaemia. Catalase cytochemistry. Endoplasmic reticulum is swollen and contains densely packed, irregularly arranged tubules
reflecting the accumulation of fibrinogen (Pfeifer et al 1981). Peroxisomes show a normal electron
density but their shape is strongly irregular, an observation not made in control livers. Bar = 1 μm

ones (25% of the peroxisomes) was 0.19 μm, while the mean radius of the larger organelles
was 0.43 μm. The volume density of the peroxisomal compartment was not different from
that in control livers. Peroxisomes often contained an electron-lucent matrix. The enlarged
peroxisomes were more frequently surrounded by cisternae of the rough endoplasmic
reticulum. This was interpreted as autophagocytosis of the peroxisomes. Prenecrotic cells
were almost completely deprived of peroxisomes (Landrieu et al 1982).

Alterations of the peroxisomal matrix were repeatedly reported in several disorders of
bilirubin metabolism. In a patient with combined Gilbert syndrome (McKusick 143500)
and alkaptonuria, peroxisomes with electron-dense crystalloids were observed (Brown and
Smuckler 1970). The liver of a patient with idiopathic recurrent cholestasis contained
several nucleated microbodies (de la Iglesia 1969). In a patient with faecal porphyrin

abnormalities and with features of Rotor syndrome (McKusick 237450), several peroxisomes had a flocculent matrix and contained paracrystalline inclusions and marginal plates (Evans et al 1981); the number of peroxisomes was increased by visual evaluation. In familial cholestatic cirrhosis the predominant profile of the peroxisomes was round; the mean maximum diameter of the organelles was increased when compared to three control human livers (Sternlieb and Quintana 1977). This enlargement of the peroxisomes should be interpreted with caution as only 30 peroxisomes in routinely stained liver sections of a single patient were measured.

In Wilson disease (McKusick 277900), a disorder of copper metabolism, peroxisomal alterations included the presence of amorphous, centrally located electron-dense inclusions (Sternlieb and Quintana 1977) and marginal plates (Sternlieb 1970); a rarified, less osmiophilic matrix (Sternlieb and Quintana 1982); and often irregularly shaped profiles (Sternlieb and Quintana 1977). By visual evaluation, a strong variation in number has been reported (Sternlieb and Quintana 1977, 1982; Phillips et al 1987). In 3 of 5 patients an increase in maximum diameter (up to $1.54\,\mu$m) was found (Sternlieb and Quintana 1977); this is not unexpected in view of the presence of irregularly shaped peroxisomes. In 2 patients with Menkes disease (McKusick 309400), another disorder of copper metabolism, the density of the peroxisomal matrix was increased (Sternlieb and Quintana 1977). The mean maximum diameter of the predominant round to ellipsoid peroxisomes was significantly decreased ($0.39\,\mu$m and $0.49\,\mu$m); and in one case, peroxisomal number 'seemed' to be increased in some cells (Sternlieb and Quintana 1977).

In Reye syndrome (McKusick 228100) an increased number of peroxisomes was suggested by several authors, but their photographs sometimes display clusters of peroxisomes that are present in most livers (Thaler et al 1970; Partin et al 1971, 1984; Iancu et al 1972; Phillips et al 1987). Others reported that the size and number of peroxisomes were normal (Svoboda and Reddy 1980). Quantitative data are needed to resolve this contradiction. Only Svoboda and Reddy (1980) reported the presence of several peroxisomes with a dense eccentrically located core that was of an amorphous, non-crystalline composition.

In the liver of a patient in a hyperthyroid state, an increase in number of peroxisomes (catalase-positive organelles) was evaluated by light microscopy (Kerckaert et al 1989). In contrast, peroxisomes were reported normal in an ultrastructural study of the liver of patients with hyperthyroidism (Klion et al 1971). No quantitative investigation of the peroxisomes was performed. Morphometric and enzymological data on hepatic peroxisomes are only available in rodents with an experimentally induced change of thyroid hormones. Administration of thyroid hormone causes an increase in peroxisomal β-oxidation capacity (Just and Hartl 1983). Catalase reaction in the peroxisomes was decreased (Kerckaert et al 1989). Concomitant morphological alterations of the peroxisomes include an increase in number and a simultaneous reduction in size (Fringes and Reith 1982; Just et al 1982; Just and Hartl 1983; Kerckaert et al 1989). In contrast, in thyroidectomized rats the population of peroxisomes mainly consisted of low numbers of enlarged peroxisomes ('mega-peroxisomes') (Riede et al 1978).

Alcoholic liver diseases

Hepatic alterations induced by alcohol abuse include fatty liver, hepatitis, fibrosis and cirrhosis. In several patients with alcoholic fatty liver, alcoholic hepatitis or ethylic cirrhosis, a decrease in electron density of the peroxisomal catalase reaction product was observed (Roels et al 1983; De Craemer et al 1991a, 1993b, 1995). Occasionally, nucleoids and amorphous cores are found in the peroxisomal matrix (De Craemer et al 1993b).

By light microscopic evaluation of liver sections stained for catalase activity, an arrangement of peroxisomes in a perinuclear configuration in several hepatocytes of patients with an alcoholic liver disease was observed (Roels et al 1983; De Craemer et al 1991a, 1995).

Peroxisomal size was considered to be increased in alcoholic liver diseases (Rubin and Lieber 1967; Phillips et al 1987). Using morphometric analysis, we found no changes in mean peroxisomal diameter in alcoholic hepatitis, although a subpopulation of swollen organelles was evident (De Craemer et al 1991a). In cirrhosis and in steatosis of different origin including alcohol abuse, mean peroxisomal size was significantly smaller (De Craemer et al 1993b, 1995).

Alcohol often induces an increase in peroxisomal number in the human liver, as was visually evaluated by several groups of investigators (Rubin and Lieber 1967; Horvath et al 1973; Phillips et al 1987). Peroxisomal proliferation was morphometrically confirmed in all 4 patients with alcoholic hepatitis (De Craemer et al 1991a), in all 11 patients with ethylic cirrhosis (De Craemer et al 1993b) and in 10 of 12 patients with alcoholic fatty liver (De Craemer et al 1995). In addition, an inverse correlation between number and size of the peroxisomes was found in fatty liver (De Craemer et al 1995). Uchida et al (1983) reported a decreased number of peroxisomes in alcoholic foamy degeneration, an observation that was not confirmed by morphometric analysis.

Mean volume density and mean surface density of the peroxisomes was not altered in patients with alcoholic liver diseases (Oudea et al 1973; Chen et al 1987; De Craemer et al 1991a, 1993b, 1995); in a few patients, volume density and/or surface density may fall outside the control range. No significant differences in peroxisomal parameters were found between patients with alcoholic fatty livers and controls (Oudea et al 1973), or between patients with different stages of alcoholic disease (Grases et al 1987; Ryoo and Buschmann 1989).

Malignancy

Peroxisomes with amorphous inclusions and small lipid droplets were reported in routinely stained sections of hepatomas; in one hepatoma, peroxisomes revealed a condensed matrix (Ruebner et al 1967). Sometimes peroxisomes were frequently arrayed along the periphery of Mallory bodies (Keeley et al 1972). In 2 of 5 human hepatocarcinomas studied by Ma and Blackburn (1973), peroxisomes were absent. In one hepatoma, 'large' numbers of peroxisomes were observed (Keeley et al 1972). Phillips et al (1987) stated that the peroxisomes may be abundant in well-differentiated tumours and absent in those poorly differentiated. In liver cell adenoma, peroxisomes are decreased in number or are even absent (Phillips et al 1987).

By light-microscopic evaluation of the peroxisomal staining after catalase cytochemistry, a decrease in catalase activity in the liver of several but not all patients with extrahepatic malignant diseases was observed (Roels et al 1983; De Craemer et al 1993a). A perinuclear configuration of the peroxisomes was sometimes obvious (Roels et al 1983; De Craemer et al 1993a).

The normal hepatocytes in 1 of 2 patients with primary hepatocellular carcinoma

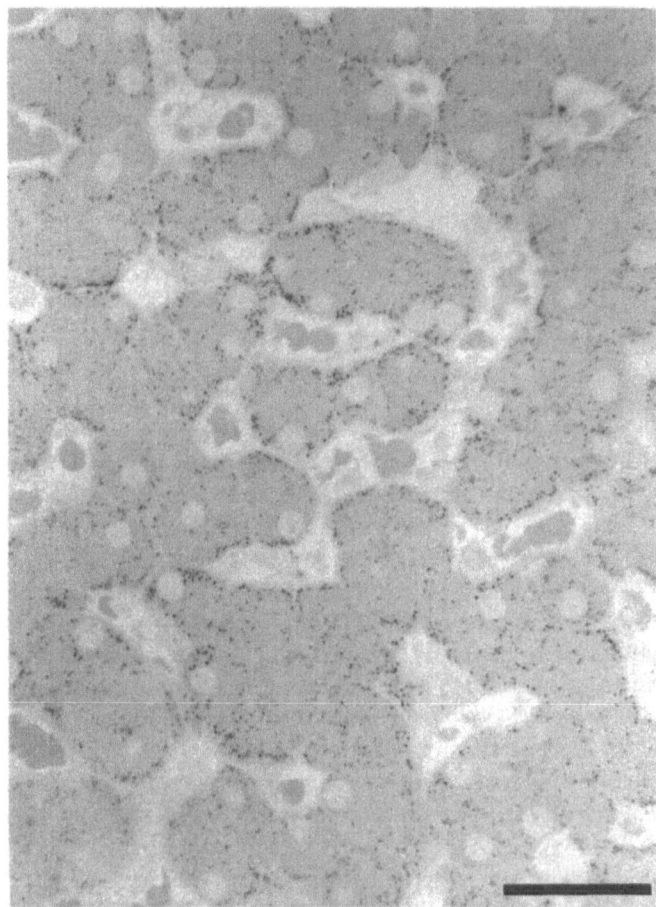

Figure 10 Light micrograph of the liver parenchyma of a patient with X-linked adrenoleuko-dystrophy, stained for catalase activity. The peroxisomes have a preferential localization along the sinusoidal membrane of the hepatocyte. This peculiar distribution is not seen in human control livers (see Figure 2) and was recently reported for the first time in hepatocytes of a patient with nodular regenerative hyperplasia (De Craemer and Roels 1994). It was not seen in other patients with X-linked adrenoleukodystrophy (Roels et al 1988). Bar=40 μm. (Liver sections obtained from F. Roels, Gent)

contained significantly more peroxisomes than controls (De Craemer et al 1993a). In the unaffected liver parenchyma of patients with extrahepatic carcinoma with or without detectable liver metastases, peroxisomal number was increased in 18 of 20 cases, the mean size of the organelles being decreased in 13 of 20 patients (De Craemer et al 1993a).

In the liver of one patient with nodular regenerative hyperplasia of the liver, a benign alteration of the liver parenchyma, normal peroxisomal staining based on catalase activity was observed; in a restricted number of hepatocytes, peroxisomes showed a preferential localization near the cell membrane facing sinusoidal cavities (Figure 10) or were arranged around the nucleus (De Craemer and Roels 1994).

Inflammatory processes

Using catalase cytochemistry, Roels et al (1983) found that peroxisomal catalase staining was decreased in several patients with an extrahepatic inflammatory process. This is in agreement with data on catalase activity in animal experiments (Canonico et al 1976). Peroxisomes often revealed a perinuclear distribution (Roels et al 1983).

PEROXISOMAL FEATURES DURING ADAPTATION

For the sake of the (electron) microscopist we now classify and discuss alterations of each morphological and morphometric feature (matrix, inclusions, distribution, size, shape, number, volume and surface density) of peroxisomes.

Peroxisomal matrix

An increased electron density of the peroxisomal matrix was observed in routinely stained sections of liver of patients with several diseases (Ruebner et al 1967; Sternlieb and Quintana 1977; Lake et al 1987). In long- and medium-chain fatty acid oxidation defects, the dense peroxisomal matrices were equated with an increased peroxisomal activity (Lake et al 1987). The opposite alteration, a less osmiophilic, less electron-dense aspect of the peroxisomal matrix, was also reported (de la Iglesia 1969; Landrieu et al 1982; Sternlieb and Quintana 1982). In line with the hypothesis of Lake et al (1987), this phenomenon may be interpreted as an indication of a decrease in peroxisomal activity. It is unclear whether a change in the activity of one single peroxisomal enzyme is sufficient to induce an alteration of the electron density of the peroxisomal matrix, and whether each of the more than 50 peroxisomal enzymes identified so far is able to induce changes in electron density of the peroxisomes when its activity is altered. Changes in peroxisomal size associated with the decrease of a single peroxisomal enzyme activity have been reported: a reduction of catalase synthesis was accompanied by a decrease in mean volume of the peroxisomes in the liver of guinea pigs (Geerts et al 1984). In the liver of 2 patients with a deficiency of alanine glyoxylate aminotransferase, smaller peroxisomes were observed than in controls (De Craemer et al 1989).

In liver of patients with acquired liver diseases or with cancer, a decrease in peroxisomal catalase staining was often observed when compared to the staining intensity of the organelles in control liver (Roels et al 1983; De Craemer et al 1991a,b, 1993a,b). This observation is in line with the well-known decrease in hepatic catalase activity in patients with cancer (Blumenthal and Brahn 1910; Ohnuma et al 1966; Wittrin et al 1975; Bellisola et al 1987). A decrease in catalase activity or in staining of the peroxisomes may be directly linked to a reduction of the protein synthesis: the short half-life (± 1.5 days) of catalase (Price et al 1962) makes this enzyme a sensitive marker for alterations in protein synthesis. In nude mice bearing xenografts of human pancreatic adenocarcinomas, hepatic catalase activity and peroxisomal staining (after catalase cytochemistry) were decreased when compared to control mice (De Craemer et al 1993c). In rats injected with tumour necrosis factor α, peroxisomal β-oxidation was decreased (Beier et al 1992). Recently it was shown that the catalase gene expression is depressed in the liver of tumour-bearing mice (Yamaguchi et al 1992).

After catalase cytochemistry, clumped electron-dense reaction product leaving electron-lucent, transparent spaces in the peroxisomal matrix was frequently observed in the liver of patients with malignant diseases, viral (Figure 5), alcoholic, amitriptyline and aprindine hepatitis and cirrhosis (De Craemer et al 1991a,b, 1993a,b). This morphological feature is related to swelling as the subgroup of peroxisomes with clumped matrix material revealed a higher mean diameter than the subgroup with a homogeneously dense matrix (De Craemer et al 1991a,b).

Matrix inclusions

Matrix inclusions in the hepatocellular peroxisomes of patients with several diseases include amorphous cores (Figure 4), crystalline nucleoids and marginal plates (Table 1). The precise nature of these types of inclusions in human liver peroxisomes is unknown. In 2 of 9 patients with colorectal cancer, crystalloid cores in hepatic peroxisomes were observed; they were associated with urate oxidase activity (De Netto et al 1991), an enzyme that has otherwise not been demonstrated in human liver. In rat liver, a crystalline core is a constant finding in peroxisomal profiles and corresponds to the localization site of urate oxidase and xanthine oxidase (Angermüller et al 1987). Marginal plates are constant findings in the peroxisomes of bovine and sheep kidney and liver. Immunocyto-chemical studies have elucidated that the L-α-hydroxyacid oxidase B is exclusively localized in this part of the peroxisomal matrix (Zaar et al 1992).

Other inclusions in human liver peroxisomes are lipid droplets and small incomplete myelin figures; they were reported in tyrosinaemia (Phillips et al 1987) and in a hepatoma with trabecular and tubular areas (Ruebner et al 1967), respectively.

In peroxisomal diseases, membrane-bound inclusions in the peroxisomal matrix containing cytoplasmic material such as glycogen were reported in several patients and interpreted as invaginations of the peroxisomal membrane (Roels et al 1988; Espeel et al 1991, 1993; Roels 1991). Peroxisomal membrane invaginations were also observed in patients with alcoholic or viral hepatitis (Figure 8), total lipodystrophy, hypofibrino-genaemia (Figure 9) and extrahepatic cancer (De Craemer et al 1991a, 1992, 1993a).

Peroxisomal distribution

An aberrant distribution of peroxisomes has been described in several pathologies of the human liver. Besides a relatively homogeneous dispersion of single or clustered peroxisomes over the cytoplasm of several hepatocytes (Figure 2), a concentration of organelles around the nucleus may be obvious in a variable number of cells (Figure 3). This was observed in sections of the liver of patients with several acquired intrahepatic and extrahepatic diseases (Roels et al 1983; De Craemer et al 1991a,b, 1993a,b; De Craemer and Roels 1994). In most cases this type of peroxisomal distribution was accompanied by an increase in number morphometrically (De Craemer et al 1991a, 1993a,b).

An arrangement of the hepatocellular peroxisomes around Mallory bodies, glycogen bodies and cytoplasmic crystals was found, respectively, in a hepatoma (Ruebner et al 1967), after treatment with trimethoprim/sulphamethoxazole (Phillips et al 1987) and in cerebrotendinous xanthomatosis (Salen et al 1978). The concentration of peroxisomes around these inclusions may be linked to the size or the abundance of these inclusions: the cytoplasmic space is restricted to a limited area surrounding these inclusions so that a concentration of peroxisomes becomes a peculiar observation. The same holds true for the perinuclear configuration of the peroxisomes: in light-microscopic liver sections stained for catalase activity, numerous peroxisomes surround the centrally located nucleus.

In the liver of one patient with nodular regenerative hyperplasia, the peroxisomes lined the sinusoidal membrane in several hepatocytes (De Craemer and Roels 1994). It was hypothesized that in this configuration the organelles were less influenced by the lowered supply of nutrients from blood in the affected regions of the liver parenchyma of this patient. This type of peroxisomal distribution was also observed in one patient with X-linked adrenoleukodystrophy (McKusick 300100) (Figure 10). However, no biochemical data supported a decreased blood supply to the liver of this patient.

Peroxisomal size

Based on visual impression as well as on morphometric results, changes in peroxisomal size have been reported in several non-peroxisomal diseases. Sometimes, the subjective (visually evaluated) data are contradictory to the objective data obtained by morphometric analysis (Table 1).

We have observed only a single case of borderline increase in mean diameter of the peroxisomes in more than 80 patients with acquired or congenital non-peroxisomal diseases studied by morphometric analysis to date. The smallest hepatocellular peroxisomes were found in a patient with phenytoin hepatitis and in a patient with alcoholic fatty liver: mean peroxisomal d-circle was $0.364 \mu m$ and $0.359 \mu m$, respectively, being 69% and 68% of the mean diameter in 7 control livers (De Craemer et al 1991a; De Craemer, unpublished observation).

Morphometric analysis of the true peroxisomal diameter (i.e. the diameter of the circle with the same area as the measured profile = d-circle) reveals that peroxisomes are smaller or normal-sized in acquired and congenital non-peroxisomal diseases (Table 1); only one study reported an enlargement of the organelles (in hyperlipoproteinaemia type IIA) but no cytochemical marker was used (Gariot et al 1984). In the liver of patients with an inborn error of peroxisomal β-oxidation, peroxisomes are enlarged or are extremely small (maximum diameter $<0.25 \mu m$) (Roels 1991; Roels et al 1991, 1993; Espeel et al 1995).

An inverse correlation between electron density of the catalase reaction product in the peroxisomal matrix and the size of the peroxisomal profile has been observed in human livers stained for catalase activity (Roels 1991; Roels and Cornelis 1989). Similar observations were made in peroxisomes in routinely stained liver sections investigated at the ultrastructural level (de la Iglesia 1969; Oudea et al 1973). In mice fed a high-fish-oil diet, we recently found an inverse correlation between peroxisomal size and catalase activity measured in liver homogenates (De Craemer et al 1994). Larger peroxisomes have a lower membrane area to volume ratio; when the exchange over the peroxisomal membrane becomes the limiting factor, peroxisomal metabolism will be unfavourably influenced by an increase in size of the peroxisomes (Roels and Cornelis 1989; De Craemer et al 1991; Roels et al 1991).

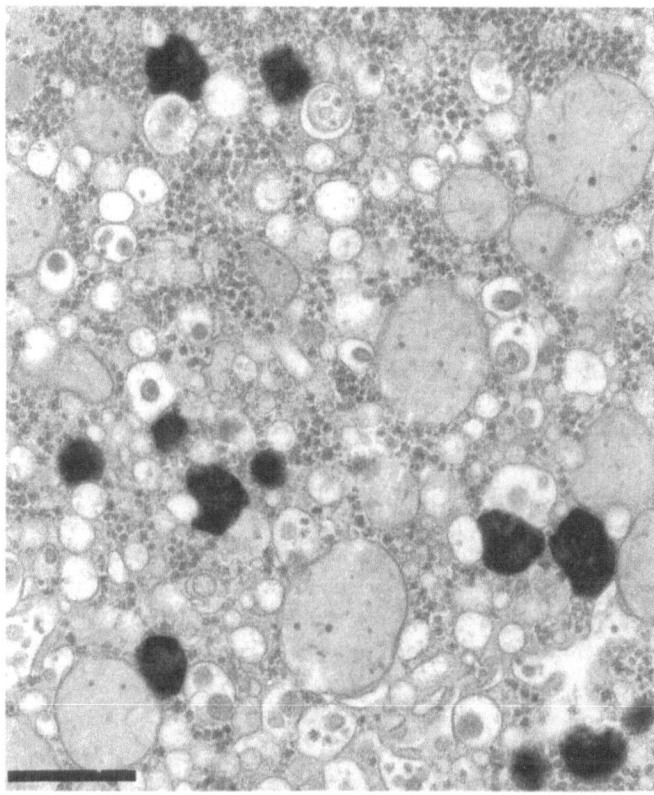

Figure 11 Part of a hepatocyte of a patient with liver cirrhosis. Catalase cytochemistry. Endoplasmic reticulum is strongly vesiculated and provokes one or more dents in the membrane of a majority of peroxisomes. Bar = 1 μm

Shape of the peroxisomes

In the liver of patients with a variety of 'non-peroxisomal' diseases, pleomorphic, elongated and irregularly shaped ('misshapen') peroxisomes with cytoplasmic invaginations, gastruloid cisternae and protrusions were frequently observed: in viral (Figs 6–8), alcoholic and cholestatic hepatitis (Sternlieb and Quintana 1977; Phillips et al 1987; De Craemer et al 1991a), after treatment with acetylsalicylic acid, clofibrate (see references in Phillips et al (1987)), amitriptyline, aprindine, methimazole (De Craemer et al 1991a) or isoniazid (Sternlieb and Quintana 1977), in malignant diseases (Sternlieb and Quintana 1977; De Craemer et al 1993a), in various types of cirrhosis (Figure 11) (Sternlieb and Quintana 1977; De Craemer et al 1993b), in abetalipoproteinaemia (Collins et al 1989), in hypofibrinogenaemia (Figure 9), and in Wilson disease (Sternlieb and Quintana 1982). As these irregular peroxisomal profiles are often observed in livers with an increased peroxisomal number (De Craemer et al 1993a,b), they may represent intermediary stages in the proliferation of peroxisomes. According to the current concept of peroxisomal biogenesis, peroxisomes originate by division (Fahimi et al 1993). Sections through a dividing peroxisome may render profiles with small protrusions or with cytoplasmic invaginations. Apart from being indications of proliferation, the presence of abnormally shaped peroxisomes is also favourable for an increase in peroxisomal metabolism as more membrane (the place of import and export of molecules) surrounds the same volume of matrix containing most enzymes (Roels 1991).

Peroxisomal shape may also be imposed by alterations in other cellular compartments such as vesiculation of the endoplasmic reticulum (Figure 11), dilatation of the endoplasmic reticulum, hyperplasia of mitochondria and accumulation of lipid droplets (De Craemer 1993), changes which may represent an increase in volume density of these subcellular components. As endoplasmic reticulum often partially or completely encloses peroxisomes (a morphological characteristic often used to identify peroxisomal profiles in routinely stained sections), a dilation of its cisternae may force the peroxisomes to generate the most resistant shape (a sphere). Consequently, only circular-shaped peroxisomes are often observed in sections through cells with a strong dilatation of the cisternae of the endoplasmic reticulum, a striking uniformity not found in hepatocytes with undilated endoplasmic reticulum. Peroxisomes closed in by several mitochondria may reveal angular profiles with flat sides, a shape that conforms to the shape of the mitochondria. Small subcellular structures such as vesicles of endoplasmic reticulum in the near vicinity of peroxisomes may provoke dents in the peroxisomal membrane, rendering them irregularly shaped (Figure 11). Peroxisomes apposed to large structures such as lipid droplets are often reniform or flat-sided (see Fig. 8 in Roels et al (1988)).

Number of peroxisomes

An alteration in number of peroxisomes has been reported in many non-peroxisomal diseases (Table 1). Sometimes, contradictory results were obtained by visual evaluation of the organelles: in Reye syndrome, some authors reported a normal peroxisomal number (Svoboda and Reddy 1980) while others observed an increase in number (Thaler et al 1970; Partin et al 1971, 1984; Iancu et al 1972; Phillips et al 1987). In congenital total lipodystrophy, an increase (Harbour et al 1981; Klar et al 1987) as well as a decrease (Phillips et al 1987) in peroxisomal number was evaluated visually. Using morphometric analysis, the peroxisomal number was nearly doubled in one of 2 patients with this disorder; in the other patient peroxisomal number was normal (De Craemer et al 1992). These contradictory results are probably due to differences in methods used to evaluate the number of organelles, although they may also indicate heterogeneity between patients with the same disease. Additional quantitative studies are needed to resolve this problem.

As shown in Table 1, morphometric data nearly always indicate a proliferation of peroxi-somes in acquired or congenital non-peroxisomal diseases. Only in a single patient with partial ornithine transcarbamylase deficiency and acute hepatic failure was a decrease in peroxisomal number observed in routinely stained liver sections (Landrieu et al 1982). This observation has to be interpreted with caution as no peroxisome-specific marker was used.

A decrease in peroxisomal number leads to a diminution of the total peroxisomal volume unless, at the same time, the size of the organelles is increased. Peroxisomal metabolism will be influenced unfavourably in this situation, as was discussed above. In contrast, an increase in number of normal-sized peroxisomes or even of smaller-sized organelles (on condition that the total peroxisomal volume is not decreased), is favourable for the activity of the peroxisomal enzymes. This was recently illustrated in mice fed a high-fish-oil diet: a positive linear correlation between the number of (normal-sized) peroxisomes on the one hand and the catalase activity and the peroxisomal β-oxidation capacity on the other was found (De Craemer et al 1994).

A decrease in peroxisomal number in chronic hepatitis B combined with immuno-suppressive therapy was the result of lysis as well as of autophagocytosis (Reinke et al 1988). Autophagocytosis was also the basis for the peroxisomal disappearance in the liver of a patient with partial ornithine transcarbamylase deficiency (Landrieu et al 1982). Recently, rats whose dietary supply of dioctyl phthalate (a peroxisome proliferator) had been stopped were injected with leupeptin during the recovery process in order to block the breakdown of the excess of peroxisomes: an accumulation of autophagosomes was indeed observed (Yokota 1993). This phenomenon was not studied in steady-state liver.

Volume and surface density

A normal volume density of peroxisomes in a patient's liver does not always indicate a normal peroxisomal population; it can also be the result of an increased number of smaller organelles, or of a decreased number of larger peroxisomes. As discussed above, each of these situations may have a different effect on the metabolic activity of the peroxisomal enzymes. Therefore, data on the size and/or the number of peroxisomes should also be included each time the volume density is determined in a group of patients.

The surface density expresses the amount of membrane area per reference volume (cell, organ, cytoplasm) and is correlated with the number, shape and size of the organelles. As is the case with the volume density, the surface density may be normal in some liver pathologies while significant changes in number, shape or size of the individual organelles are observed.

CONCLUSIONS

The morphology of the peroxisomes in control human liver may be summarized as a regularly shaped cell organelle with a diameter of approximately $0.5-0.6\,\mu$m surrounded by a single tripartite membrane and filled with a homogeneous matrix without inclusions. Identification of these organelles is improved by use of cytochemical and immunocyto-chemical techniques. Peroxisomes have a rather homogeneous distribution in the hepato-cytes but they often form clusters that also show a homogeneous dispersion throughout the cells.

Alterations of peroxisomes cover a large spectrum of changes in their morphological and morphometric characteristics. When compared to the peroxisomal changes that are only found in primary peroxisomal disorders (Roels et al 1991, 1993), the following seem to be specific secondary changes: a proliferation of the peroxisomes that is *never* accompanied by an increase in size (confirmed by morphometry); a minor reduction in mean size; and proliferation-related alterations in shape of at least part of the peroxisomes. In contrast, enlarged peroxisomal size and the presence of very small (diameter $<0.3\,\mu$m) organelles appear to be specific for primary peroxisomal disorders, in addition to vesicles devoid of one or more enzymes or the complete absence of organelles of the peroxisome family (Espeel et al 1995).

Alterations in peroxisomal characteristics in non-peroxisomal diseases are not disease-specific: a heterogeneous spectrum of secondary peroxisomal alterations may be observed in patients with the same disease; in addition, similar sets of peroxisomal changes may be found in the liver of patients suffering from unrelated diseases. The morphological and morphometric changes in the peroxisomal compartment may have a significant effect on the activity of the peroxisomal enzyme activities: as peroxisomal enzymes intervene in several essential metabolic pathways, alterations in the peroxisomal compartment may be responsible for some of the 'side-effects' observed in human pathology. To clarify the actual effect of changes in peroxisomal morphology on the physiology *in vivo* of several peroxisomal enzymes and on the different liver functions in patients with non-peroxisomal diseases, peroxisomal metabolites (very long-chain fatty acids, plasmalogens, abnormal bile acids, etc.) and enzyme activities should be studied. At present, little attention has been paid to this aspect when compared to the literature available for peroxisomal diseases.

NOTE ADDED IN PROOF

Recently, it was found that dehydroepiandrosterone (DHEA), a naturally occurring C_{19} steroid in mammals, acts as a peroxisome proliferator when added to the diet of rats and mice: it induces a 10-fold increase in mRNA level and activity of peroxisomal β-oxidation enzymes (Wu et al 1989; Sakuma et al 1992; Rao et al 1993), a doubling of catalase mRNA and activity (Wu et al 1989; Rao et al 1993), and a 5 to 6-fold increase in peroxisomal volume density (Rao et al 1992; 1993) in the liver of treated animals. By visual evaluation the number of peroxisomes was increased (Yamada et al 1991; Bellei et al 1992; Rao et al 1993). A morphometric study on size and number of peroxisomes has not been performed yet. In contrast to these *in vivo* findings, no effect of DHEA was observed in cultured primary rat hepaocytes (Ram and Waxman 1994). However *in vitro* treatment with DHEA 3β-sulphate stimulated major increases in peroxisomal acyl-CoA oxidase mRNA (Ram and Waxman 1994). No data on the effect of exogenous DHEA or its analogues on the hepatocellular peroxisomes in humans is available.

ACKNOWLEDGEMENTS

The author thanks all clinicians who provided liver biopsy samples and clinical data, Professor Dr F. Roels (Gent) for the invitation to write this paper, and Marina Pauwels and René Stien for technical assistance. This work was supported by grants of the Research Council (OZR) of the Vrije Universiteit Brussel, and by grants of the National Fund for Scientific Research (NFWO).

REFERENCES

Angermüller S, Bruder G, Völkl A, Wesch H, Fahimi HD (1987) Localization of xanthine oxidase in crystalline cores of peroxisomes: a cytochemical and biochemical study. *Eur J Cell Biol* **45**: 137–144.

Beier K, Fahimi HD (1986) Application of automatic image analysis for morphometric studies of peroxisomes stained cytochemically for catalase. *Cell Tissue Res* **246**: 635–640.

Beier K, Völkl A, Fahimi HD (1992) Suppression of peroxisomal lipid β-oxidation enzymes by TNF-α. *FEBS Lett* **310**: 273–276.

Bellei M, Battelli D, Fornieri C, et al (1992) Changes in liver structure and function after short-term and long-term treatment of rats with dehydroepiandrosterone. *J Nutr* **122**: 967–976.

Bellisola G, Casaril M, Gabrielli GB, Caraffi M, Corrocher R (1987) Catalase activity in human hepatocellular carcinoma (HCC). *Clin Biochem* **20**: 415–417.

Bentley P, Calder I, Elcombe C, Grasso P, Stringer D, Wiegand HS (1993) Hepatic peroxisome proliferation in rodents and its significance for humans. *Fd Chem Toxic* **31**: 857–907.

Bichet N, Cahard D, Fabre G, Remandet B, Gouy D, Cano J-P (1990) Toxicological studies on a Benzofuran derivative. III. Comparison of peroxisome proliferation in rat and human hepatocytes in primary culture. *Toxicol Appl Pharmacol* **40**: 521–528.

Blaauboer BJ, van Holsteijn CW, Bleumink R, et al (1990) The effect of beclobric acid and clofibric acid on peroxisomal β-oxidation and peroxisome proliferation in primary cultures of rat, monkey and human hepatocytes. *Biochem Pharmacol* **10**: 521–528.

Blumcke S, Schwartzkopff W, Lobeck H, Edmondson NA, Prentice DE, Blane GF (1983) Influence of fenofibrate on cellular and subcellular liver structure in hyperlipidemic patients. *Atherosclerosis* **46**: 105–116.

Blumenthal F, Brahn B (1910) Die Katalasewirkung in normaler und in carcinomatöser leber. *Z Krebsforsch* **8**: 436–440.

Brown NK, Smuckler EA (1970) Alkaptonuria and Gilbert's syndrome: report of two affected siblings and hepatic ultrastructure in one sibling. *Am J Med* **48**: 759–765.

Canonico PG, Rill W, Ayala E (1976) Effects of inflammation on peroxisomal enzyme activities, catalase synthesis, and lipid metabolism. *Lab Invest* **37**: 479–486.

Chen TS, Murphy DP, Marquet G, et al (1987) Morphometric study of hepatic ultrastructure in alcoholic hepatitis. *Histol Histopathol* **2**: 429–432.

Christiansen EN, Flatmark T, Kryvi H (1981) Effects of marine oil diet on peroxisomes and mitochondria of rat liver. A combined biochemical and morphometric study. *Eur J Cell Biol* **26**: 11–20.

Collins JC, Scheinberg IH, Giblin DR, Sternlieb I (1989) Hepatic abnormalities in abetalipoproteinaemia. *Gastroenterology* **97**: 766–770.

Conway JG, Cattley RC, Popp JA, Butterworth BE (1989) Possible mechanisms in hepatocarcinogenesis by the peroxisome proliferator di(2-ethylhexyl)phthalate. *Drug Metab Rev* **21**: 65–102.

Cooper PJ, Danpure CJ, Wise PJ, Guttridge KM (1988) Immunocytochemical localization of human hepatic alanine: glyoxylate aminotransferase in control subjects and patients with primary hyperoxaluria type I. *J Histochem Cytochem* **36**: 1285–1294.

Cooper PJ, Danpure CJ, Simpson KJ (1989) Peroxisomal and mitochondrial proliferation and increased alanine: glyoxylate aminotransferase activity in human liver after chlorpromazine-induced cholestasis. *Biochem Soc Trans* **17**: 1071–1072.

Danpure CJ, Cooper PJ, Wise PJ, Jennings PR (1989) An enzyme trafficking defect in two patients with primary hyperoxaluria type 1: peroxisomal alanine: glyoxylate aminotransferase rerouted to mitochondria. *J Cell Biol* **108**: 1345:1352.

Danpure CJ, Guttridge KM, Fryer P, Jennings PR, Allsop J, Purdue PE (1990) Subcellular distribution of hepatic alanine: glyoxylate aminotransferase in various mammalian species. *J Cell Sci* **97**: 669–678.

De Craemer D (1993) Changes in subcellular organelles may influence the peroxisomal shape in human hepatocytes. *Biol Cell* **77**: 127.

De Craemer D, Roels F (1994) A peculiar distribution of the peroxisomes in a patient with nodular regenerative hyperplasia of the liver. *J Hepatol* **20**: 394–397.

De Craemer D, Rickaert F, Wanders RJA, Roels F (1989) Hepatic peroxisomes are smaller in primary hyperoxaluria type 1 (PHI), cytochemistry and morphometry. *Micron Microsc Acta* **20**: 125–126.

De Craemer D, Bingen A, Langendries M, Martin JP, Roels F (1990a) Alterations of hepatocellular peroxisomes in viral hepatitis in the mouse. *J Hepatol* **11**: 145–152.

De Craemer D, Espeel M, Langendries M, Schutgens RBH, Hashimoto T, Roels F (1990b) Postmortem visualization of peroxisomes in rat and in human liver tissue. *Histochem J* **22**: 36–44.

De Craemer D, Kerckaert I, Roels F (1991a) Hepatocellular peroxisomes in human alcoholic and drug-induced hepatitis: a quantitative study. *Hepatology* **14**: 811–817.

De Craemer D, Kerckaert I, Roels F (1991b) Alterations of hepatocellular peroxisomes in human viral hepatitis. In Schiraldi O, Pastore G, Dentico P, eds. *Progress and Prospects in Viral Hepatitis*. San Severo: Gerni Editori, 461–467.

De Craemer D, Zweens MJ, Lyonnet S, et al (1991c) Very large peroxisomes in distinct peroxisomal disorders (rhizomelic chondrodysplasia punctata and acyl-CoA oxidase deficiency): novel data. *Virchows Arch (A)* **419**: 523–525.

De Craemer D, Van Maldergem L, Roels F (1992) Hepatic ultrastructure in congenital total lipodystrophy with special reference to peroxisomes. *Ultrastruct Pathol* **16**: 307–316.

De Craemer D, Pauwels M, Hautekeete M, Roels F (1993a) Alterations of hepatocellular peroxisomes in cancer patients: catalase cytochemistry and morphometry. *Cancer* **71**: 3851–3858.

De Craemer D, Pauwels M, Roels F (1993b) Peroxisomes in cirrhosis of the human liver: a cytochemical, ultrastructural and quantitative study. *Hepatology* **17**: 404–410.

De Craemer D, Pauwels M, Vergeylen A, Roels F, Van den Branden C (1993c) Peroxisomes in liver, kidney and duodenum of nude mice bearing xenografts of human pancreatic adenocarcinomas. *Virchows Arch (B)* **64**: 7–12.

De Craemer D, Roels F, Van den Branden C (1993d) Rapid effects of dietary fish oil on peroxisomes in mouse liver. *Eur J Morphol* **30**: 331–335.

De Craemer D, Vamecq J, Roels F, Vallée L, Pauwels M, Van den Branden C (1994) Peroxisomes in liver, heart, and kidney of mice fed a commercial fish oil preparation: original data and review on peroxisomal changes induced by high-fat diets. *J Lipid Research* **35**: 1241–1250.

De Craemer D, Pauwels M, Van den Branden C (1995) Alterations of peroxisomes in steatosis of the human liver: a quantitative study. *Hepatology*, **22**: 744–752.

de Duve C, Baudhuin P (1966) Peroxisomes (Microbodies and related particles). *Physiol Rev* **46**: 323–357.

de la Iglesia FA (1969) Comparative analysis of hepatic microbodies (A review). *Acta Hepatosplenol* **3**: 141–160.

de la Iglesia FA, Lewis JE, Buchanan RA, Marcus EL, McMahon G (1982) Light and electron microscopy of liver in hyperlipoproteinemic patients under long-term gemfibrozil treatment. *Atherosclerosis* **43**: 19–37.

De Netto LA, Tappia PS, Malik ZA, et al (1991) Human hepatic peroxisomes with crystalloid cores associated with urate oxidase activity. In Harkness RA, ed. *Purine and Pyrimidine Metabolism in Man*, 7th edn. New York: Plenum Press, 373–376.

Espeel M, Hashimoto T, De Craemer D, Roels F (1990a) Immunocytochemical detection of peroxisomal β-oxidation enzymes in cryostat and paraffin sections of human post mortem liver. *Histochem J* **22**: 57–62.

Espeel M, Jauniaux E, Hashimoto T, Roels F (1990b) Immunocytochemical localization of peroxisomal β-oxidation enzymes in human fetal liver. *Prenat Diagn* **10**: 349–357.

Espeel M, Roels F, De Craemer D, Dacremont G, Wanders RJA, Hashimoto T (1991) Peroxisomal localization of the immunoreactive β-oxidation enzymes in a neonate with a β-oxidation defect. *Virchows Arch (A)* **419**: 301–308.

Espeel M, Heikoop J, Smeitink JAM, et al (1993) Cytoplasmic catalase and ghostlike peroxisomes in the liver from a child with atypical chondrodysplasia punctata. *Ultrastruct Pathol* **17**: 623–636.

Espeel M, Roels F, Giros M, et al (1995) Immunolocalization of a 43 kDa peroxisomal membrane protein in the liver of patients with generalized peroxisomal disorders. *Eur J Cell Biol*, **67**: 319–327.

Evans J, Lefkowitch J, Lim CK, Billing B (1981) Fecal porphyrin abnormalities in a patient with features of Rotor's syndrome. *Gastroenterology* **81**: 1125–1130.

Fahimi HD, Baumgart E, Völkl A (1993) Ultrastructural aspects of the biogenesis of peroxisomes in rat liver. *Biochimie* **75**: 201–208.

Flatmark T, Nilsson A, Kvannes J, et al (1988) On the mechanism of induction of the enzyme systems for peroxisomal β-oxidation of fatty acids in rat liver by diets rich in partially hydrogenated fish oil. *Biochim Biophys Acta* **962**: 122–130.

Fringes B, Reith A (1982) Time course of peroxisome biogenesis during adaptation to mild hyperthyroidism in rat liver. A morphometric/stereologic study by electron microscopy. *Lab Invest* **47**: 19–26.

Gariot P, Pointel JP, Barrat E, Drouin P, Debry G (1984) Etude morphométrique des péroxysomes hépatiques chez des malades hyperlipoprotéinémiques traités par fénofibrate. *Biomed Pharmacother* **38**: 101–106.

Gariot P, Genton P, Barrat E, et al (1986) The hepatocyte and type IIA and IIB hyperlipoproteinemia: a morphometric study. *J Hepatol* **2**: 495–503.

Geerts A, De Prest B, Roels F (1984) On the topology of the catalase biosynthesis and degradation in the guinea pig liver. *Histochemistry* **80**: 339–345.

Goldfischer S, Sobel HJ (1981) Peroxisomes and bile acid synthesis. *Gastroenterology* **81**: 196–197.

Grases PJ, Millard PR, McGee JO'D (1987) The ultrastructure of alcoholic liver disease: a review and analysis of 100 biopsies. *Histol Histopathol* **2**: 19–29.

Hanefeld M, Kemmer C, Kadner E (1983) Relationship between morphological changes and lipid-lowering action of *p*-chlorphenoxyisobutyric acid (CPIB) on hepatic mitochondria and peroxisomes in man. *Atherosclerosis* **46**: 239–246.

Harbour JR, Rosenthal P, Smuckler EA (1981) Ultrastructural abnormalities of the liver in total lipodystrophy. *Hum Pathol* **12**: 856–862.

Holmer G, Hoy CE, Kirstein D (1982) Influence of partially hydrogenated vegetable and marine oils on lipid metabolism in rat liver and heart. *Lipids* **17**: 585–593.

Horvath E, Kovacs K, Ross RC (1973) Alcoholic liver lesion. Frequency and diagnostic value of fine structural alterations in hepatocytes. *Beitr Path* **148**: 67–85.

Hughes JL, Poulos A, Crane DI, Chow CW, Sheffield LJ, Sillence D (1992) Ultrastructure and immunocytochemistry of hepatic peroxisomes in rhizomelic chondrodysplasia punctata. *Eur J Pediatr* **151**: 829–836.

Hughes JL, Bourne AJ, Poulos A (1993) Establishment of a normal range of morphometric values for peroxisomes in paediatric liver. *Virchows Arch (A)* **423**: 453–457.

Iancu TC, Mason WH, Neustein HB (1972) Ultrastructural abnormalities of liver cells in Reye's syndrome. *Hum Pathol* **8**: 421–431.

Ishii H, Fukumori N, Horie S, Suga T (1980) Effects of fat content in the diet on hepatic peroxisomes of the rat. *Biochim Biophys Acta* **617**: 1–11.

Jezequel AM, Librari ML, Mosca PG, Novelli G (1983) The human liver in extrahepatic cholestasis: ultrastructural morphometric data. *Liver* **3**: 303–314.

Just W, Hartl F (1983) Rat liver peroxisomes. II. Stimulation of peroxisomal fatty-acid β-oxidation by thyroid hormones. *Hoppe-Seyler's Z Physiol Chem* **364**: 1541–1547.

Just W, Hartl F, Schimassek H (1982) Rat liver peroxisomes. I. New peroxisome population induced by thyroid hormones in the liver of male rats. *Eur J Cell Biol* **26**: 249–254.

Kamei A, Houdou S, Takashima S, Suzuki Y, Becker LE, Armstrong DL (1993) Peroxisomal disorders in children: immuno-histochemistry and neuropathology. *J Pediatr* **122**: 573–579.

Keeley AF, Iseri OA, Gottlieb LS (1972) Ultrastructure of hyaline cytoplasmic inclusions in a human hepatoma: relationship to Mallory's alcoholic hyalin. *Gastroenterology* **62**: 280–293.

Kerckaert I, Claeys A, Just W, Cornelis A, Roels F (1989) Automated image analysis of rat liver peroxisomes after treatment with thyroid hormones: changes in number, size and catalase reaction. *Micron Microsc Acta* **20**: 9–18.

Klar A, Livni N, Gross-Kieselstein E, Navon P, Shahin A, Branski D (1987) Ultrastructural abnormalities of the liver in total lipodystrophy. *Arch Pathol Lab Med* **111**: 197–199.

Klion F, Segal R, Schaffner F (1971) The effect of altered thyroid function on the ultrastructure of the human liver. *Am J Med* **20**: 317–324.

Lake BD, Clayton PT, Leonard JV, Bhuiyan AKM, Bartlett K, Aynsley Green A (1987) Ultrastructure of liver in inherited disorders of fat oxidation. *Lancet* **1**: 382–383.

Landrieu P, François B, Lyon G, Van Hoof F (1982) Liver peroxisome damage during acute hepatic failure in partial ornithine transcarbamylase deficiency. *Pediatr Res* **16**: 977–981.

Litwin JA, Völkl A, Stachura J, Fahimi D (1988) Detection of peroxisomes in human liver and kidney fixed with formalin and embedded in paraffin: the use of catalase and lipid β-oxidation enzymes as immunohistochemical markers. *Histochem J* **20**: 165–173.

Ma MH, Biempica L (1971) The normal human liver cell: cytochemical and ultrastructural studies. *Am J Pathol* **62**: 353–390.

Ma MH, Blackburn CRB (1973) Fine structure of primary liver tumors and tumor-bearing livers in man. *Cancer Res* **33**: 1766–1774.

Neat CE, Thomassen MS, Osmundsen H (1980) Induction of peroxisomal β-oxidation in rat liver by high-fat diets. *Biochem J* **186**: 369–371.

Neat CE, Thomassen MS, Osmundsen H (1981) Effects of high-fat diets on hepatic fatty acid oxidation in the rat. *Biochem J* **196**: 149–159.

Ohnuma T, Maldia G, Holland JF (1966) Hepatic catalase activity in advanced human cancer. *Cancer Res* **26**: 1806–1818.

Oudea MC, Collette M, Dedieu P, Oudea P (1973) Morphometric study of the ultrastructure of human alcoholic fatty liver. *Biomedicine* **19**: 455–459.

Partin JC, Schubert WK, Partin JS (1971) Mitochondrial ultrastructure in Reye's syndrome (encephalopathy and fatty degeneration of the viscera). *N Engl J Med* **285**: 1339–1343.

Partin JS, Daugherty CC, McAdams AJ, Partin JC, Schubert WK (1984) A comparison of liver ultrastructure in salicylate intoxication and Reye's syndrome. *Hepatology* **4**: 687–690.

Pfeifer U, Ormanns W, Klinge O (1981) Hepatocellular fibrinogen storage in familial hypofibrinogenemia. *Virchows Arch (B)* **36**: 247–255.

Phillips MJ, Poucell S, Patterson J, Valencia P (1987) *The Liver: An Atlas and Text of Ultrastructural Pathology.* New York: Raven Press.

Price VE, Stirling WR, Tarantola VA, Hartley RW Jr, Rechcigl M Jr (1962) The kinetics of catalase synthesis and destruction in vivo. *J Biol Chem* **237**: 2468–2475.

Ram PA, Waxman DJ (1994) Dehydroepiandrosterone 3β-sulphate is an endogenous activator of the peroxisome-proliferation pathway: induction of cytochrome P-450 4A and acyl-CoA oxidase mRNAs in primary rat hepatocyte culture and inhibitory effects of Ca^{2+}-channel blockers. *Biochem J* **301**: 753–758.

Rao MS, Musunuri S, Reddy JK (1992) Dehydroepiandrosterone-induced peroxisome proliferation in the rat liver. *Pathobiology* **60**: 82–86.

Rao MS, Ide H, Alvares K et al (1993) Comparative effects of dehydroepiandrosterone and related steroids on peroxisome proliferation in rat liver. *Life Sci* **52**: 1709–1716.

Reddy JK, Mannaerts GP (1994) Peroxisomal lipid metabolism. *Annu Rev Nutr* **14**: 343–370.

Reddy JK, Warren JR, Reddy MK, Lalwani ND (1982) Hepatic and renal effects of peroxisome proliferators: biological implications. *Ann NY Acad Sci* **386**: 81–110.

Reinke P, David H, Uerlings I, Decker T (1988) Pathology of hepatic peroxisomes in chronic hepatitis B and immunosuppression. *Exp Pathol* **34**: 71–77.

Riede UN, Riede PR, Horn R, Batthiany R, Kiefer G, Sandritter W (1978) Mechanisms of adaptation of hepatocytes to a chronical hypothyroidism (a cytophotometrical and morphometrical study). *Pathol Res Pract* **162**: 398–419.

Roels F (1991) *Peroxisomes, A Personal Account.* Brussels: VUB Press, 1–151. [ISBN 90-70289-94-6].

Roels F, Cornelis A (1989) Heterogeneity of catalase staining in human hepatocellular peroxisomes. *J Histochem Cytochem* **37**: 331–337.

Roels F, Goldfischer S (1979) Cytochemistry of human catalase: the demonstration of hepatic and renal peroxisomes by a high temperature procedure. *J Histochem Cytochem* **27**: 1471–1477.

Roel F, Pauwels M, Cornelis A, et al (1983) Peroxisomes (microbodies) in human liver. Cytochemical and quantitative studies of 85 biopsies. *J Histochem Cytochem* **31**: 235–237.

Roels F, Pauwels M, Poll-The BT, et al (1988) Hepatic peroxisomes in adrenoleukodystrophy and related syndromes: cytochemical and morphometric data. *Virchows Arch (A)* **413**: 275–285.

Roels F, Espeel M, De Craemer D (1991) Liver pathology and immunocytochemistry in congenital peroxisomal diseases: a review. *J Inher Metab Dis* **14**: 853–875.

Roels F, Espeel M, Poggi F, Mandel H, Van Maldergem L, Saudubray JM (1993) Human liver pathology in peroxisomal diseases: a review including novel data. *Biochimie* **75**: 281–292.

Rohr HP, Lüthy J, Gudat F, Oberholzer M, Gysin C, Bianchi L (1976) Stereology of liver biopsies from healthy volunteers. *Virchows Arch (A)* **371**: 251–263.

Rubin E, Lieber CS (1967) Early fine structural changes in the human liver induced by alcohol. *Gastroenterology* **52**: 1–13.

Ruebner BH, Gonzalez-Licea A, Slusser RJ (1967) Electron microscopy of some hepatic hepatomas. *Gastroenterology* **53**: 18–30.

Rustan AC, Christiansen EN, Drevon AC (1992) Serum lipids, hepatic glycerolipid metabolism and peroxisomal fatty acid oxidation in rats fed ω-3 and ω-6 fatty acids, *Biochem J* **283**: 333–339.

Ryoo JW, Buschmann RJ (1989) Morphometry of liver parenchyma in needle biopsy specimens from patients with alcoholic liver disease: preliminary variables for the diagnosis and prognosis of cirrhosis. *Mod Pathol* **2**: 382–389.

Sakuma M, Yamada J, Suga T (1992) Comparison of the inducing effect of dehydroepiandrosterone on hepatic peroxisome proliferation-associated enzymes in several rodent species: a short-term administration study. *Biochem Pharmacol* **43**: 1269–1273.

Salen G, Zaki G, Sabesin S, Boehme D, Shefer S, Mosbach EH (1978) Intrahepatic pigment and crystal forms in patients with cerebrotendinous xanthomatosis (CTX). *Gastroenterology* **74**: 82–89.

Schaffner F (1966) Intralobular changes in hepatocytes and the electron microscopic mesenchymal response in acute viral hepatitis. *Medicine* **45**: 547–552.

Staubli W, Hess R (1975) Lipoprotein formation in the liver cell. Ultrastructural and functional aspects relevant to hypolipidemic action. In Kritchevsky D, eds. *Hypolipidemic Agents*. Springer-Verlag, Berlin, 248–289.

Sternlieb I (1970) Marginal plate in human hepatic microbody. *N Engl J Med* **283**: 1290.

Sternlieb I (1979) Electron microscopy of mitochondria and peroxisomes of human hepatocytes. In Popper H, ed. *Progress in Liver Diseases*, vol. 6. New York: Grune and Stratton, 81–104.

Sternlieb I, Quintana N (1977) The peroxisomes of human hepatocytes. *Lab Invest* **36**: 140–149.

Sternlieb I, Quintana N (1982) Abnormalities of human hepatocellular peroxisomes. *Ann NY Acad Sci* **386**: 530–533.

Svoboda DJ, Reddy JK (1980) Pathology of the liver in Reye's syndrome. *Lab Invest* **32**: 571–579.

Thaler MM, Bruhn FW, Applebaum MN, Goodman J (1970) Reye's syndrome in twins. *J Pediatr* **77**: 638–646.

Thomassen MS, Christiansen EN, Norum KR (1982) Characterization of the stimulatory effect of high-fat diets on peroxisomal β-oxidation in rat liver. *Biochem J* **206**: 195–202.

Thomassen MS, Norseth J, Christiansen EN (1985) Long-term effects of high-fat diets on peroxisomal β-oxidation in male and female rats. *Lipids* **20**: 668–674.

Uchida T, Kao H, Quispe-Sjogren M, Peters RL (1983) Alcoholic foamy degeneration: a pattern of acute alcoholic injury of the liver. *Gastroenterology* **84**: 683–692.

Vamecq J, Roels F, Van den Branden C, Draye JP (1987) Peroxisomal proliferation in heart and liver of mice receiving chlorpromazine, ethyl-2-(5-(4-chlorophenyl)pentyl)oxiran-2-carboxylic acid or high fat diet: a biochemical and morphometrical comparative study. *Pediatr Res* **22**: 748–754.

Vamecq J, Vallée L, Lechêne de la Porte P, et al (1993) Effect of various n-3/n-6 fatty acid ratio contents of high fat diets on rat liver and heart peroxisomal and mitochondrial β-oxidation. *Biochim Biophys Acta* **1170**: 151–157.

Vanhole G, de Zegher F, Casaer P, et al (1994) A new peroxisomal disorder with fetal and neonatal adrenal insufficiency. *Arch Dis Child* **71**: F55–F56.

Wittrin G, Horstmann HG, Hunger J, Schnieder G (1975) Die bestimmung der Katalaseaktivität in Lebergewebe. *Klin Wschr* **53**: 723–725.

Wu HQ, Masset-Brown J, Tweedie DJ, et al (1989) Induction of microsomal NADPH-cytochrome P-450 reductase and cytochrome P450IVA1 (P-450LAω) by dehydroepiandrosterone in rats: a possible peroxisomal proliferator. *Cancer Res* **49**: 2337–2343.

Yamada J, Sakuma M, Ikeda T, Fukuda K, Suga T (1991) Characteristics of dehydroepiandrosterone as a peroxisome proliferator. *Biochim Biophys Acta* **1092**: 233–243.

Yamaguchi Y, Sato K, Endo H (1992) Depression of catalase gene expression in the liver of tumor bearing nude mice. *Biochem Biophys Res Commun* **189**: 1084–1089.

Yamazaki RK, Shen T, Schade GB (1987) A diet rich in (n-3) fatty acids increases peroxisomal β-oxidation activity and lowers plasma triacylglycerols without inhibiting glutathione-dependent detoxication activities in the rat liver. *Biochim Biophys Acta* **920**: 62–67.

Yokota S (1993) Formation of autophagosomes during degradation of excess peroxisomes induced by administration of dioctyl phthalate. *Eur J Cell Biol* **61**: 67–80.

Zaar K (1992) Structure and function of peroxisomes in the mammalian kidney. *Eur J Cell Biol* **59**: 233–254.

J. Inher. Metab. Dis. 18 Suppl. 1 (1995) 214–222
© SSIEM and Kluwer Academic Publishers.

Diagnostic work-up of a peroxisomal patient

J.G. LEROY[1], M. ESPEEL[2], J.F. GADISSEUX[3], H. MANDEL[4], M. MARTINEZ[5],
B.T. POLL-THE[6], R.J.A. WANDERS[7] and F. ROELS[2]
PANEL DISCUSSION, 21 MAY 1994, UNIVERSITY HOSPITAL GENT; UPDATED
AUGUST 1995
[1]*Departments of Pediatrics and Genetics, University of Gent, Gent, Belgium;*
[2]*Department of Human Anatomy, Embryology and Histology, University of Gent,
Belgium;* [3]*Service de Neurologie Pédiatrique, Université Catholique de Louvain,
Belgium;* [4]*Department of Pediatrics, Rambam Medical Centre, Haifa, Israel;*
[5]*Biochemical Research Unit, Laboratory of Lipids, Nutrition and Brain Development,
Maternity–Children's Hospital, Barcelona, Spain;* [6]*Wilhelmina Children's Hospital,
Utrecht, The Netherlands;* [7]*Department of Pediatrics, University Hospital Amsterdam,
AMC, Amsterdam, The Netherlands*

Chaired by J.G. Leroy, the panel discussed three main topics:

- Which clinical symptoms prompt suspicion of a peroxisomal disorder?
- Is there a logical flow chart of laboratory tests?
- Which other diagnoses should be considered?

Treatment of peroxisomal disorders was also examined.

CLINICAL SYMPTOMS PROMPTING SUSPICION OF A PEROXISOMAL DISORDER

Poll-The stressed the age-dependence of clinical presentation, as summarized in Table 1. Most symptoms are not specific for peroxisomal disorders but combinations of symptoms may be. Atypical CDP has the biochemical abnormalities of classical RCDP, but there is no shortening of the limbs (Poll-The et al 1991; Smeitink et al 1992). Many other neurological disorders, chromosomal aberrations and several malformation syndromes share symptoms with peroxisomal disorders. When laboratory tests such as plasma VLCFA are performed routinely in all doubtful cases, peroxisomal function may be normal in more than 80%. A karyotype is always very useful. Zellweger syndrome and infantile Refsum disease have been mistaken for Down syndrome. Dysmorphic features due to chromosomal aberrations remain constant, whereas in metabolic diseases they become more severe as the disease progresses (**J.F. Gaddisseux**). The clinical picture of neuronal ceroid lipofuscinosis may appear similar to a peroxisomal disorder and is comparatively common; diagnosis of the former is by electron microscopy of a rectal biopsy for inclusion bodies (**B.T. Poll-The, P. Augoustides, J. Leroy**).

Hanna Mandel first recalled the clinical guidelines of Theil and colleagues (1992) based upon a retrospective study of 40 patients. They proposed that biochemical

Table 1   Clinical onset of peroxisomal disorders with neurological involvement

Symptoms	*Disorder*
Neonatal period	
	ZS, ZS variants
	Neonatal ALD
Hypotonia, areactivity, seizures	Pseudo-neonatal ALD
Craniofacial dysmorphia	(acyl-CoA oxidase deficiency)
Skeletal abnormalities	Bifunctional enzyme deficiency
Conjugated hyperbilirubinaemia	Thiolase deficiency
	RCDP (typical/atypical)
	THC acidaemia
	Pipecolic acidaemia
	Mevalonic aciduria
First six months of life	
Hepatomegaly, prolonged jaundice	IRD, pseudo-IRD
Digestive problems, hypocholesterolaemia	Pipecolic acidaemia, neonatal ALD
Vitamin E deficiency	Milder forms of ZS
Failure to thrive	Atypical chondrodysplasia
Visual abnormalities	Mevalonic aciduria
Six months – four years	
Neurological presentation	IRD, pseudo-IRD
Psychomotor retardation	Pipecolic acidaemia
Visual and hearing impairment (ERG, BAEP)	Neonatal ALD
	Milder forms of ZS
Failure to thrive	Atypical chondrodysplasia
Osteoporosis	Di- and THC acidaemia
Beyond four years of age	
Behaviour changes	X-linked ALD
Deterioration intellectual functions	
White matter demyelination	
Visual and hearing impairment	Classical Refsum disease
Peripheral neuropathy, gait abnormality	

investigation of peroxisomal functions is warranted by the combined presence of at least *3 major* clinical characteristics (present in >75% of the affected patients), namely *psychomotor retardation, hypotonia, impaired hearing, low/broad nasal bridge, abnormal ERG and hepatomegaly, and one or more minor characteristics* (present in 50%–75% of the patients), e.g. *large fontanelles, shallow orbital ridges, epicanthus, anteverted nostrils and retinitis pigmentosa.*

Experience in Haifa demonstrated that these guidelines may not be sufficient in some instances, and that some symptoms may be absent in the neonatal period (Table 2: 9 patients with X-ALD and 21 with primary hyperoxaluria are excluded). The large number of patients is explained by the frequency of intermarriage in Israel, peroxisomal disorders being rare recessive diseases.

Table 2 shows four patients with straightforward ZS, presenting more than three major symptoms. However the following five patients did not fulfil the criteria and the correct diagnosis of a β-oxidation defect might have been missed. Patient 5 was born with a low

Table 2 Patients with peroxisomal disorders seen at **Rambam Medical Center, Haifa (1989–1994)**

Patient no.	Age (months)	Features at presentation — Major clinical criteria					Hepatomegaly	Special clinical presentation
		MR	Hypotonia	Deafness	ERG ↓	Facial dysm.		
Zellweger syndrome								
1	Birth		+	+	+	+		
2	Birth		+	+	+	+	+	
3	Birth		+	+	+	+	+	
4	Birth		+	+	+	+	+	
β-Oxidation defect								
5[a]	Birth		+					Neonatal asphyxia and seizures
6[a]	Birth		+					Neonatal asphyxia and seizures
7	Birth		+					
8	Birth		+					Neonatal seizures
9	Birth		+			+		
Infantile Refsum disease								
10[b]	1		+					FTT, chronic diarrhoea
11[b]	4							Hypo-β-lipoproteinaemia, malabsorption
12[c]	1							Jaundice, suspected BDA
13[c]	1							Jaundice, CMV infection
14	5		+	+	+			Normal FPF
15	5		+	+	+			Normal FPF
16	7		+	+	+	+		
Liver/tissue mosaicism								
17[d]	13	+	+					Normal FPF
18[d]	17	+	+					Normal FPF

MR = mental retardation, FTT = failure to thrive, BDA = bile duct atresia, FPF = fibroblast peroxisomal functions
[a]Mandel et al 1992a, [b]Mandel et al 1992b, [c]Goez et al 1995, [d]Mandel et al 1994.

Apgar score. The early seizures and brain damage were thus attributed to perinatal asphyxia (Mandel et al 1992a). Lacking a diagnosis, genetic counselling was not provided to the family and prenatal diagnosis was not performed on a subsequent pregnancy, which resulted in the birth of an affected baby (Mandel et al 1992a). Patients 8 and 9 also presented in the neonatal period with only hypotonia and seizures, but facial dysmorphism, deafness and abnormal ERG later. In both families, subsequent affected fetuses were diagnosed prenatally and the pregnancies were terminated. Patient 10 presented at the age of 1 month with a bleeding diathesis due to a vitamin K-responsive coagulation defect. She was repeatedly hospitalized during the following years with failure to thrive, diarrhoea, blindness and severe psychomotor retardation, and died at the age of 3.5 years (Mandel et al 1992b). Her brother presented at the age of 4 months with clinical and laboratory manifestations suggestive of β-hypolipoproteinaemia. A diagnosis of IRD was not made until the patient was 7 years old (Mandel et al 1992b). Patient 12 was referred with hepatomegaly and jaundice at the age of 4 weeks (Goez et al 1995). In view of the inconclusive investigations, surgical exploration was performed to rule out biliary atresia. At the age of 13 months, hypotonia, psychomotor retardation, retinitis pigmentosa, abnormal ERG and deafness suggested the diagnosis of IRD. Patient 13, who presented with a similar clinical constellation at 4 weeks of age, was diagnosed as having CMV infection. IRD was diagnosed only at the age of 4 months, after it was realized that patients 12 and 13 are second-degree cousins (Goez et al 1995). Patients 17 and 18 were healthy siblings and developed normally initially. The onset of neurological regression after the first year of life was the sole manifestation of their disorder, and this did not arouse suspicion of peroxisomal disease (Mandel et al 1994).

The clinical heterogeneity of peroxisomal disorders is becoming increasingly evident, and warrants periodic revision of the criteria for their detection. Seizures could be the sole presentation of a peroxisomal disorder during the neonatal period (Mandel et al 1992a) and IRD should be included in the differential diagnosis of a gastrointestinal disease mimicking hypo-β-lipoproteinaemia (Mandel et al 1992b). There is a need for increased awareness of the hepatic presentation, especially in the neonatal period, when cholestatic jaundice could be the first and only manifestation of IRD (Goez et al 1995), to be followed only later by the evolution of classic clinical signs of peroxisomopathy. Evaluation of plasma VLCFA should be considered in patients of any age with unexplained neurological regression (Mandel et al 1994).

The clinician should be aware that not all peroxisomal disorder patients will fulfil the criteria as originally listed, and that others may display features that have not been hitherto considered as indicative. The clinical manifestations may initially be very mild but early diagnosis becomes crucial for genetic counselling, and for prenatal diagnosis of the disorder in future pregnancies.

During the ensuing discussion other clinical features were mentioned. The nose in RCDP is hypoplastic and similar to the nose in babies of mothers treated with coumarin. A possible connection between plasmalogen biosynthesis and vitamin K abnormality should be investigated (**H. Mandel, J. Leroy**). The dysmorphic features in peroxisomal patients are not malformations but deformations and hypoplasias (**J. Leroy**). Hypoplastic earlobes in Zellweger babies were observed by some but not all physicians. Prenatal diagnosis by echography is easy in RCDP, but rather late in pregnancy. Zellweger

Table 3   Biochemical classification of peroxisomal disorders

I   *Disorders of peroxisome biogenesis*

Zellweger syndrome
neonatal adrenoleukodystrophy
infantile Refsum disease
hyperpipecolic acidaemia[a] (Gatfield et al 1968; Burton et al 1981; Thomas et al 1975)

II   Rhizomelic chondrodysplasia punctata (classic and milder variants)[b]
Zellweger-like syndrome

III   X-linked adrenoleukodystrophy
acyl-CoA oxidase deficiency
bifunctional enzyme deficiency
thiolase deficiency
isolated DHAPAT deficiency (Wanders et al 1992)
alkylDHAPsynthase deficiency (Wanders et al 1994)
primary hyperoxaluria type I
glutaryl-CoA oxidase deficiency (Bennett et al 1991)
mevalonate kinase deficiency
acatalasaemia[c]

problematic: pristanic and trihydroxycoprostanoic acidaemia (Christensen et al 1990)
pristanic and VLCFA acidaemia

adult Refsum: not clear

[a]The classification of hyperpipecolic acidaemia (HPA) as a peroxisome biogenesis disorder comparable to ZS is challenged by Roels. In two of the original patients hepatic peroxisomes were described by Challa et al (1983); the classification as biogenesis disorder is based on fibroblast studies (Wanders et al 1988). However, Wiemer et al (1991) reported catalase-containing granules in the cultured fibroblasts from the same HPA patient. Poll-The has seen two patients with HPA, the liver of which contained peroxisomes. Roels mentions other cases of discrepancy between fibroblasts and liver and he emphasizes that liver results are more reliable (Roels 1991; Espeel et al 1993; Mandel et al 1994; Schutgens et al 1994; Roels et al 1995). Cultured cells can display heterogeneity (Wiemer et al 1991)
[b]Patients showing the tetrad of biochemical defects but without rhizomelia of the limbs were described (Poll-The et al 1991; Smeitink et al 1992; Espeel et al 1993)
[c]It is mainly erythrocyte catalase which is inactive. The enzyme deficiency in other tissues is variable; hepatocellular peroxisomes have not been investigated

syndrome with patellar calcifications and polycystic kidneys was well recognized at 20 weeks of gestation by Powers et al (1985) (**J.F. Gadisseux**). It was queried whether the echographic detection of any dysmorphism should be followed by a search in amniotic fluid for peroxisomal metabolites (bile acids, VLCFA) (**C. Jakobs**).

LABORATORY TESTS TO BE PERFORMED

R.J.A. Wanders presented the biochemical classification of the peroxisomal disorders (Table 3).

The number of reported cases, reflecting their frequency, was as follows (1994): Zellweger, at least 400 patients; NALD and IRD, both more than 50 cases; X-ALD, more than 1000 patients; RCDP, more than 50 cases are known in Amsterdam but few were published; DHAPAT deficiency, five cases; alkylDHAP synthase deficiency, one patient.

Table 4 Flow chart of laboratory tests

VLCFA

(i) Plasma VLCFA increased:
ZS, NALD, IRD, ZS-like
X-ALD/AMN (very mild increase in some)
isolated β-oxidation defects

(ii) Plasma VLCFA normal:
in RCDP
    isolated THCA acidaemia
    assay of:
        erythrocyte plasmalogens[d]
        fibroblasts plasmalogen biosynthesis
        plasma phytanic, pristanic, pipecolic acid, bile acids
        urinary organic acids[c]

Complete postnatal diagnosis by

* tests listed sub (ii)
* docosahexaenoic acid (DHA) in erythrocyte membranes
* liver biopsy for immunolocalization and immunoblotting (except in X-ALD[a])
* cultured fibroblasts[b]: *de novo* plasmalogen synthesis
                DHAPAT activity
                alkylDHAP synthase activity
                β-oxidation of C26, phytanic, pristanic acid
                catalase immunofluorescence
                complementation analysis

First trimester prenatal diagnosis:

Native chorionic villi:
* Immunoblotting and microscopy of peroxisomes (ZS, NALD, IRD, RCDP)
* DHAP AT activity (ZS, NALD, IRD)
* plasmalogens (RCDP)

Cultured CV-fibroblasts or cultured amniotic fluid cells:
* VLCFA, β-oxidation of C26 and pristanic acids (X-ALD, β-oxidation defects)

[a]As yet VLCFA acyl-CoA synthase, which is functionally impaired in X-ALD, cannot be visualized microscopically. The ALD protein has not been localized cytochemically in normal human liver (Mosser et al 1994) but studies on fibroblasts and blood cells show the antigen to be present in 22 out of 106 ALD patients (Aubourg et al, Watkins et al, Jorge and Brites, International Symposium on Peroxisomes, Aspen, Colorado, USA, June 28 –July 2, 1995). [Editor's note]
[b]Roels underlines that in cases of discrepancy the results from blood assays and liver should be preferred over results from cultured fibroblasts (see footnote in Table 3)
[c]emphasized by Dr C Jakobs
[d]Björkhem et al 1991.

Deficiencies of acyl-CoA oxidase, bifunctional protein and thiolase form a significant group.

When clinical symptoms indicate a possible peroxisomal disorder, which biochemical tests should be performed and in what order? This is summarized in Table 4.

The classification of peroxisomal disorders was again discussed. Clinical syndromes given historical names may not correspond to biochemical defects nor to complementation groups, of which there are now 16 (Moser AB et al 1995). In the same large group one finds Zellweger, NALD and IRD patients. Only a few mutations have been identified (Shimozawa et al 1992, 1993; Gärtner et al 1992; Dodt et al 1995; see also Editors' notes nos. 2, 3 and 4). However, the clinical classification does give indications on the severity of the illness and prognosis (**Poll-The, Molzer** referring to **Saudubray, Wanders**).

Several panel members stressed the variability within families. In X-ALD the childhood cerebral form, adult adrenomyeloneuropathy and Addison disease only occur within the same family. In addition a few remain asymptomatic, and older female heterozygotes may manifest the disease. **H. Mandel** described the different ages of onset in two brothers where one has peroxisome mosaicism in liver but normal fibroblast functions (Mandel et al 1994; Espeel et al 1995a), as well as a family with IRD where the first child died at 4 months of age, and her brother is now 12 years and walking (see also Editors' note no. 1). **M. Espeel** suggested that in mosaic patients, the residual population of normal peroxisomes in part of the liver contributes to the clinical and biochemical variability. If mosaicism exists also in skin, this might contribute to the discrepancies between liver and fibroblasts.

SOME REMARKS ON TREATMENT

M. Martinez urged the inclusion of DHA assay in erythrocytes as a standard procedure, and to treat patients with a low level with oral DHA ethylester, without restriction of the diet. PUFA concentrations should be monitored during treatment but irreversible changes such as in Zellweger brain cannot be expected to respond. With respect to those patients with normal erythrocyte DHA levels, **H. Mandel** drew attention to lactate levels which are sometimes normal in blood but increased in CSF. DHA levels in CSF should be investigated as this may provide a more accurate assessment of levels in the brain; however this is technically more difficult because concentrations are very low. It is extremely important to investigate DHA in brain tissue from patients with normal erythrocyte levels, such as X-ALD patients; data are lacking.

Treatment with GTO-GTE (Lorenzo's oil) is recommended by H.W. Moser (Baltimore) in *asymptomatic* boys but not in symptomatic cerebral ALD and AMN. The effectiveness of this treatment is still under investigation. Bone marrow transplantation has favourably influenced the course of X-ALD in a limited number of cases (Moser et al 1992; Moser H 1995; Krivit et al 1992, 1995) and is still recommended by Krivit in 1995. However it requires an HLA-identical donor, is a risky and costly procedure and must be done as soon as neuropsychological and MRI markers of cerebral disease are detected (Shapiro et al 1995).

Pentoxifylline, a TNF-α antagonist (Noel et al 1991; Alegre et al 1991) and β-interferon are experimentally applied in a few ALD patients in Europe and the United States.

Transfection of haematopoietic cells of the ALD patient with wild type cDNA of the ALD protein could be the therapy of the future. The first *in vitro* steps leading to this procedure are presently being tried out by several laboratories. Prenatal prevention should also be intensified (Editors' notes).

EDITORS' NOTES

1. A brother and a sister with IRD, who differ strikingly in clinical presentation, have been reported by Van Maldergem et al (1993). Both siblings have no hepatic peroxisomes, but the liver of the relatively mild case shows peroxisomal membrane ghosts; the liver of the other child does not (Espeel et al 1995b).
2. Shimozawa et al (1992) described mutations in the *peroxisomal assembly factor-1* (PAF1), gene symbol PXMP3, of a patient with typical Zellweger syndrome (MIM 170993). In a subsequent paper, Shimozawa et al (1993) reported that a Dutch patient (Brul et al 1988) was homozygous for the same C-T transition at nucleotide 355 as their original patient.

3. Gärtner et al (1992) described two point mutations in the PMP70 gene (gene symbol PXMP1) in two patients of the same complementation group. PMP70 is a member of the ATP-binding cassette transporter family (MIM 170995).
4. Recently, a third gene (PXR1) encoding the receptor for PTS-1 proteins has been isolated; mutations in PXR1 define a second complementation group of the generalized peroxisomal disorders (Dodt et al 1995). PTS-1 proteins are peroxisomal proteins which use the carboxy terminal SKL-motif as peroxisomal targeting signals.

REFERENCES

Alegre M-L, Gastaldello K, Abramowicz D et al (1991) Evidence that pentoxifylline reduces anti-CD3 monoclonal antibody induced cytokine release syndrome. *Transplantation* **52**: 674–679

Bennett MJ, Pollitt RJ, Hale DE, et al (1991) Atypical riboflavin-responsive glutaric aciduria and deficient peroxisomal glutaryl CoA oxidase activity: a new peroxisomal disorder. *J Inher Metab Dis* **14**: 165–173.

Björkhem I, Sisfontes L, Boström B, et al (1986) Simple diagnosis of the Zellweger syndrome by gas–liquid chromatography. *J Lipid Res* **27**: 786–791.

Brul S, Westerveld A, Strijland A, et al (1988) Genetic heterogeneity in the cerebrohepatorenal (Zellweger) syndrome and other inherited disorders with a generalized impairment of peroxisomal functions: a study using complementation analysis. *J Clin Invest* **81**: 1710–1715.

Burton BK, Reed SP, Reny WT (1981) Hyperpipecolic acidemia: Clinical and biochemical observations in two male siblings. *J Pediatr* **108**: 729–734.

Challa, VR, Geisinger KR, Burton BB (1983) Pathologic alterations in the brain and liver in hyperpipecolic acidemia. *J Neuropathol Exp Neurol* **42**: 627–638.

Christensen E, Van Eldere J, Brandt NJ, et al (1990) A new peroxisomal disorder: di- and trihydroxycholestanoyl-CoA oxidase deficiency. *J Inher Metab Dis* **13**: 363–366.

Dodt G, Braverman N, Wong C, et al (1995) Mutations in the PTS1 receptor complementation group 2 of the peroxisome biogenesis disorders. *Nature Genetics* **9**: 115–125.

Espeel M, Heikoop JC, Smeitink JAM, et al (1993) Cytoplasmic catalase and ghost-like peroxisomes in the liver from a child with atypical rhizomelic chondrodysplasia punctata. *Ultrastruct Pathol* **17**: 623–636.

Espeel M, Mandel H, Poggi F, et al (1995a) Peroxisomal mosaicism in the livers of peroxisomal deficiency patients. *Hepatology* **22**: 497–504.

Espeel M, Roels F, Giros et al (1995b) Immunolocalization of a 43 kDa peroxisomal membrane protein in the liver of patients with generalized peroxisomal disorders. *Eur J Cell Biol* **67**: 319–327.

Gärtner J, Moser H, Valle D (1992) Mutations in the 70 kDa peroxisomal membrane protein gene in Zellweger syndrome. *Nature Genetics* **1**: 16–23.

Gatfield PD, Taller E, Hinton GG, et al (1968) Hyperpipecholemia: A new metabolic disorder associated with neuropathy and hepatomegaly. *Canad Med Assoc* **99**: 1215–1233.

Goez H, Meiron D, Horwitz J, et al (1995) Infantile Refsum disease: Neonatal cholestatic jaundice presentation of a peroxisomal disorder. *J Pediatr Gastroenterol Nutr* **17**: 98–101.

Krivit W, Shapiro E, Lockman L, et al (1992) Recommendations for treatment of childhood cerebral form of adrenoleukodystrophy. In: Hobs JR, Riches PG, eds. *Correction of certain genetic diseases by transplantation.* London: COGENT Trust.

Krivit W, Lockman LA, Watkins PA, Hirsch J, Shapiro EG (1995) The future for treatment by bone marrow transplantation for adrenoleukodystrophy, metachromatic leukodystrophy, globoid cell leukodystrophy and Hurler syndrome. *J Inher Metab Dis* **18**: 398–412.

Mandel H, Berant M, Aizin A, et al (1992a) Zellweger-like phenotype in two siblings: A defect in peroxisomal β-oxidation with elevated very long chain fatty acids but normal bile acids. *J Inher Metab Dis* **15**: 381–384.

Mandel H, Meiron D, Schutgens RBH, et al (1992b) Infantile Refsum disease: gastrointestinal presentation of a peroxisomal disorder. *J Pediatr Gastroenterol Nutr* **14**: 83–85.

Mandel H, Espeel M, Roels F, et al (1994) A new type of peroxisomal disorder with variable expression in liver and fibroblasts. *J Pediatr* **125**: 549–555.

Moser AB, Rasmussen M, Naidu S, et al (1995) Phenotype of patients with peroxisomal disorders subdivided into sixteen complementation groups. *J Pediatr* **127**: 13–22.

Moser HW (1995) Adrenoleukodystrophy: natural history, treatment and outcome. *J Inher Metab Dis* **18**: 435–447.

Moser HW, Moser AB, Smith KD, et al (1992) Adrenoleukodystrophy: phenotypic variability and implications for therapy. *J Inher Metab Dis* **15**: 645–664.

Mosser J, Lutz Y, Stoeckel ME, et al (1994) The gene responsible for adrenoleukodystrophy encodes a peroxisomal membrane protein. *Hum Mol Genet* **3**: 265–271.

Noel P, Nelson S, Bokulic R, et al (1991) Pentoxifylline inhibits lipopolysaccharide-induced serum tumor necrosis factor and mortality. *Life Sci* **47**: 1023.

Poll-The BT, Lombes A, Lenoir, et al (1988) Joubert's syndrome associated with hyperpipecolatemia. Three siblings. *Doctoral Thesis*, Amsterdam, Chap 8: 201–219.

Poll-The BT, Marotaux P, Narcy C, et al (1991) A new type of chondrodysplasia punctata associated with peroxisomal dysfunction. *J Inher Metab Dis* **14**: 361–363.

Powers JM, Moser HW, Moser AB, et al (1985) Fetal cerebro-hepato-renal (Zellweger) syndrome. Dysmorphic, radiologic, biochemical and pathological findings in four affected fetuses. *Hum Pathol* **16**: 610–620.

Roels F, Verdonk V, Pauwels M, et al (1987) Light microscopic visualization of peroxisomes and plasmalogens in first trimester chorionic villi. *Prenat Diagn* **7**: 525–530.

Roels F (1991) *Peroxisomes: a personal account*. Brussels: VUB Press, ISBN: 90-70289-94-6, 151p.

Roels F, Espeel M, De Craemer D (1991) Liver pathology and immunocytochemistry in peroxisomal disorders: A review. *J Inher Metab Dis* **14**: 853–875.

Roels F, Espeel M, Poggi F, et al (1993) Human liver pathology in peroxisomal diseases: A review including novel data. *Biochimie* **75**: 281–292.

Roels F, Espeel M, Mandel H, et al (1995) Cell and tissue heterogeneity in peroxisomal patients. In: Wanders RJA, Schutgens RBH, Tabak HF, eds. *Peroxisomal disorders in relation to functions and biogenesis of peroxisomes*. Amsterdam: Royal Acad of Arts and Science, Elsevier (In press).

Shapiro EG, Lockman LA, Balthazor M, Krivit W (1995) Neuropsychological outcomes of several storage diseases with and without bone marrow transplantation. *J Inher Metab Dis* **18**: 413–429.

Schutgens RBH, Wanders RJA, Jakobs C, et al (1994) A new variant of Zellweger syndrome with normal peroxisomal functions in fibroblasts. *J Inher Metab Dis* **17**: 319–322.

Shimozawa N, Tsukamoto T, Suzuki Y, et al (1992) A human gene responsible for Zellweger syndrome that affects peroxisome assembly. *Science* **255**: 1132–1134.

Shimozawa N, Suzuki Y, Orii T, et al (1993) Standardization of complementation grouping of peroxisome-deficient disorders and the second Zellweger patient with peroxisomal assembly factor-1 (PAF-1) defect. *Am J Hum Genet* **52**: 843–844.

Smeitink JAM, Beemer FA, Espeel M, et al (1992) Bone dysplasia associated with phytanic acid accumulation and deficient plasmalogen synthesis: A peroxisomal entity amenable to plasmapheresis. *J Inher Metab Dis* **15**: 377–380.

Theil AC, Schutgens RBH, Wanders RJA, et al (1992) Clinical recognition of patients affected by a peroxisomal disorder: a retrospective study of 40 patients. *Eur J Pediatr* **151**: 117–120.

Thomas GH, Haslam RHA, Batshaw ML, et al (1975) Hyperpipecolic acidemia associated with hepatomegaly, mental retardation, optic nerve dysplasia and progressive neurologic disease. *Clin Genet* **8**: 376–382.

Van Maldergem L, Ureel D, Roels F, et al (1993) Clinical variability within sibship with infantile Refsum disease. In Abstracts of 31st Annual SSIEM Symposium, ISBN: 1 870617-06-1, Sheffield; SSIEM, p. W23.

Wanders RJA, van Roermund CWT, van Wijland MJA, et al (1988) Peroxisomes and peroxisomal functions in hyperpipecolic acidemia. *J Inher Metab Dis* **11 Suppl 2**: 161–164.

Wanders RJA, Schumacher H, Heikoop J, et al (1992) Human dihydroxy-acetonephosphate acyltransferase deficiency: a new peroxisomal disorder. *J Inher Metab Dis* **15**: 389–391.

Wanders RJA, Schutgens RBH, Barth PG, et al (1993) Postnatal diagnosis of peroxisomal disorders: A biochemical approach. *Biochimie* **75**: 269–279.

Wanders RJA, Dekker C, Hovarth VAP, et al (1994) Human alkyldihydroxyacetonephosphate synthase deficiency: A new peroxisomal disorder. *J Inher Metab Dis* **17**: 315–318.

Wiemer EAC, Out M, Schelen A, et al (1991) Phenotypic heterogeneity in cultured skin fibroblasts from patients with disorders of peroxisome biogenesis belonging to the same complementation group. *Biochem Biophys Acta* **1097**: 232–237.

J. Inher. Metab. Dis. 18 Suppl. 1 (1995) 223–226
© SSIEM and Kluwer Academic Publishers.

Subject Index

Journal of Inherited Metabolic Disease

Aims and Scope of the Journal

The *Journal of Inherited Metabolic Disease* is the official scientific and clinical journal of the Society for the Study of Inborn Errors of Metabolism. The aim of this international and multidisciplinary journal is to provide otherwise unavailable information on inherited metabolic disorders covering clinical (medical, dental and veterinary), biochemical (including molecular genetics), genetic, experimental (including cell biology), theoretical, epidemiological, ethical and counselling aspects. Widespread and efficient communication between professional workers should improve the handling and understanding of inherited disorders.

The journal publishes papers, case and short reports, short communications based on work presented at the annual symposium of the SSIEM, letters, invited articles which are generally reviews, and book reviews. *Papers for submission* and correspondence should be written in English and sent to:

> The Editor-in-Chief
> Journal of Inherited Metabolic Disease
> Kluwer Academic Publishers
> PO Box 55, Lancaster, LA1 1PE, United Kingdom

Acceptance of articles for publication is at the discretion of the editors. Authors are advised to consult a current issue of the journal before submission. It is a condition of acceptance that articles have not been and will not be published elsewhere in substantially the same form. The submitting author must have circulated the article to and have secured agreement from all co-authors before submission of the article. The absence of previous similar or simultaneous publications, their inspection of the manuscript, their substantial contribution to the work and their agreement to submission must be confirmed by all authors in a signed letter or letters on submission. It should be noted that the conditions are later confirmed by the corresponding author in a copyright transfer form when the paper is accepted.

Detailed *Instructions to Authors* are published in issue no. 1 of each volume of the journal, and in other issues as space permits, and are listed in the Contents.

Ethical considerations

The editors reserve the right to reject for publication work which, even though scientifically sound, they consider should not have been undertaken on ethical grounds. Where there is a difference of opinion, authors may be asked to explain in their publication why the work was undertaken. There are no absolute ethical standards, and studies on children present special problems. In this last respect the editors agree with the policy stated by the *Archives of Disease in Childhood* 42 (1967), 109; 48 (1973) 751–752; 54 (1978) 441–442.

The purely legal aspect of studies in children are discussed in *Lancet* 2 (1977) 754–755; *Archives of Disease in Childhood* 53 (1978) 443–446.

Books for Review should be sent to The Editor-in-Chief at the above address.

Offprints. 25 offprints are provided free of charge. Authors will receive a form for ordering an extra 125 offprints with the proofs of their paper.

No page charges are levied on authors or their institutions.

Advertisements: As well as commercial advertisements, announcements of forthcoming scientific meetings and other material relevant to the journal can be included, at rates available from the publishers. All advertising is subject to the discretion of the editors.

Consent to publish in this journal entails the author's irrevocable and exclusive authorization of the publisher to collect on behalf of the copyright owners, SSIEM and the publishers, any sums or considerations for copying or reproduction payable by third parties (as mentioned in article 17, paragraph 2, of the Dutch Copyright act of 1912 and in the Royal Decree of June 20, 1974 (S.351) pursuant to article 16b of the Dutch Copyright act of 1912) and/or to act in or out of court in connection therewith.

Publication programme, 1995: Volume 18 (6 issues)
Subscriptions should be sent to: **Kluwer Academic Publishers Group, PO Box 322, 3300 AH Dordrecht, The Netherlands,** or at **PO Box 358, Accord Station, Hingham, MA 02018-0358, USA,** or to any subscription agent.
Changes of mailing address should be notified together with your latest label.
Subscription prices, per volume (6 issues): NLG 577.- plus postage: NLG 54.- (NLG 631.- per annum).
Subscription prices, per volume (bimonthly): USD 361.00
Second Class Postage paid at Rahway, NJ. USPS No. 757-750.
US Mailing Agent: Mercury Airfreight International Ltd., 2323 Randolph Ave., Avenel, NJ 07001, USA.
Published by Kluwer Academic Publishers, PO Box 55, Lancaster, LA1 1PE, UK.
Postmaster: please send all address corrections to: *Journal of Inherited Metabolic Disease,* c/o Mercury Airfreight International Ltd., 2323 Randolph Ave., Avenel, NJ 07001, USA